Cradle to Grave

CRADLE TO GRAVE

Life, Work, and Death at the Lake Superior Copper Mines

LARRY LANKTON

OXFORD UNIVERSITY PRESS
New York   Oxford

Oxford University Press

Oxford New York Toronto
Delhi Bombay Calcutta Madras Karachi
Kuala Lumpur Singapore Hong Kong Tokyo
Nairobi Dar es Salaam Cape Town
Melbourne Auckland Madrid

and associated companies in
Berlin Ibadan

Copyright © 1991 by Larry Lankton

Published by Oxford University Press, Inc.,
198 Madison Avenue, New York, New York 10016

First issued as an Oxford University Press paperback, 1993

Oxford is a registered trademark of Oxford University Press

All rights reserved. No part of this publication may be reproduced,
stored in a retrieval system, or transmitted, in any form or by any means,
electronic, mechanical, photocopying, recording, or otherwise,
without the prior permission of Oxford University Press.

Library of Congress Cataloging-in-Publication Data
Lankton, Larry
Cradle to grave : life, work, and death at
the Lake Superior copper mines./
Larry Lankton.
p. cm. Includes bibliographical references and index
ISBN 0-19-506263-9 ISBN 0-19-508357-1 (pap)
1. Copper industry and trade—Michigan—Keweenaw Peninsula—History.
2. Copper miners—Michigan—Keweenaw Peninsula—History.
I. Title. HD9539.C7U548 1991
338.7′662343′0977499—dc20 90-40520

8 10 9 7

Printed in the United States of America
on acid-free paper

To John, Martha, and Jim

Preface

> It is quite possible that the Keweenaw was in the long run the most profitable of any American nonferrous mining district. Perhaps the reason for its lack of fame outside Michigan itself is that its annals, like those of happy peoples, are relatively dull.
>
> Otis E. Young, Jr., 1984

Otis E. Young, Jr., the noted historian of western mining, is certainly right in one regard: little historical attention has been paid to the Lake Superior copper district of Upper Michigan. The reasons for its "lack of fame" might be several.

When it comes to mining and the rise of industrialism, historians have paid particular attention to minerals and metals sold by the ton or by the ounce; they have paid less notice to metals, like copper, sold by the pound. Iron and coal quite properly have attracted historians' attention, because together they were the two key materials that fueled and built the Industrial Revolution. On the other end of the scale, historians have been captured by the romantic allure of the precious metals—gold and silver. Tracking the human pursuit of gold and silver has proved a seductive task, just as the metals themselves proved so seductive to the forty-niners and to other nineteenth-century prospectors who went in search of them.

Copper, then, remains an important metal, but it is certainly not *the* most important metal, nor is it the dearest one. In the case of Michigan copper in particular, another factor might help account for its lack of historical recognition: it was taken from the upper Midwest, not from the "real" West, and it was taken not from mountains but from near the shore of a lake. For some unexplained cultural reason, Americans seemingly prefer that their nonferrous metals come from high, western regions, and the higher the elevation, and the more remote the mountain, the more interesting the site, the more fascinating its history.

Young attributes this mining district's lack of notoriety to the fact that it was populated by "happy peoples" who just went about their business in a steadfast way, without fanfare, and presumably without the rowdiness, lawlessness, strikes, and violence that occasionally punctuated day-to-day life in other mining regions. In short, the Michigan miners were dull in comparison with the fractious men found in eastern coal fields, or with the union men found in

Butte or Cripple Creek. This is the accepted view of the Lake Superior copper district—that it was home to a society that remained remarkably harmonious as the industrialization of mining progressed. It remained at peace, even while new, large companies and new machines obliterated traditional forms of organization and work within the mining industry.

There certainly is some truth to this interpretation of the region's past, particularly if attention is focused on its history prior to about 1890. But in the subsequent twenty to thirty years, through 1910 and up to 1920, this smoothly run society and its single supporting industry seriously deteriorated. After a long run of success beginning in the mid-1840s, this district turned the corner and headed toward decline and failure.

This book, after a fashion, tries to tell the broad history of the Lake Superior copper district from its discovery until its final demise about 1970. But within that expansive framework, its focus remains on the tensions and conflicts, on the social and economic problems, that surfaced during the critical 1890 to 1920 period, and which the local industry and its attendant community never successfully erased or resolved.

My intent has been to write a "social history of technology"—one that works that particular space where machine and man meet, and where industry and society connect. I placed special emphasis on the social, economic, and political ramifications of technological change; on mine safety and fatal accidents as a social issue; and on the companies' use of paternalism to enhance and direct development, to contain costs, and to effect social control. I attempted to create a text that is business, labor, social, and technological history all at the same time. If, at the end, readers still think that the "annals" of Lake copper are "relatively dull," then the fault lies with this historian, and not with the people he was writing about.

Acknowledgments

Late in the summer of 1977, Charlie Hyde, an historian at Wayne State University, called me at my office in Washington, D.C., where I worked for the Historic American Engineering Record (HAER). Up near the very top of Michigan, he said, he had run across a very interesting historic site: the remains of the Quincy copper mine. Charlie said that I should look the site over, with a view to having HAER document it. Taking his advice, in September 1977 I flew to the Keweenaw Peninsula of Upper Michigan, and with Charlie as my guide I toured the Lake Superior copper district for the first time. Now, in 1990, I am finally finishing this book. Needless to say, in all the intervening years, numerous institutions and individuals have helped me out along the way.

I indeed headed up a HAER recording project at the Quincy Mine in the summer of 1978, and much I learned then has worked its way into this book. I am grateful to the organizations that sponsored the project that got me involved with the history of Michigan copper: HAER, the Quincy Mining Company, the Quincy Mine Hoist Association, Michigan's Bureau of History, and Michigan Technological University.

As the years rolled by and my research changed in purpose and scope, other institutions rendered assistance. I benefited greatly from a post-doctoral fellowship awarded by the Smithsonian Institution, which allowed me to spend nearly a year within the Division of Civil and Mechanical Engineering at the National Museum of American History. After moving from HAER to my faculty position at Michigan Tech, the University supported me with summer research and creativity grants. Finally, I was very happy to receive an advanced research fellowship from the Hagley Museum and Library, which was funded by the National Endowment for the Humanities and the Mellon Foundation. That fellowship, coupled with sabbatical leave suport from Michigan Tech, gave me the time needed to write most of this volume.

Organizations, of course, whether they are old mining companies or modern research libraries, are made up of individuals, and fortunately I have encountered many helpful persons in many different settings. Thanks go to the members of the original HAER recording project, especially Charles O'Connell, Richard Anderson, and Charlie Hyde. Charlie, for well more than a decade now, has always shared ideas and sources with me—and when necessary, he has applied some corrective scolding to make sure that in studying technological change, I always paid adequate attention to the economics of mining.

At the Smithsonian, curator Robert Vogel proved a gracious host during my stay there in 1980–81. It was fitting that I spent some time in Vogel's domain, for about a decade earlier he and Pat Malone had been two of the persons principally responsible for kindling my interest in historic sites and industrial archeology. I am also grateful to the Smithsonian's Robert Post, who doubles as the editor of *Technology and Culture*. Bob encouraged me to keep going by always telling me he liked my work—and he backed up his word by publishing two articles on the history of mechanization and mine safety at Lake Superior in *T&C*.

At the Hagley Museum and Library in Delaware, the professional staff was always receptive to this researcher's needs, and the grounds crew gave me a beautiful place to work overlooking the Brandywine. I want to thank Liz Kogan, who coordinated the advanced research fellowship program while I was there. I am particularly indebted to Ed Lurie, of Hagley and the University of Delaware, who read my early drafts, who shared my fascination with the autocratic mine boss and Harvard scientist Alexander Agassiz, and who steered my manuscript in the direction of Oxford University Press. Hagley's director, Glenn Porter, also read some early draft chapters and did some important steering of his own, by encouraging me to think about the life cycles of industries, and about how competition from new, young producers can threaten the viability of older ones.

Numerous individuals on the Keweenaw Peninsula have assisted me in important ways. Louis Koepel, who spent nearly his entire career with the Quincy Mining Company, quite literally opened up all the vaults filled with company records—thereby making a marvelous collection of business records available to historians for the first time. Theresa Spence and her staff at the Michigan Tech Archives always made my work days there both fruitful and pleasant. And Ed Koepel personally took the time to tutor me in the arrogance inherent in the old-time paternalism of the mining companies.

My colleagues within the Science, Technology and Society Program at Michigan Tech have offered both general encouragement and specific advice and help. Bruce Seely, Carol MacLennan, Susan Martin, and Terry Reynolds have suffered through my innumerable tales about life and work in the mines; they have never turned me away, but instead always graciously heard me out. Mark Rose taught me to pay more attention to a company's organizational form, and to the internal politics of decision making. Two other colleagues in particular must be acknowledged. Jack Martin, a sociologist, rendered invaluable assistance when it came to collecting and analyzing data regarding fatal mine accidents in the copper district. And Pat Martin, an archaeologist interested in copper mining before the Civil War, kept prodding me, and rightfully so, to remember that small and failed companies were a part of this story, and that I should not write a history only of big successes.

Michigan Tech students, most of them engineering majors, moved this work along by putting down their calculators and marching off to the MTU Archives or to local county courthouses. I would especially like to thank Jay Lukkarila, Pam Moravec, Joe Sarazin, Kevin Carpenter, Lauren Hutson, Karen Keeley, and Brian Taivalkoski. Each owns a piece of this work. The same can be said of Claire Chaput and Annette Coburn, staff with MTU's Department of Social Sciences. Whenever called upon, they helped push these words with dispatch through chips, circuits, wires, and fast printers.

Finally, a deep bow to my family, which has increased by two since all this began. In pursuit of this research and the final product you hold in your hands, I spent two summers away from home; then I moved us from Washington, D.C., to Houghton, Michigan; and then I dragged wife Rachel and our two native-born "Yoopers," David and Laura, back east to Delaware for my sabbatical leave at Hagley. ("Yoopers" are people from the "U.P."—the Upper Peninsula.) Rachel has "processed" more than her share of these words, while David and Laura have kept their games off the 'puter whenever Dad needed it. At different (and difficult) times, I have been absent, absent-minded, possessed, and cranky (such as when the p.c. ate my list of illustrations and then, its appetite whetted, my bibliography). Through it all, the rest of the family has weathered this experience far better than I have, and I always appreciated their good humor and support.

After acknowledging the help of others, I now free them from all liability regarding any erroneous or murky sections remaining in this book. Indeed, I alone am responsible for any such failings that a quick-witted reader might find. I would point out, however, that Jack, Pat, Ed, Glenn, Charlie, Bob, Carol, Susan, Terry, Mark, Bruce, et al., do bear some responsibility for the length of this tome, for they all helped to head me off in the several major directions that this book tries to take.

Houghton L.L.
May 1990

Contents

1. Company and Community: Copper Mining on Lake Superior, **3**

2. The Underground: Change and Continuity, **26**

3. The Surface: A Celebration of Steam Power, **41**

4. Men at Work: Agents, Captains, and Contract Miners, **58**

5. Change and Consensus: Miners, Managers, and Two-Man Drills, **78**

6. Change and Conflict: Nitro, Trolleys, and One-Man Drills, **93**

7. The Cost of Copper: One Man Per Week, **110**

8. Weeping Widows, Generous Juries, **126**

9. Homes on the Range, **142**

10. Cradle to Grave, **163**

11. The Social Safety Net: Patients, Paupers, and Pensioners, **181**

12. Social Control at Calumet and Hecla, **199**

13. Showdown: The Strike of 1913–14, **219**

14. Shutdown: Death of an Industry, **244**

Abbreviations Used in Notes, **269**
Notes, **271**
Bibliography, **301**
Index, **311**

Illustrations follow page 162

Cradle to Grave

1

Company and Community: Copper Mining on Lake Superior

> When it is considered that nearly the entire copper region is an unreclaimed wilderness, the miners' settlements appearing like mere dots on its surface, covered with a dense growth of trees, through which the explorer with difficulty forces a path; and that, except where the streams have worn their beds in the rock, or the hills terminate in bold and craggy ledges, the ground is covered with a thick carpet of mosses and lichens, effectually concealing every trace of veins,—it is surprising that such an amount of mineral wealth has been revealed within so short a time.
>
> J. W. Foster and J. D. Whitney,
> *Report on the Geology . . . of the
> Lake Superior Land District,* 1850

Social critics, particularly since the Second World War, have chided Americans for homogenizing our culture. Thanks largely to the automobile, the highway, and to standardized roadside architecture—as well as to the mass media, particularly television—we are looking, sounding, and acting more alike. Metropolitan spaces seem cut from the same cloth. They exhibit the same gas stations, convenience stores, malls, motels, and fast-food outlets.

Certainly, local color and regional variation have taken a beating in modern America. But all has not been lost. When travelers leave the interstates and head down the smaller highways and byways, they can still find curious survivals, pockets of culture rendered unique, colorful, and interesting by their setting and history. The Keweenaw Peninsula of Upper Michigan is such a place.

Located on the western end of Upper Michigan, the Keweenaw is a narrow, jagged finger of land about 70 miles long. It juts northeastward into Lake Superior, the greatest body of fresh water in the world. Three counties occupy the peninsula: Ontonagon is the base; Houghton the middle; and Keweenaw the tip. Today the region is still remote. Some 550 miles separate it from

Detroit; it is considerably nearer Milwaukee and Minneapolis, "only" 300 to 350 miles away. When travelers to the northwoods wander onto the Keweenaw, it is common (and understandable) for them to ask: "What do people do here for a living?"

Many of them work for Michigan Technological University, a school of 6500 students that started out, more than a century ago, as the Michigan College of Mines. Michigan Tech, located in the small city of Houghton, is the biggest employer in Houghton County. Near its campus, modern America is quite evident, having finally arrived about 1980. A modest mall, anchored by a K-Mart and a J.C. Penneys, contains the obligatory mini-theatres and an oversupply of shoe stores. Fast-food restaurants peddle their nationally advertised hamburgers, pizzas, or chicken. Houghton even has a few elevators, parking decks, and a host of ranch-style houses in suburban developments. But aside from the Michigan Tech/Houghton anomaly, the rest of the Keweenaw looks as though it has missed much of the twentieth century, as though time has stood still since 1910.

Curiosities prevail that set this place apart from others. Restaurant booths are fashioned from real oak, not Formica, and menus feature the Cornish pasty and the Finnish pannukakku, or pancake. The two-lane highway down the center spine of the Keweenaw threads its way through forests of birch, spruce, cedar, pine, and maple, occasionally through low-lying swamps, and every 5 to 10 miles or so through a small village or ghost town. At the merest of hamlets, Gay, Michigan, the only active business seems to be a tavern (the "Gay Bar," of course), but nearby towers a massive, reinforced concrete smokestack, a marker of a dead industry. Many other settlements are marked by these abandoned smokestacks; by empty industrial buildings of stone, their roofs caved in by too many winters of 200 inches of snow; by modest, uniform dwellings; and by piles of rock, lying against the base of a bluff, or rising as strange hillocks on otherwise level terrain. In winter, warm, moist air leaves a deep hole next to the road (an old mine shaft) and becomes a tall plume of steam standing up in the sky.

Other roads follow the Lake Superior coastline, which is often rugged, sometimes treacherous. They pass the villages that once had been vital ports of call and supply centers for the peninsula: Copper Harbor, Eagle Harbor, Eagle River, and Ontonagon. Still other roads run in the margin of the Keweenaw's inland lakes. Here the scenery, as it is along Lake Superior, is often gorgeous. The forest line meets the shoreline; occasionally a quaint nineteenth-century town sits next to the water.

Then more smokestacks and ramshackle industrial buildings come into view. Occupying the shore, they disturb the otherwise picturesque scene. Monolithic concrete foundations, 10 to 20 feet tall, litter the lakeside. A strange, flat beach—usually composed of a coarse, gray-black sand—spreads out over acres and acres. It is an unnatural, unearthly landscape. Nothing grows on the black sand, but out in the water, worrisome, cancerous tumors grow on the fish.[1]

The Keweenaw is not the lush wilderness that much of it appears to be.

Human industry has applied a hand—a heavy hand—to its land and water. It has its peculiar history that explains its present condition. It is the history of a mining district having unique copper deposits, which were first worked by primitive technologies, and then by very sophisticated ones—which were first worked by aboriginal miners and then, thousands of years later, by some of the earliest, large-scale mining companies in America.

What made the Keweenaw so unusual—and what provided its greatest allure over the centuries—was the fact that it held the world's largest deposit of native copper. Here, almost pure copper existed naturally in its metallic state, unalloyed with other elements. Some of this copper came in the form of tiny specks or flakes, almost indiscernible in the rock matrix that held them. Some came in massive pieces, weighing up to several hundred tons. And some of the metal, called "float" copper, was found right on top of the ground, where it had been dropped off by glaciers.

The rock that forms the Keweenaw Peninusula is extremely old. From 880 to 1,280 million years ago, some 400 basaltic lava flows overtopped the peninsula's original sandstone base. These flows, which reached a cumulative depth of thousands of feet, were interbedded with 20 to 30 layers of sedimentary rock. The lava and sedimentary beds gave rise to two distinct types of copper-bearing lodes: amygdaloids and conglomerates. Also, cracks or fissures in these rock beds allowed for the formation of mineralized veins, which tended to be charged with large pieces of native copper, called mass copper. In all cases, the deposition of copper depended upon the permeability of the rock and upon the existence of cracks or voids to receive the copper.

The successive volcanic flows spread out over large areas. Each lava flow included immense quantities of hot gas. The gas collected into bubbles that migrated toward the top of the flow. As the lava cooled, became more viscous, and finally solidified, it captured the gas bubbles, which became cavities or "vesicles" in the top part of the flow. The denser rock at the bottom lacked vesicles and was largely impermeable to later copper-bearing solutions. While it remained unmineralized, the solutions flowed into and occupied the voids in the less-dense portions of the lava flows. Geologists termed the permeable rock "amygdaloid," a term taken from the Greek word meaning "almond"—which described the shape of the cavities. The red metal gleaned from such rock was known as amygdaloid copper.

Volcanic activity was not constant during the formation of the Keweenaw geological series. During some long interruptions, sedimentary rock formed over earlier lava flows as erosion took place along the Lake Superior basin. Conglomerate copper came from these deposits, from "conglomerations" of boulders, sand, water-worn pebbles, and sandstone fragments. The interstices or open spaces in conglomerate rock—like the gas-bubble voids in amygdaloid rock—later could receive mineralizing, copper-rich solutions.

After their original formation, the once-horizontal amygdaloid and conglomerate rock formations tilted. They bent into a basin, with one upper rim, about 100 miles long, running from the northern tip of the Keweenaw southwestward into Wisconsin. The rock strata included in this portion of the rim angled into the ground at from 15 to 80 degrees. The basin dipped beneath Lake Superior, and its opposite rim appeared in Minnesota, along the Canadian shore, and on Isle Royale, which lay some 45 miles northwest of the Keweenaw. All this movement created longitudinal and cross-fractures on the Keweenaw, creating still more voids in the rock.

Either during or following this tilting, copper-bearing solutions mineralized select rock formations. These solutions probably originated deep underground. They were expelled under tremendous pressures and forced to flow upward through porous amygdaloid and conglomerate rock and through fractures or fissures. It was a distinct geological novelty that the mineralizing solutions, whatever their exact constituency, apparently reacted with ferric compounds in the rock in such a way that metallic copper—and not copper compounds—precipitated out.[2]

The copper-rich solutions did not flow uniformly throughout the Keweenaw rock formations, so the native copper left behind was unevenly distributed. Much porous rock remained barren of copper because it was bounded by an impermeable layer of rock, or because no fissure channeled solution into it. Many rock beds were reached by solution, but were poorly charged with copper because they were too large and permeable. If no barriers existed to stop and contain the solution, it spread throughout too great a volume of rock, meaning that the copper deposit was not concentrated enough to support commercial mining. Of the hundreds of lodes and fissures on the Keweenaw, very few presented real opportunities for mining, and even the best ones tended to be copper-rich only for short distances, rather than along their entire length.

Nature, in the form of glaciation, first "mined" Keweenaw copper. Glaciers laid bare the upturned edges of the lodes, where they outcropped at the surface. They snagged pieces of mass copper, transported them across the ground, and left them behind as tantalizing pieces of float copper—metallic copper that could be had without resort to mining or smelting. Indians availed themselves of this copper, starting at least 5,000 years ago.

The Lake Superior basin provided almost all copper used by prehistoric American Indians in the eastern portion of the United States.[3] American Indians did not merely glean the nuggets and masses of copper found on the forest floor and along beaches. They actually mined for it, using primitive tools to sink pits or even shafts into the ground. They used hammerstones— hard, dense igneous stones—which they bashed against the softer bedrock. They laboriously chipped away at the rock matrix, until they exposed the corners or jagged appendages of a piece of mass copper. Then they hammered on these points, until the copper yielded and tore away from the mass. Later, other Native Americans traded for the copper and hammered the malleable metal into tools and ornaments.

When early European explorers and missionaries arrived in the Lake Superior region in the 1600s, they learned of the Keweenaw's legendary deposits of native copper. The French launched several expeditions to the region between 1660 and 1740, but made no significant progress in tapping its resources. After the British took control of French Canada in 1763, they too demonstrated an interest in Lake copper. In 1766 an expedition to the banks of the Ontonagon River confirmed the existence of a piece of float copper weighing nearly two tons. The band of explorers failed to remove the "Ontonagon Boulder." (Indeed, it remained in situ for another three-quarters of a century, before it was taken to Washington, where it now sits in the Smithsonian's Museum of Natural History.) But the British expedition of 1766, coupled with the work of a second party of miners sent to the Ontonagon River in the early 1770s, did a great deal to publicize this wondrous copper boulder.[4] For decades it served as the Keweenaw Peninsula's main attraction and claim to fame.

The early French and British attempts to exploit Lake Superior copper were doomed to failure. The exploring and mining parties, understaffed and underfunded, lacked the transportation and mining technologies needed to launch an industry. Nearly two centuries passed between the time Europeans first searched for Lake copper and the time when Americans first turned a dollar by wrenching native copper from the ground, shipping it to the East Coast, smelting it, and selling it.

The chain of events that led directly to the successful taking of Lake copper began about 1820, before Michigan was a state. Territorial governor Lewis Cass, along with Henry Schoolcraft, mounted an expedition to the Keweenaw that included a visit to the Ontonagon Boulder. The region seemed so promising that Schoolcraft recommended that the U.S. government mine the deposits. The government declined, but in 1826 concluded a treaty with the Ojibwa that allowed for federal explorations for minerals on Indian lands. In 1831, Schoolcraft launched another Keweenaw excursion; this time a young geologist named Douglass Houghton accompanied him.[5]

Michigan became a state in 1837, after the federal government had settled a border dispute between it and neighboring Ohio. Both new states had wanted the area around Toledo. When Ohio received that piece of territory, Michigan was due some compensation—so it received the Upper Peninsula, even though it nowhere touched the state's Lower Peninsula. Douglass Houghton became Michigan's first state geologist, and serving in the role, he became an important contributor to the opening up of the copper district, ten years after he had first visited it.

Douglass Houghton was a geologist with a practical bent. He scoured Michigan for natural resources that could be exploited and turned into wealth and jobs. In 1840, he returned to the Keweenaw, and the next year he reported on its minerals and economic potential to the state legislature. Houghton's report was only guardedly optimistic. He was bothered by the fact that this was native copper—and no successful mining industry had ever been based on such a deposit. In fact, in other parts of the world, where native copper had been found in small quantities, its existence had been taken as contraindicative of mining potential. Geologists took it as a sign that an aged deposit had weathered and broken down, had lost its mineral value. Finally, Houghton feared that any speculative rush to claim the Keweenaw's copper would produce far more losers than winners:

> While I am fully satisfied that the mineral district of our state will prove a source of eventual and steadily increasing wealth to our people, I cannot fail to have before me the fear that it may prove the ruin of hundreds of adventurers, who will visit it with expectations never to be realized. . . . I would by no means desire to throw obstacles in the way of those who might wish to engage in the business of mining this ore . . . , but I would simply caution those persons who would engage in the business in the hope of accumulating wealth suddenly and without patient industry and capital, to look closely before the step is taken, which will most certainly end in disappointment and ruin.[6]

Douglass Houghton doubted that a profitable mining industry soon could be founded on the Keweenaw. Nevertheless, his confirmation of the copper deposits, coupled with the newsworthy, belated removal of the Ontonagon Boulder, generated a great deal of interest in the copper's economic potential. The federal government negotiated a treaty with the Ojibwa, whereby the Indians ceded to the United States their lands in the western Upper Peninsula, including the copper district. In that same year, 1843, the U.S. Secretary of War established a land office at Copper Harbor, far out on the Keweenaw's tip. The land office initially offered lease permits to mine for copper on the Keweenaw and on Isle Royale; soon it switched to outright sales of mineral lands. In any event, by the mid-1840s the first real mine rush in American history was on.

Participants in the California gold rush, which came a half-decade after the rush to Lake Superior, could hope to strike it rich as lone prospectors equipped with a modest array of tools, because the taking of relatively little gold provided them with a sizable fortune. Such was not the case with Michigan copper, which did not have the value of precious gold. Copper was used to sheath the hulls of wooden ships and to roof buildings. It went into pots and pans. Alloyed with zinc to make brass, or with tin to make bronze, copper went into buttons, candlesticks, machinery bearings, weapons, decorative items, and hardware. To make much money from this versatile metal, a producer had to take large quantities to market. And just getting it out of the ground from the Michigan wilderness required considerable expenditures for wages, tools, equipment, forest clearing, shaft-sinking, and road and building construction.

The journalist Horace Greeley, noted for his advisory "Go West, young man," did so himself in 1847–48, when he visited the Lake copper district. Greeley was not a romantic prospector, but a shareholder and director of the Pennsylvania Mining Company. Incorporated mining companies, not individuals or partnerships, developed this region. It took at least $50,000 in capital investment, Greeley reasoned early on, to try to open a copper mine.[7] It also took patience and luck. Investing in these mines was like purchasing a ticket in a "subterranean lottery." Even if a company made no mistakes—if it secured a copper-rich parcel of land and had adequate skill and capital to develop it—it did remarkably well if it ran in the black and paid dividends within five years. And most companies never paid any dividends at all based on copper production, because they were speculative "paper" companies or because they went bust trying to develop an unprofitable property.

Investors from scattered midwestern locales and from Detroit, Pittsburgh, and New York did the most to promote the development of the copper district during its formative, high-risk years from the 1840s through the end of the Civil War. At least 300 ventures were launched over this time, including some 94 joint-stock companies. Fifty-two of these 94 companies produced no copper by 1865, and another dozen produced fewer than 30 tons each. Altogether, two-thirds of the companies achieved little or no production, and in the first 20 years of mining, only 8 paid any dividends. Investing about $25 million in these 94 firms had realized a dividend return of only $5.6 million.[8]

The nascent industry consumed over four times more money than it paid out. As Douglass Houghton had feared, the failure rate was high. Still, the Lake Superior copper district was deemed an overall success. It carried a part of America's western frontier onto the Keweenaw, and established a substantial population there (over 18,000 by 1865). Its record of copper production rose impressively, from 51 tons in 1848 to 7,200 tons annually by the close of the Civil War. From its opening through 1865, the Keweenaw, with a small contribution from Isle Royale, accounted for three-fourths of America's copper production. Its production levels dwarfed those achieved by other early copper-producing districts in Connecticut, New Jersey, Vermont, and Tennessee.[9]

The notoriety of two successful companies encouraged investors to keep putting capital into the industry, despite its high failure rate. In the 1840s and early 1850s, mining companies largely ignored amygdaloid and conglomerate deposits, seeing them as too hard to work and as too poor in copper. They searched instead for fissure veins charged with mass copper. Two mass copper mines in particular struck it rich: the Cliff, located on the northern end of the Keweenaw, and then Minesota [sic], located at the base of the peninsula.

In 1844 and 1845, the Pittsburgh and Boston Mining Company tried unsuccessfully to develop mines at two properties near Copper Harbor. Then it moved south and opened the Cliff mine over an encouraging fissure deposit. The Cliff indeed yielded impressive quantities of native copper. Production reached more than a million pounds per year in 1849, when the company paid out $60,000 in dividends, the first to be received by any investors in Lake Superior copper. Before attaining this success, the firm had spent six years in developing various properties; stockholders had paid in $110,000 in assessments to fund the work; and the company had plowed another $250,000 of early revenues back into its development effort. But before the company was through, it handsomely rewarded its stockholders with dividends totaling $2.5 million.[10]

The second major success on the Lake was the Minesota mine. Like many early ventures, the Minesota Mining Company searched for a profitable fissure deposit on property previously worked by Native Americans. The mining companies wisely used the ancient diggings as signposts leading them to copper. For the Minesota mine, started in 1847, the discovery and removal of one six-ton mass of copper in 1848 came easy—the mine's agent discovered it at the bottom of a 26-foot deep shaft, supported on some timber cribbing. American Indian miners had robbed the mass of its corners and appendages before leaving the rest behind. This welcome and sizable discovery soon paled, however, in comparison with the Minesota's total production, which reached nearly four million pounds per year after a decade's operation.[11] To launch the mine, stockholders paid in $366,000 in assessments. By the end of the Civil War, they had received $1.8 million in dividends.

In the 1850s the Cliff and Minesota mines served as important and encouraging models. Profits could be made on the Lake, if a company had adequate capital investment to sustain development, coupled with a modicum of patience. But at the same time, these two mines were false harbingers of the district's future. They worked the mass copper found in fissure veins, but such veins were not well distributed along the Keweenaw, and even the best of them tended to peter out suddenly. The Pittsburgh and Boston Company closed its Cliff mine in 1870, after its production fell off by two-thirds in only two years. The Minesota mine, too, was basically defunct by 1870. From 1855 through 1864, the mine averaged 2.8 million pounds of refined copper per year. From 1865 through 1870, production fell to only 360,000 pounds annually.

The long-lived mines, and the majority of the successful ones, would work not fissure deposits but the district's amygdaloid or conglomerate lodes. They would not depend on discoveries of mass copper to carry their production. Instead, they would mine and mill vast tonnages of rock charged with small flecks of copper that made up as little as 1 percent of the whole.

The Cliff and Minesota mines steered early development to the opposite ends of the Keweenaw, first north to Keweenaw County and then south to Ontonagon County. But after the mid-1850s, the middle of the Keweenaw—Houghton County—proved home to the best lodes, the richest companies, and the largest settlements. As the fissure mines waned, the fortunes of other mines rose. Over 95 percent of the district's total native copper production came from the stretch of mineral range running 15 miles north and 10 miles south of Houghton County's Portage Lake.[12]

Close to the shores of Portage Lake, and near the new towns of Houghton and Hancock, a cluster of mines formed between the mid-1850s and the early 1870s to work amygdaloid lodes. South of Houghton stood the Huron, Isle Royale, and Atlantic mines, plus several lesser operations. Just north of Hancock, the Quincy, Pewabic, and Franklin mines stood in a row atop a single lode. Ten to 15 miles northeast of these operations, another string of important mines developed between the late 1860s and the early 1880s. These included the Osceola, Tamarack, and Calumet and Hecla mines in the northern end of Houghton County, and the Allouez and Ahmeek mines, just over the border into Keweenaw County. These mines worked several amygdaloid lodes, plus the Allouez and Calumet conglomerates. About 10 miles south of

Portage Lake, a final major group of Houghton County mines opened at the turn of the century—the Baltic, Trimountain, and Champion mines, which all exploited the Baltic amygdaloid lode.

The companies owning property along the richest stretch of the mineral range were advantaged on a number of counts. Their lodes carried commercial quantities of copper near the surface. By tapping it, they could use revenues from early copper sales to help pay for developing the mine. This meant that investors did not have to risk so much capital at the outset. But just as important, the lodes ran deep and supported full-bore mining for decades. Several mines exceeded five decades of operation, and a few nearly reached the century mark. In addition to copper, the companies here were well endowed with another important natural resource: water.

On the mining end of metals production, water was a distinct nuisance. After gravity carried water to a mine's bottom, some mechanical contrivance had to lift it up and out of the way. But in the Lake Superior copper district, mining was only the first of three distinct steps in the production of marketable copper. The second and third steps, milling and smelting, both used water, either for processing or transporting material. In the center of the Keweenaw, close to the largest mines, Torch Lake connected to Portage Lake, which connected to Lake Superior. Companies filled these shorelines with docks, stamp mills, and copper smelters.

The Lake mines yielded various products. The early fissure mines extracted large pieces of mass copper, as well as smaller, fist-sized pieces known as "barrel copper," (named after their shipping container). Mass and barrel copper required no milling. Since they were essentially pure, they went straight into a furnace to be melted down.

The mines working amygdaloid and conglomerate lodes encountered some mass and barrel copper, but these formed only a small fraction of their production. The vast bulk of their output came from "copper rock." (Although geologists argued for using the term "copper ore," local mining men ignored the scientists and embraced the term "copper rock." They preferred this term, because it emphasized that the included copper was not a compound, but was in its unalloyed, metallic state.) Even at one of the wealthiest mines, Quincy, this copper rock was only about 2 percent copper and 98 percent waste. This material could not go directly from mine to smelter. It would have choked a furnace, producing too much slag and too little metal. Before smelting, the rock headed to a stamp mill that separated and concentrated its copper.

Heavy stamping machines first broke and abraded the copper rock, reducing it to the size of a peppercorn or smaller. Mechanically breaking the rock liberated its copper particles. A steady stream of water flushed the copper and rock out of the stamps and onto other types of esoteric mill equipment, such as hydraulic separators, jigs, wash tables, and slime buddles.

This machinery, much of it "homegrown" to work Lake Superior's peculiar copper deposits, used the difference in the specific gravities of copper and rock to separate the two materials.[13] In the jigs, for instance, plungers agitated a watery suspension of copper and rock particles over a sieve. When the agitation stopped, and the water drained out of the box, the materials settled out on the sieve plate, with the copper on the bottom. Mill hands skimmed off the rock, then collected the copper. At the buddles, a slimy mixture of minuscule particles of rock and copper flowed out very slowly from the center of a large, rotating disc. The heavier copper stayed on the disc; water carried the lighter rock particles across the disc's surface and then tailed them over its edge.

Many steps in the milling process used water to help separate materials into three categories: copper concentrate (locally called "mineral"), which was processed no further because it was suitable for smelting; middlings, a copper and rock mixture that required additional separation; and tailings, or waste rock, that was run out of the mill and into its nearby lake. Day after day, the stamp sand tailings flowed out of the mills in vast quantities, filling lake bottoms and redrawing shorelines. Since the stamp rock that arrived at the mills was far richer in rock than in copper, the mills always tailed out more in waste sands than they produced in the form of concentrates, which at 60 to 80 percent copper were ready to be smelted.

Some early attempts to smelt the mines' products on the Keweenaw and on Isle Royale failed, so companies initially shipped their mass copper and mineral to faraway smelters in Boston or Baltimore. There, furnacemen solved one technical problem—how to get large pieces of mass copper into a reverberatory furnace—by developing one with a removable top. As the industry expanded, new smelters located closer to the copper source. They sprang up in Pittsburgh, Cleveland, and Detroit, and by as early as 1860, on the shores of Portage Lake.

The key structure at any Lake Superior smelter was a reverberatory furnace building, rendered distinctive by the tall brick smokestack penetrating the roof near each of its corners. Inside sat four furnaces built of fireclay and brick. At one end of each furnace the heat source, coal, burned in a fire box. A low bridge wall kept the fuel from mixing with and chemically contaminating the mineral and mass, which rested in a shallow hearth. Access doors and a flue and chimney stood at the other end of the furnace. Heat and flame swept over the bridge wall and reverberated along the hearth before going up the stack.

Smelting native copper was basically a melting and casting operation; it required little refining, due to the purity of the metal put in the furnace. It was a batch process; men produced a furnaceful of copper at a time. After the copper reached its melting point, furnacemen opened an access door and skimmed off the slag of molten rock. They splashed the copper around with a paddlelike "rabble." Rabbling mixed air into the melt to oxidize any impurities. Then they eliminated excess oxygen by plunging hardwood poles into the copper. The rapid combustion of the poles agitated the melt, and the wood's carbon combined with oxygen to produce gases that passed up the stack. Furnacemen again skimmed off the slag. Then they ladled the copper into molds to produce ingots and cakes of various sizes and shapes for different markets.[14]

Every mine of consequence working an amygdaloid or conglomerate lode built its own stamp mill. A mill was nearly obligatory, just as soon as a company moved from an exploratory phase into actual production. But while mills were ubiquitous features on the Keweenaw landscape, smelters remained relatively few. Only the firms running the largest mines, including Calumet and Hecla, Quincy, and Copper Range, produced enough copper to justify the cost of building and operating their own smelter. Most companies sent their concentrates to these firms for custom smelting, or they sent them to one of the region's few independent smelters, such as the one operated by the Detroit and Lake Superior Copper Company.

The region's key smelters occupied waterfront sites on Portage Lake and Torch Lake. They located near the largest stamp mills, in order to minimize the distance that mineral concentrates had to be shipped. Also, the smelter docks could receive, straight off Great Lakes freighters, incoming supplies such as coal. The same docks also stockpiled copper ingot, until a freighter drew alongside to take on cargo. From late autumn to late spring, the wait for the arrival of the steamers was considerable, due to the storms and ice on the upper Great Lakes that kept the ships off Superior for nearly five months each year. Companies mined, milled, and smelted copper year round, but they shipped it to market only from about late April to mid-November.

By the time winter receded and navigation commenced, the residents of the Keweenaw were more than ready for it. Tired of looking at piles of snow and of being confined indoors, they welcomed the arrival of "robins with gay plumage and song" and "the boats with bright colors." In the spring, citizens of Houghton and Hancock massed on the opposite sides of Portage Lake to shout "hurrah!" at the sight of the first ship's prow, coming to deliver much needed supplies and to take away their copper. They rushed shouting to the docks, waving hats and handkerchiefs, and indulging in "a thousand extravagent antics." All the steam whistles in the towns blew in celebration. The ships were a physical tie to "a life which has been withheld from us for a long time, and we love them as among the dearest objects." The opening of navigation was an emotional event: "To those who have never been shut out from the world half of the time, all these manifestations of joy and delight may seem wild and absurd, but let them take our place only one season, and see how wild they will be."[15]

In the winter, residents of the Keweenaw endured months of continuous below-freezing temperatures. They scooped and shoveled snow, prayed that their woodpile would last for the duration, and fought off bouts of depression or cabin fever, often with the help of alcohol. In the too-short summer, they suffered blood-sucking mosquitoes and, even worse, black flies. At least mosquitoes were big enough to be seen, and they buzzed a warning while circling their pray. Black flies, so tiny as to be nearly invisible, buzzed not at all. Preferring to do their damage while under cover (up a sleeve, beneath a collar, or burried in a person's hair), female black flies bored holes in their victims. The blood often flowed freely, even after the gorged guest took its leave. A red rivulet, trickling down the wrist, ankle or neck, signaled that an unwanted visitor had come and gone.

Over time, life on the Keweenaw became decidedly less harsh, isolated, and bleak. Railroads offered an alternative connection to the outside world, one that operated throughout the winter months. Stores, schools, and churches multiplied. Communities erected theatres and opera houses. In the summer, streetcars and boats carried picnickers out to parks. Circuses came to town, and the sons and daughters of miners, excitedly chattering in more than a dozen languages while they lined the streets, thoroughly enjoyed the gross incongruity of a parade of pachyderms and camels in northern Michigan.

The mining frontier first pushed back the wilderness, and then after the Civil War the rough-and-tumble frontier settlements grew more urban-like and refined. They came to wear the appearance of commercial and industrial success. They looked like permanent fixtures on the landscape—at least some of them did, such as Houghton, Hancock, Calumet, and Laurium. But thriving mines, surrounded by bustling, growing communities, were not always found up and down the Keweenaw. Failed or failing companies, attended by small, struggling communities, remained the rule, even late in the nineteenth century and early into the twentieth.

From the 1840s through the 1920s, of the hundreds of ancient lava flows uncovered and explored in this region, only six amygdaloid lodes developed into major producers: the Kearsarge, Baltic, Pewabic, Osceola, Isle Royale, and Atlantic. Of the region's conglomerate lodes, only two—the Calumet and the Allouez—supported intensive, profitable mining.[16] Because so few lodes were well charged with copper, it followed that at any given time, even if 30 companies operated, the region did well if half a dozen ran deep in the black. Stockholders—and indeed, workers, too—were lucky if they had tied into one of the dozen or so most fortunate companies. Each of these accounted for 2 to 10 percent of the region's 7.5 billion pounds of copper produced through 1925, and for similar shares of its $300 million paid out as dividends. And stockholders were particularly blessed if they had invested early in the giant success, Calumet and Hecla, one of the world's foremost mines. Sited atop the Calumet conglomerate lode—the Keweenaw's richest copper deposit by far—C&H alone accounted for 43 percent of all Lake copper produced through 1925, and for half of all dividends.[17]

The historical importance of this one company to the post–Civil War development of the Copper Country can hardly be overstated. Take away its highly profitable performance, and the district only broke even. The investments squandered on failed mines would have cancelled out the dividends paid by the successful ones. Calumet and Hecla made the district a clear winner and a producer of wealth.

Calumet and Hecla dominated the Lake Superior copper industry. Until the rise of Copper Range Consolidated in the early twentieth century, C&H had no legitimate rivals, and it ruled its region with a haughty self-assuredness that the only way to mine for copper, or to run a mining community, was the C&H way. It became the principle magnet for immigrants. It set the standards for wages, for company paternalism, for technologies. And because Boston investors had launched C&H, after 1870 that eastern city became the most important home of money still to be invested in the Lake Superior copper district—and of money taken out of that district in dividends. As one wag wrote in 1928: "The four greatest words in the annals of New England are: Concord and Lexington, and Calumet and Hecla. The first two made New England history and the last two made New England fortunes."[18]

Exercising the advantage of hindsight, it is easy to tell where the mining entrepreneurs should have gone to strike it rich; easy to tell which properties they should have left alone, to save financial ruin. But in real time, as the history of the region unfolded year by year, the participants—miners and capitalists alike—had no way of divining its future. They had no way of peering at the outcropping of a lode in a forest and of knowing if, in ten years, the site would be scene to a booming, vital community or a ghost town. At the very start of operations, future successes and future failures looked exactly the same.

The Norwich mine in Ontonagon County was one of the failures. To put it more accurately, Norwich was several of the failures all rolled up into one, because different parties tried to make a go of this mine, and none succeeded. Norwich was one of the most devilish types of properties. It produced enough copper to be encouraging, but never enough to pay. It teased investors, but never rewarded them.

Between 1849 and the end of 1854, the American Mining Company, chartered in Vermont, launched at least 15 to 20 new mines in the diverse locales of Cuba, North Carolina, Pennsylvania, and Michigan. Five of these were copper mines on the Keweenaw, and the most promising was the Norwich. Initial exploration of its site took place in 1846, when a small band of prospectors, traversing lands near the west branch of the Ontonagon River, discovered a copper-bearing lode. From it, they carried away two samples of mass copper weighing 12 and 110 pounds.

Between 1846 and 1850, very little work went forth at the Norwich location. A small crew erected a long hut for their living quarters and sank a single exploratory shaft to a depth of 60 feet. Then in 1850, the American Mining Company decided to test the site more fully. It organized the Norwich Mine as one of its interests and issued 10,000 shares of stock in the venture. As was usually the case in capitalizing Michigan copper mines, the takers of the American Mining Company's Norwich stock did not immediately pay in full the $5 per share. Instead, the company periodically assessed stockholders 25 to 50 cents per share to support ongoing development.

Opening the Norwich was particularly difficult, because it was even more remote and isolated than most mines on the Keweenaw. Men, trudging through 12 miles of dense forest, carried in tools and supplies on their backs. The mine site itself—dominated by a 400-foot-tall bluff—was a hard one to clear and to work. A visitor to Norwich remarked that its pioneers had "met and surmounted difficulties and dangers . . . which it is difficult for those at a distance, and residents of a 'settled country' to realize."

Between 1853 and 1855, employment at the Norwich peaked at 120 to 140 men. This work force cleared 100 acres of land at the foot of the bluff and put

65 acres into cultivation. They built a 10½-mile-long road from the mine to a landing on the Ontonagon River. Workers erected at least 20 to 25 buildings through 1855. Included were two large boarding houses for single men, at least 10 log cabins (presumably for married men), and a finer, larger dwelling for the mine's superintendent. Several other buildings stood next to the houses at the base of the bluff: a company office, a storehouse, barn and powderhouse; a company store; a change or wash house for the men; blacksmith and carpenter shops, along with a sawmill; a couple of kilnhouses for dressing and breaking copper rock; and a stamp mill and steam-engine house.

Far more important than these surface improvements was the work done underground. By 1855, mining teams had driven three horizontal tunnels, or adits, into the side of the bluff. They sank five shafts a total of 665 feet. Branching off from the shafts, and tunneling through the copper lode, miners drifted a distance of 1,480 feet. This work produced some 220 tons of copper mineral and mass copper, which the company shipped to Detroit for smelting.

The company sold its ingot copper and plowed the proceeds back into the mine. It assessed the entire $5 on 10,000 shares of stock. But after calling in and using up all its investment capital, the American Mining Company could not get the Norwich mine to support itself. Its operation continued to produce losses and not profits, so the original Norwich went under early in 1855.[19]

Some investors, however, still hoped that with an infusion of new capital, and with another burst of development work, the mine would break through into richer ground. They resurrected the operation by incorporating a new Norwich Mining Company under the laws of Michigan. This company purchased the Norwich property from the American Mining Company and paid off its debts, including back wages owed to miners from throughout the previous winter. The new company improved the mine's technologies and pushed production, to see if the mine could be made to pay.

This effort was doomed by the same problem that had driven the earlier company into closure: the Norwich lode, in places only a foot wide, simply was not rich or large enough. In 1856, the R. B. Dun & Co. credit-rating firm had said of the revitalized Norwich: "This mine is good." Two years later, in 1858, it said: "This comp'y is abt. closing up. They have not been successful in mining operations. It would not be safe to sell them many goods on long time." Indeed, after producing about 280 tons of mineral and mass since 1855, in 1858 the Norwich Mining Company suspended work.[20] To get some income from the property, without risking any more of its own capital, the company leased the mine to tributors. In return for providing the use of its property, the company received one-fourth of all copper the tributors produced. That did not amount to much, however, for the tributors, too, failed to make a go of it. But the Norwich was not done yet.

With the onset of the Civil War, copper prices rose, and this encouraged the reopening of many dormant Lake Superior mines. In 1863, after another corporate reorganization, the Norwich was unwatered and work started anew. Miners exploited new lodes on the property, as well as the original one, and in 1865 the mine produced 92,000 pounds of ingot copper.[21] But only wartime, inflationary copper prices sustained the operation. When the war ended, the Norwich Mining Company again stopped mining on the company account and let the mine out to tributors. They, too, soon gave up.

In 1885, the Norwich Mining Company failed to renew its incorporation charter, so it passed out of existence. The property, however, still seduced other investors. Between 1885 and 1917, three more companies—the Essex, the Copper Crown, and the Cass—tried (and failed) to make the mine pay. Eventually, the abandoned buildings tumbled down and the forest regenerated itself. Its floor, however, is still littered with archaelogical remains, the leavings of unrewarded enthusiasm. Littered with abandoned shafts and adits. With exploratory trenches. With cellar pits, stamp shoes, blacksmith's equipment, and pieces of cast-iron stoves.

Failed companies did not have a monopoly on hard times, bad luck, or poor decisions. Even the most successful companies often had troubled starts. The Quincy Mining Company, organized in 1846, spent a decade searching in vain for copper on a single square mile of land it owned just above Portage Lake. Ten years of trenching, driving adits and sinking exploratory shafts produced several false starts and no profits. Stockholders became very discouraged when the first $42,000 invested in the firm seemed utterly wasted. Then Quincy's neighbor, the Pewabic mine, uncovered the Pewabic amygdaloid lode. Quincy traced the strike of the lode onto its property—and suddenly its future was secured.[22] Between 1856 and 1860 the company levied heavy assessments on its stock, totaling $156,000, to fund development work on the Pewabic lode. This investment paid off handsomely. The company paid its first dividends in 1861, and except for two years at the end of the Civil War, it continued to pay dividends every year through 1920.

Quincy was extremely fortunate. A paying lode first discovered on another company's property continued onto its own, with no diminishing of copper values. Other companies were not so lucky, particularly those in the vicinity of Calumet and Hecla. Because of the great Calumet conglomerate lode, C&H returned fabulous profits. But other firms just northeast and southwest of Calumet and Hecla failed to develop paying mines on the same lode. A 4,200-foot stretch along the outcropping of the conglomerate was prime for mining—and Calumet and Hecla happened to own all of it.

Northeast of C&H, the Schoolcraft and then the Centennial mines turned no profit on the conglomerate. Centennial did extensive work on the lode over 29 years, sinking seven shafts, one to a depth of 3,200 feet, but did not find commercial mineralization. To the southwest, the Osceola mine—which became the region's third or fourth largest producer in the nineteenth century—initially failed just the same.

Organized in 1873, Osceola worked only the Calumet conglomerate lode for its first four or five years. But the company became successful and profitable only after it wrote off its original works and commenced operations on the nearby Osceola amygdaloid lode. Between 1873 and 1880, Osceola produced 12.5 million pounds of refined copper from the conglomerate. Between 1879 and 1920, its amygdaloid lode yielded 191 million pounds of ingot copper.[23]

Even the great Calumet and Hecla Mining Company experienced myriad problems at the start. In his annual report for 1899, Michigan's commissioner of mineral statistics allowed himself a colorful bit of hyperbole when he wrote that C&H "sprang, fully panoplied, like Minerva from the head of Jove, to a commanding position among the wonderous treasure vaults of man." In fact, C&H stumbled to its commanding position in the mining industry. It faltered in the face of several challenges early on, but succeeded in spite of itself.

Finding stable, competent, and compatible leadership proved difficult—and C&H succeeded only after a second group of investors and experts took over from the mine's original backers. The earliest attempt to open the conglomerate lode left the mine site in a shambles that impeded subsequent work. Major errors were made in thinking that the conglomerate rock was so rich that it could be smelted directly, without first milling or concentrating it. Then, when the need for milling became obvious, capital was wasted on a poorly sited mill that had to be abandoned, and on the novel technology of crushing rock with rollers, instead of using tried and true stamps. For a period of about five years, it seemed that nothing could be done right. The mine, quite literally, had trouble getting on track: it was embarrassing, for instance, to discover that the gauge of the mine's freshly-built railroad did not match the gauge of its just-arrived steam locomotive.[24]

Calumet and Hecla's origins traced back to Edwin Hulbert, a surveyor who was always on the lookout for new deposits of Keweenaw copper. In 1858, as Hulbert ran the line for a new road from Portage Lake out to the tip of the Keweenaw, he discovered a curious pit that contained a cache of copper in the bottom. He expected to find evidence that the pit was an old Indian mineworks, situated atop a lode that was right underfoot. He found no tools to confirm his idea, but, intrigued by the site, he returned to it in 1859, and in 1860 he started purchasing mineral lands in the area.

Hulbert sought the outside investment capital needed to acquire, explore, and develop all the desired properties. The onset of the Civil War caused delays, but by the mid-1860s he had confirmed the existence of the Calumet conglomerate lode near the pit, and he had found some backers—a group of Bostonians associated with Horatio Bigelow, a financier who had had his hands in more than a dozen Lake mines since the 1850s.

The Hulbert Mining Company, formed in 1864, finished the task of acquiring the desired lands. Then a bit later, Hulbert and his backers spun off two more new companies in order to raise the capital needed to put these lands into production. The new Calumet Mining Company and the new Hecla Mining Company purchased their adjacent mineral lands from the original Hulbert company. They had the same stockholders, and the same administrative heads. The Calumet mine was created to develop one end of the conglomerate lode, while the Hecla mine would work the other.[25]

In 1866 Edwin Hulbert thought himself on the verge of becoming an extremely wealthy and important figure in the annals of Michigan copper. But just when the new companies seemed poised for a quick take-off, Hulbert made some poor decisions and quickly started losing control of the lode he had discovered. In 1866, he pushed development work at the Calumet mine, instead of at the Hecla—but it was the Hecla end of the conglomerate lode that later proved to be wider and richer at grass-roots, and less covered by overburden. Operating as a tributor in opening up the Calumet mine, Hulbert did not quickly sink shafts into the conglomerate. Instead, he mined it through large open pits that filled with water and suffered collapses of the pit walls.

Hulbert sent glowing reports back to Boston about the richness of the Calumet conglomerate, but other experts on the Keweenaw, especially Henry D'Aligny, sent back letters "full of doubt and misgiving."[26] D'Aligny, an engineer, held Calumet Mining Company stock and knew many of the company's chief backers. The discouragement spread by D'Aligny caused many of the original investors to pull out, selling their stock short. One man who was not so discouraged was the Bostonian Quincy A. Shaw, who with his friends and relatives bought up much of this stock and began to direct operations. In the summer of 1866 and again in the winter of 1866–67, Quincy Shaw sent his brother-in-law, Alexander Agassiz, out to the Keweenaw to determine the conglomerate lode's value and to assess Hulbert's performance in working it. These visits proved of absolutely critical import to the region. Because of them, the Shaw-Agassiz team, for as long as either of the two principals lived, dominated the Michigan copper industry.

Long before he ever left Boston for Lake Superior, 31-year-old Alexander Agassiz was a well-traveled man who had covered a great deal of geographical and intellectual terrain. Born in Neuchatel, Switzerland, in 1835, he emigrated to America in 1849 to live and complete his schooling in Cambridge, Massachusetts. A decade after arriving on the Atlantic coast, he found himself traveling up and down the Pacific coast of the Americas, from Panama to Puget Sound.

Agassiz's intellectual odyssey was no less sweeping than his journeys across the globe.[27] He was born into a remarkable family. His father, Louis, studied several different scientific fields at various European universities, and his subsequent work in ichthyology, geology, and paleontology cemented his reputation as one of the leading naturalists in the nineteenth century. His mother, Cecile, was an excellent artist, who illustrated some of her husband's works. On his mother's side, Alexander also claimed some notable uncles: Alexander Braun, a botanist, and Maximilian Braun, an eminent mining engineer.

It seems that Alexander Agassiz pulled together different strands from his family's experience to create a very successful and uncommon life for himself. Few would qualify for the description he received in the Dictionary of American Biography, where he is hailed, at one and the same time, as having been a "zoologist, oceanographer, and mine operator." Although this combination of careers or professions may seem odd, in fact Agassiz melded them nicely. The profits taken from the business side of his life paid for his passion for scientific investigation on the other side.

Alexander came to the United States after his father had accepted a professorship at Harvard. The son moved naturally into scientific studies, and like his father, he pursued several different fields. For a time, he made a career out of his education. He took a bachelor of arts degree from Harvard in 1855, a bachelor of science degree in engineering from Lawrence Scientific School in 1857, and a degree in zoology from Lawrence in 1862. Before heading out to the copper fields, Agassiz taught at a school for girls; worked on the U.S. Coast Survey, which took him to the Pacific; spent some time in Pennsylvania coal mining; and wrote and illustrated studies of starfish and jellyfish.

On his first trips to Lake Superior, Alexander Agassiz became a staunch believer in the wealth held by the conglomerate lode, but he was appalled by Hulbert's ineptness in opening it up. On 3 February 1867, he wrote:

> The value of the mines, both Calumet and Hecla, is beyond the wildest dreams of copper men, but with the kind of management many of the mines have had, then even if the pits were full of gold, it would be of no use.[28]

In March of 1867, the directors of the Calumet and the Hecla mining companies discharged Edwin Hulbert. They entrusted Alexander Agassiz with the job of resident superintendent, and perhaps as an incentive for Agassiz to accept this post in the hinterland of Michigan, Quincy Shaw presented him with a substantial gift of stock to accompany his regular salary. Agassiz's decision to go to Lake Superior certainly proved wise in the long run. After a few years at the mines, he lived out the rest of his days as a wealthy scientist ensconced in the more comfortable confines of Boston and of Harvard's Museum of Comparative Zoology.

In the course of correcting Hulbert's mistakes, Agassiz made several errors of his own. But he worked through the difficult times, brought order to the endeavor, and it paid off handsomely. Of course, it did not hurt matters that when first opened, the conglomerate lode was as much as 15 percent copper. Even after that figure declined to about 5 percent, the conglomerate ran far richer than other paying lodes, which did well if they were as rich as 2 percent copper.

By 1870, both the Calumet and the Hecla mines were running in the black, and the conglomerate lode's superiority was established. In that year, Michigan accounted for 87 percent of the nation's production of new copper, and the Calumet and Hecla mines, combined, accounted for over half of Michigan's share.[29] In the spring of the next year, the two closely associated companies consolidated into one dominant firm: the Calumet and Hecla Mining Company. Alexander Agassiz became president of C&H on 1 August 1871, and he did not relinquish that post until his death in 1910. After a fitful start, Calumet and Hecla prospered under consistent, predictable leadership for a great many years thereafter.

Consistent leadership was a hallmark of many of the region's profitable mines. They tended to be formed—and subsequently directed for decades—by stable groups of major stockholders who knew each other as friends or relatives. Key "families" of investors developed and held onto their own mines. This trend started early, when a group of investors lead by Thomas

Howe and Curtis Hussey, encouraged by their success with the Cliff mine, went out and formed another seven mines by the 1860s. T. Henry Perkins and Thomas F. Mason headed another of the early, dominant financial groups. Their single best investment by far proved to be the Quincy mine, but during the Civil War they served as directors for eight companies paying out half the district's dividends. Shortly thereafter, the Shaw-Agassiz group gained control of Calumet and Hecla, while Horatio Bigelow, Albert S. Bigelow, and their group of Bostonian investors corralled the Tamarack and Osceola mines, plus several others.[30]

Another important investors' group, headed by John Stanton and William A. Paine, came together in the 1890s to develop a host of new mines. Stanton's interests in Lake Superior copper dated back to the mid-1860s, when he became secretary of the Central mine. Later he was involved with the Atlantic and Allouez mines, and in the 1890s he and his associates, including Paine, established the Winona, Mohawk, Wolverine, and Michigan mines.

Of particular importance, Stanton and Paine cooperated to develop the new Baltic, Trimountain, and Champion mines atop the Baltic lode, one of the last high-producing lodes to be discovered. Paine, a founder of the investment firm of Paine, Webber & Company, used his copper stocks to catapult himself to the top of the list of the richest men in New England. In 1901, Paine presided over, and Stanton served as a director of, a newly incorporated holding company, known as Copper Range Consolidated. Copper Range owned nearly 100 percent of the stock of the Baltic and Trimountain mines and of the Copper Range Railroad. It also owned half the stock of the Champion mine, and 60 percent of a newly erected smelter at Portage Lake. Paine and Stanton had built a fine copper empire for themselves out in Michigan.[31]

As this mining district matured, and as dozens of marginal or poor mines came and went from the scene, control over the industry became concentrated in the hands of a small number of investors' groups, each of which had struck it rich with at least one major mine. By 1904, four such groups controlled 96 percent of Michigan's production. The Shaw-Agassiz group still held only a single company, C&H, but it produced 39 percent of all Lake copper. The Paine-Stanton group, heading up Copper Range Consolidated, plus another six or so mines, produced 30 percent of Michigan's copper. The Bigelow group, directing the affairs of four mines, and the Mason group, heading up only Quincy, accounted for 18 and 9 percent of production, respectively.[32]

The chaotic years of the "subterranean lottery" had given way to more orderly investments in the industry. The early era of many small mining firms had passed, making way for an era dominated by fewer, but larger, companies. The Lake copper mines, because they started so early, pioneered in the development of large-scale mining operations, employing upwards of several hundred men, arranged into a complex hierarchy of work. They pioneered in the development of an integrated metals industry, whereby the mining firm also owned the mill, the smelter, and the connecting railroad. Working relatively narrow, inclined lodes, the mines year after year went down further and further to get copper—which made them the deepest mining district in the

United States. They innovated—they quite literally broke new ground—in working at depths that other mines had not achieved.

When the Lake mines opened, they used very labor-intensive technologies. Manual labor, assisted by simple tools, performed the work. Brawn counted—swinging a sledge hammer was hard physical labor—but so did skill and knowledge. If the miner did not know what he was doing, he had trouble extracting enough copper to make a living. But over time, virtually all technologies changed as the industrialization of mining proceeded. Machines drilled the rock, not sledges and hand drills. Dynamite blew it up, not black powder. Electric locomotives, not men, delivered the rock to the shafts. Massive steam hoists, not windlasses or horsewhims, raised it to the surface. As the technologies became more powerful and sophisticated, the general skill level of miners declined, as did the status and role of traditional bosses, those practically trained men who had worked their way up the ranks. New experts, college-educated engineers, started directing a mine's business and technological affairs.

The communities surrounding the mines changed as much as the mines themselves. In the beginning, men made up nearly the entire population at the new settlements. While the work force stayed wholly male, local society became more gender balanced when, with the companies' full blessing and encouragement, women and children arrived in large numbers. The companies hastened evolutionary improvements within their communities that made them more stable and comfortable. They built housing and engaged in many other paternalistic practices: they helped build churches and schools, opened up farmland, encouraged storekeepers to set up shop, and hired company doctors. Indeed, the companies practiced as broad a paternalism as could be found in American industry.

Moving from a miniscule population in the 1840s, the three-county Copper Country boasted 100,000 residents by 1910. The population had ebbed and flowed with the discoveries of copper, the growth of select mines, and the depletion of others. The population had moved north, then south, and had ended up concentrated along the richest part of the mineral range in Houghton County. The population had also changed considerably in its ethnic constituency over the decades, as successive waves of immigrant groups washed up on the Keweenaw's shores. The early immigrants were largely Cornish, Irish, German, and French Canadian; the later arrivals more and more came from Finland, Italy, and eastern Europe.

Wholesale technological and social changes kept altering established patterns of work and life on the Keweenaw, and yet its largest mines, led by Calumet and Hecla, continued to thrive. Through the 1870s, they averaged three-fourths of the entire nation's output of new copper each year. As late as 1886, the Keweenaw mines still produced over half of America's copper, but that was just barely, and for the last time.[33] The Lake mines thereafter were surpassed as producers by western copper districts, especially in Butte, Montana, and later, in Arizona.

Still, Lake copper mining remained a vital and expanding industry. It reached its highest levels of production, employment, and profitability a quarter-century after the rival western copper districts had opened up, and not before. The market for copper expanded in the 1890s with the rise of the electrical industry, and the Michigan mines increased their annual output by about 50 percent over that decade. The mines ended the nineteenth century with a string of their most profitable years ever. They also ended it with an important tradition still intact: they remained known for being among the most enlightened, fair, humane, and paternalistic employers in the American mining industry.

The Copper Country was a one-industry region, and the mining men who credited themselves with building and populating it also relegated to themselves the task of ruling it. Through their company paternalism, they exerted control not only over a man's work, but over community development and the lives of workers' families. In the United States, the last quarter of the nineteenth century witnessed much strife and bitter, often violent confrontations between capital and labor. The mining industry had seen more than its share of the workers' struggle to obtain decent wages and improved living and working conditions. But by comparison with Montana, Idaho, or Colorado, the Lake Superior copper district remained remarkably peaceful, feeling only some small tremors of discontent.

For decades, Copper Country workers and bosses both had bought into paternalism, each taking from it something of value. The men got a more comfortable and secure existence; the managers got loyalty and labor peace. But early in the twentieth century, the industry, along with its accompanying social and labor relations system, began to unravel. Discontent grew far more apparent, and managers and laborers became more estranged. This happened at a time when the industry was approaching the top of its upward curve of growth, but before the onset of real and permanent economic decline.

Many different but related forces worked to create a widening breech between the mining companies and their workers. Economics was a key factor. Although the mines were still profitable, they were high-cost producers. Their managers recognized that competitive copper districts paid less to make their copper. With any downturn in copper prices, Michigan felt the pinch before anyone else. The mining companies, in search of higher productivity and lowered production costs, started to be more forceful in putting into place new technologies and work rules. In doing so, they politicized change in the workplace, making it a controversial issue.

But production cost data and diminishing profit margins caused only a part of the problems being felt in the Lake Superior copper district. The industry had long been led by men like Alexander Agassiz who were very set in their ways, and who had become very tradition-bound as their companies grew older. Once fully mature, the Lake copper industry did not cope well with change. Although it appeared modern on the surface, it remained a nineteenth-century industry and fell out of sync with the realities of the twentieth.

The industry did not cope well with the introduction of new peoples, such as Finns, Italians, and Hungarians—immigrants who were decidedly different from the industry's earlier Cornish, German, and Irish workers. It did not cope well with the introduction of new social or political movements, such as Progressivism. Workers tinged with the Progressivism spirit came to believe that too many men were dying in pursuit of copper. Meanwhile, mine owners and managers still believed that a death a week in Houghton County was an unvoidable cost of mining. And the industry did not cope well with workers' rising expectations, with their desires for a higher standard of living. Managers continued to see themselves as the standard setters. The companies would deal their workers into appropriate occupations, houses, and social settings, and the men were supposed to settle for what they got. The company was the father, and father knew best. That attitude came to rankle more and more workers having an independent, democratic bent.

The Lake copper mines, all told, passed through four eras. In the first, from the 1840s through the Civil War, they grew from mere pioneering outposts into important, productive settlements on the Upper Great Lakes. Led by Calumet and Hecla, the mines next entered their second phase, a lengthy period of robust maturity that was little marred either by economic or social problems, and that was instead characterized by great growth and successful change. But between 1900 and 1920, in their third era, the mines showed their age. They had not run out of copper, by any means, but the copper still left was harder and more expensive to get.

The mines were now deep, dangerous, and high-cost producers. In terms of employment, production, and profitability, it was during this difficult era that the mines peaked and then started their descent—during this difficult era that the major companies, even the great Calumet and Hecla, confronted questions of their own inevitable mortality. The era was marked by wrenching highs and lows: a major labor strike in 1913–14, the boom years of the First World War, and then the postwar collapse of the copper market.

The mines' fourth and final era started in 1920–21. Over several decades, the industry coped with one infirmity after the next, and no amount of technological or organizational change fully solved its major problems or reversed its long-term decline. The industry struggled to survive and did persevere (albeit in a very diminished form) until the 1960s, when Calumet and Hecla and Copper Range both shut their last native copper mines. Meanwhile, as the companies first declined and then died, they took their neighboring communities down with them.

It may be that all industries, even those with excellent leaders and fully committed workers, must one day close up shop. It may be that success can only be sustained for so long, before conditions beyond the control of the industry put it at a disadvantage, while they advantage a wholly different group. Certainly, in the history of American mining, it is a rare and special mineral range that can support an active, ongoing industry for anything approaching a century of continuous operation.

The Keweenaw Peninsula had such a mineral range, and it attracted tens of thousands of settlers to a remote region that otherwise would have remained virtually unpopulated. The copper spurred the rise of an industry, and the industry gave rise to local communities, whose fortunes were tied back to the copper. In this remote corner of Michigan, work and life and company and community were inextricably bound together. When the industry came under challenge and change became controversial, when it started to falter and contract, it was inevitable that the painful costs of competition and decline would be spread across all of society.

2

The Underground: Change and Continuity

> Nothing in the copper mines of Lake Superior is more remarkable and significant than the immense progress that has been made . . . within the last fifteen years. The rude skid shaft lining; the rickety ladder, the awkward kibble, the clumsy chains, the whim, the wheelbarrow . . . have all been superseded. . . . This progress has been made under very trying circumstances, at the cost of infinite toil, expense and discouragements. . . . If today our copper mines are prosperous, it is owing, under Providence, to their intrinsic worth, and the skill and indomitable energy of the people, aided by non-resident capitalists.
>
> A. P Swineford, *History and Review . . . of the Material Interests . . . of Lake Superior,* 1876

Between 1845 and 1910, "non-resident capitalists" and the "indomitable energy" of thousands of workers carried the copper mines of Lake Superior from the forest floor down to depths in excess of 5,000 feet. Early prospectors clear-cut the trees, grubbed out or burned the underbrush, and crisscrossed their mine sites with trenches, all to find the outcropping of a paying lode. If they succeeded, then they obliterated the nearby forest, planted a mining village in its place, and moved deeper into the earth in search of copper.

At the most successful mines, 40, 50, or 60 years later, men were still going at it, still going down. As they did so, mining techniques and the underground environment changed in many fundamental ways. Nevertheless, mining tended to be one of the more traditional industries, and change in the underground was never continuous or all encompassing. Some parts hardly changed at all over half a century. For instance, due to the nature of their copper deposits—narrow lodes that dipped steeply into the ground—the Lake mines, unlike their counterparts in Montana, never switched over to less expensive open-pit mining. They started and stayed as deep-shaft mines.

Miners sank shafts that usually ran down through the copper lode itself. The mine shafts (with few exceptions) did not run straight up and down. Instead, they were inclined at the same angle as their lode. These shafts were multipurpose thoroughfares. They opened up ground at ever greater distances from the surface. They served as key passageways for moving men and materials and for hoisting copper rock. Shafts allowed fresh air to enter and stale air to exit the underground. They also carried technological systems into a mine, such as pump rods and pipes for unwatering, pipelines that carried compressed air to drive machinery, and, later, electrical and telephone lines. Often the shafts were divided into two compartments, separated by a plank wall. The working compartment held all the appliances for mechanically transporting men, materials, and rock. The second compartment housed a ladderway for men.

Prior to about 1880, companies commonly sank numerous shafts located only 300 to 500 feet apart. This allowed them to better assess and explore the full length of their lodes, and a multitude of shafts also allowed more points for the entrance and egress of air. Also, in an early era when they had only slow drilling and hoisting technologies at their disposal, companies accelerated copper rock production by attacking their lodes through six or eight shafts, instead of through only two or four. But as mechanical systems improved and greater tonnages of rock could be handled through each shaft, companies typically abandoned several of their early openings. The mines came to use fewer shafts, spread 1,000 or even 2,000 feet apart.

In later decades, an impressive edifice for receiving, sorting, and crushing rock stood over each shaft, and each was served by a railroad siding for carrying stamp rock to the mill. Nearby stood a large engine house, a boiler house, and an air-compressor plant, all often dedicated to serving but one shaft. Each shaft became a production unit largely unto itself, with its own physical plant and its own set of mining captains, shift bosses, foremen, and workers. In the Copper Country, mine shafts served rather like different departments within a company. If quizzed about where he worked, a Quincy employee would not respond with just the company's name, but with a shaft number. He worked at Quincy No. 2, No. 6, No. 7, or No. 8. Shafts sometimes had vastly different personalities. There were good ones and bad ones. Some were hotter than others. Some were safer. Some had friendly bosses; others were run by petty tyrants.

To open a lode up along its length, miners drove tunnel-like drifts at regular intervals away from one shaft and toward another. Drifts represented the successive steps down into the underground, and all drifts the same distance beneath the surface constituted one working level of the mine. In the nineteenth century, drifts were 10 fathoms (60 feet) apart, a practice inherited directly from Cornish tin and copper mines. By the early twentieth century, mine levels moved to 100 feet apart, or even more.[1]

Miners often produced some good copper rock while drifting, but a drift's main purpose was not to recover copper but to open up underground roadways. To cut down on the cost of this development work—particularly when passing through poor rock—the mining companies made drifts relatively small: they commonly measured only six feet tall and five feet wide. Sometimes, in a highly regular lode, the drifts ran straight and true. In other mines, they zig-zagged, following the "tortuous windings" of the copper deposit.[2]

Drifts were busy places, filled with the comings and goings of numerous men and much material. Workers walked along a drift to get to their work stations, carrying drills, explosives, ladders, scaffolding, ropes, or anything else they needed. Some men struggled to transport mine timbers—often long sections of tree trunks—down these confined corridors. Still others rolled heavy cars filled with broken rock. To make this last task somewhat easier, drift floors were supposed to be graded to run slightly downhill toward the shaft where the rock would be hoisted.

Drifts also carried air across a mine. Since scant air would flow into a dead-ended drift, miners encouraged air circulation in newly developing parts of a mine by interconnecting drifts with winzes—passageways running from one drift to others above or below it. Winzes were similar to inclined shafts, except they were not used for hoisting, and they started and stopped underground, without breaking through to the surface.

To visualize how a company prepared to work its lode through shafts, drifts, and winzes, it is useful to think of a rectangular grid, like that formed by so many city streets laid out on the surface. The mines laid a rectangular grid over the face of a mineral-bearing lode dipping into the earth. On this grid, the shafts were the main up and down passages, and they were paralleled by secondary winzes, used primarily for circulating air. The drifts were the busy cross-streets. Shafts, winzes, and drifts—just like surface streets—set the lode off into blocks. It was from these blocks of ground that miners extracted copper rock by "stoping it out." Near the top of the mine one would find empty stopes—where miners had already taken out all the good copper-bearing rock. Near the bottom of the mine, stoping miners would be hard at work on several levels, getting out the mine's current production. At the very bottom of the mine, shaft-sinkers and drifters would be pressing ahead with their development work, readying blocks of ground to be exploited two or three years later.

Throughout the nineteenth century, virtually all stoping miners exploited a block of ground by working it from one level up toward the next. The chief advantage of this "overhand stoping" was that it used gravity to good effect. If men had stoped underhand, working from one level down to the next, then evey time they set off an explosive charge the broken rock would have rolled back down to settle against the working face of the stope. Before the next shift could start drilling, men would have to muck out that rock and transport it to the drift above. But with overhand stoping, broken rock rolled away from the working face and toward the drift below. Miners could start drilling almost as soon as they got to their work stations, and much broken rock could simply be stockpiled in the bottom of the stope. Since gravity moved it out of the way, no large force of men was required to muck it out expeditiously every day.[3]

The chief disadvantage of overhand stoping was that it often put mining teams in awkward positions. They did not always have solid ground to stand on, and they often worked on top of piles of broken rock, or else they created perches for themselves out of scaffold boards affixed to the mine walls. Overhand stopers also drilled many shot-holes over their heads in an upward direction. Before the advent of machine drilling, this was particularly difficult work. Swinging an eight-pound sledge upward to strike the head of a drill steel was far more laborious than swinging the same tool downward or in a horizontal plane.

When a level was ready for stoping, miners took the copper out following either the advancing or the retreating system. Advancing miners started right at or near the shaft. As they pushed their stopes upward, they also advanced them outward along the drift. Retreating miners, on the other hand, went to the the outer boundary of a drift and then stoped back along the drift toward the shaft.

Most companies, through the early twentieth century, adhered to the advancing system. It had one particular advantage: men could begin stoping out a given level earlier. They did not have to wait for the long drift to be fully done before putting the level into copper-rock production. Development work was expensive. Drifting cost money, while stoping made money. As a consequence, companies were anxious to start stoping next to the shaft as soon as possible, because the recovered copper helped pay to complete the drift. Under the advancing system, the companies could "pay as you go." Under the retreating system, all the development work had to be paid for in advance. The companies incurred all the costs of drifting, plus the costs of any timbering or tram-track laying, before the level produced copper.

Retreating had one important advantage over advancing, however. It was generally safer. Miners, in taking out copper rock, created large cavities underground. The rock beneath these cavities constituted the mine's floor or footwall. The rock arching over the openings was the mine's roof or hanging wall. The hanging wall posed the greatest danger to life and limb. The more men worked under open portions of the hanging wall, the more endangered they were by rock falls. With the advancing system men constantly passed through stoped-out ground on their way to or from newer stopes located further from the shaft. But with the retreating system, men moved closer to the shaft, leaving vacant stopes behind them. If the hanging wall collapsed in a mine using the retreating system, it was far less likely to catch any men beneath it.

Still, mining companies did not adopt the retreating system until their hanging walls proved exceptionally hazardous or costly to maintain. For half a century, and longer at some companies, the mines followed the advancing system with great abandon. As they pushed their stopes out away from their shafts, they instructed miners to take out all rock carrying a commercial grade of copper, regardless of where it was.[4] They did not leave rock pillars alongside shafts, to protect them from being crushed. They did not leave regularly spaced pillars in their stopes, or alongside their drifts. They did not lay up poor rock walls to help support the mine roof. Furthermore, the companies used scant timbering in their stopes; they stuck in an occasional stull only to support a particularly bad piece of the hanging. Calumet and Hecla had a notoriously unstable hanging wall, so it planted a "forest" of timber supports underground each year.[5] But while C&H treated its hanging wall with great caution, other companies largely ignored the need to prop up the hanging in a systematic way.

The companies hollowed out more ground than they should have. In 1876 Thomas Egleston, an eminent engineer from the Columbia School of Mines, visited the Lake mines and came away impressed by their hanging walls' ability to span large, cavernous spaces without substantial supports. Nevertheless, he faulted companies for neglecting the hanging: "The solidity of the ground is remarkable," he wrote, "yet it must one day give way."[6] That day indeed finally came, but not until after the turn of the century.

As the mines moved from 3,000 to 5,000 feet deep and beyond, their hanging walls became less stable and more demanding of attention. Companies used more timbering and left more ground unmined as pillars. The increasing costs of timbering and of leaving behind copper rock in pillars finally forced many companies to adopt the retreating system. The switch-over entailed considerable initial expense, because the mines had to defer stoping and put more men to work driving numerous long drifts to completion. But once the change was made, they could afford once again to pay less attention to the long-term support of the hanging wall. If it collapsed in an empty stope, nobody was there to get hurt. C&H, in the mid-1890s, was among the first of the major mines to adopt retreating. Quincy was one of the last, not turning to it until the late 1920s.[7]

In the Copper Country, the term "miner" did not apply to all underground workers, but was reserved for shaft-sinkers, drifters, and stopers who drilled and blasted rock. The skilled Cornish miners who arrived early at Lake Superior carved out a special niche for their occupation in the hierarchy of underground work. They wanted miners to be seen as superior to other underground laborers. In the late 1850s, Quincy's Cornishmen suddenly walked off the job one day. The perplexed mine agent first assumed they had struck for higher wages, but later discovered that they had struck over the issue of status. The Cornishmen protested that they had been demeaned when the company had handed drills and explosives to inexperienced men from another ethnic group and had called them "miners."[8]

When mining was done slowly and by hand, skilled miners had to know how to read the mine rock, how to discover the lines of least resistance, so that the fewest number of well-placed shot holes could bring down the most rock. They also needed physical skills and stamina. Their earnings depended on their ability to drive drill steels home with eight-pound sledge hammers, swung both upward and downward. Using that tool expertly was no mean feat.

The more primitive the tools, the more skill miners required to be effective. Until 1880, they battled mine rock with sledges; with octagonal drill steels (of different lengths), sharpened to a chisel point; and with black powder, an explosive of modest power. In stopes and at shaft bottoms, one man held the drill, while two others "jacked" it into the rock by rhythmically alternating hammer blows. In constricted drifts, there generally wasn't room for a second hammerman, so the drill-holder worked with a singler partner. After each hammer blow landed, the holder lifted the drill a bit and rotated it slightly, so that the next blow cut a good chip at the bottom of the shot hole. After drilling, the men cleared the shot holes of debris, charged them with black powder, set their fuses, and blasted out another few feet of mine rock. That signaled the end of their 10-hour workday. The next shift arrived two hours later. Over the break, noxious gases produced by the explosions dissipated, making it safe for fresh teams of miners to start work.

The rate of shot-hole drilling slowed considerably in particularly "coppery" stopes, rich in barrel or mass copper. Gummy, malleable copper seized the cutting edge of drill steels, causing them to stick in the hole. Miners often abandoned a shot hole for a new one when they ran into too much copper. And when they encountered a substantial deposit of mass copper, their work slowed dramatically.

Mass copper incited the mine rush to the Keweenaw, but in fact it was not the economic bonanza that people once thought. Surrounding rock firmly bound its jagged edges and appendages, making mass copper extremely difficult to liberate. Miners drilled and blasted to create a cavity under, over and around a large mass. To finally bring it out on the mine floor, they often used several kegs of powder, tucked behind the mass and surrounded by sand bags that directed the force of the explosion. Then a special subset of miners, called "copper cutters," laboriously sectioned the mass into pieces small enough to transport out of the mine. Because this work was so expensive and labor intensive, copper cutters divided large masses into as few pieces as possible. After drawing cut lines across the top of the copper, the men parted it with special chisels and hammers. They cut a long chip across the mass, then went over and over it again, until they had channeled all the way through.[9]

The mechanization of mass copper cutting eluded the mining companies until early in the twentieth century. They tried power saws to no avail, but finally succeeded with small jackhammers and also with twist drills. Men drilled a line of holes across a mass, loaded them with dynamite, and blew the sections apart.[10] The mechanization of mass copper cutting, however, was not nearly as economically important as the new machines and explosives applied to the drilling and blasting of ordinary conglomerate and amygdaloid copper rock.

Around 1880, miners put down their hand steels, sledges, and black powder and picked up nitroglycerine dynamite and rock drilling machines powered by compressed air. The new machines drilled shot holes far more rapidly, and the new high explosives brought down more rock per charge. These two innovations greatly increased worker productivity and the mines' output.

Miners broke rock. Trammers picked it up and pushed it to the shafts for hoisting. Until the 1860s, men often used wheelbarrows for this purpose, because they could be run in very narrow drifts and didn't require a flat floor.

After pushing the wheelbarrows to the shaft, men dumped their contents onto an opening called a trip plat.[11] Once enough rock had been accumulated by the shaft to justify hoisting it out, workers handled the rock again, putting it into kibbles, or wrought iron buckets, to be lifted to the surface.

By the 1860s, productive mines outstripped the capacity of wheelbarrows to move their rock. They now drove their drifts with flat floors and equipped them with crossties and lightweight, narrow-gauge rails. Men loaded rock into four-wheeled tramcars which they pushed down the tracks.[12] At the shaft, instead of dumping the rock onto a trip plat, trammers waited for the arrival of a skip, which had replaced the kibble. When companies tracked their drifts to facilitate tramming, they also laid rails down their shafts to accelerate hoisting. When the wheeled, box-like skip came to rest at the level, trammers tilted their cars to dump copper rock into the skip, and a "filler" stationed at the shaft pulled a bell cord to signal that the skip could be raised.[13]

The mines made precious little change in tramming technology for half a century following the Civil War. Very few mines replaced human trammers with horses or mules, and very few succeeded in mechanizing the work of moving heavy tramcars laden with rock. The Copper Falls mine, situated on a hillside, used a small locomotive in the 1880s to tram its rock out of the underground through an adit—a horizontal tunnel driven in from the base of the hill that intersected one of the mines' working levels.[14] Other companies, after the mid-1890s, experimented with cable-car systems or electric haulage locomotives to draw tramcars to the shaft. But prior to 1910, only Quincy succeeded with electric haulage to any extent. Companies continued to use men as "beasts of burden" to lift, fill, push, and unload vast tonnages of copper rock, and tramming remained the most physically demanding underground occupation.

Substantial numbers of timbermen joined miners and trammers underground. They erected stulls and square-set timbers in stopes, laid tram tracks in drifts, timbered shafts, laid skip roads, and erected ladders. At some mines, all men engaged in this work were called timbermen. At others, only the bosses received that moniker; other crew members were considered common underground laborers. Proper timbering required considerable knowledge on the part of bosses supervising the work, and it required hard physical labor on the part of those men who moved large timbers across and up and down a mine. A sense of the effort required can be gotten from a description of a timber crew at the Minesota mine in 1855:

> As we were passing one level we met a large log about twenty-five feet long and about two feet in diameter, that had just been let down the shaft to be taken to another part of the mine, and used as a support to the walls. A double line of men, each with a candle stuck in the front of his hat, about twenty in number, were pulling away and struggling along with the piece of wood, like as many industrious ants laboring to carry a kernal of grain through their little mines to the storehouse. There could hardly be a more impressive and singular appearance than these men bearing their lights like stars amid the darkness, and one is

forcibly reminded of the house of Cyclops, who were such a terror to the human race, and who were said to have one eye alone in the center of their forehead.[15]

Smaller occupational groups found in the mines included pumpmen, pipemen, watchmen, block-holers, track fixers, wallers, and copper pickers. Another small group of occupations had "boys" rather than "men" added as a suffix. The companies employed relatively few boys under the age of 18. At no time, probably, did boys comprise more than 10 percent of the underground labor force.[16] It is interesting that the employment of boys was generally tied to the use of machinery. Boys hand-powered small "air machines" that helped blow air into poorly ventilated drifts. "Drill boys" ran errands for miners operating compressed-air drilling machines; "puffer boys" operated small compressed-air-driven engines that performed several tasks; and "motor boys" operated the earliest electric haulage locomotives used in the district.

All underground employees had to get into and out of the mine each working day. Until the mid-1860s, men moved themselves up and down on ladders. In new, shallow mines, ladder climbing was not terribly taxing, but by the 1860s, the Cliff mine was 1,000 feet deep, and Pewabic, Quincy, and some others were over 500 feet deep. Such long ladder climbs severely stressed men's hearts and respiratory systems—and they also cut away at a company's production. Men working at deep mines started work after a long and tiring descent, and near the end of their shift they held back on their effort, so they would have enough energy left to "get to grass." Productivity also suffered because the long time men spent on ladders—which could approach an hour each way—was time lost from drilling or tramming. Finally, long ladder climbs often robbed the successful, deep mines of workers, who left to find work at shallower mines. To save men physical labor, to save time and enhance productivity, and to keep their laborers, in the mid-1860s several Lake copper mines erected "man-engines" for transporting men up and down. Such devices had been used in select German and Cornish mines since the 1840s. The ones erected at Lake Superior between 1864 and 1866 were surely among the first man-engines in the United States.

The man-engine was a mechanical ladder. The portion of the device that carried men consisted of two side-by-side wooden rods formed of timbers, about a foot square, that were bolted together end to end. The rods, operating in their own inclined shaft, rested on rollers and carried small platforms about 10 feet apart. The upper end of each rod connected to a counterweighted triangular bob on the surface, which was connected by a second rod to special gearing driven by a steam engine. The mechanical connections were such that one bob pivoted up while the other pivoted down, so that within the mine the side-by-side man-engine rods reciprocated in opposite directions. One ascended 10 feet, while the other descended 10 feet. After each stroke, there was a slight delay before the rods changed direction. During these pauses, the platforms were next to one another, and men went into or out of the mine by stepping from one rod over to the next. They descended 10 feet at a time by always stepping over to the rod ready to go down. At shift's end they reversed the procedure and reached grass by always moving to the rod set to go up.

John Rawlings, a Cornishman, installed the region's first man-engine, which began service at the Cliff mine in January 1865.[17] Others quickly followed at Pewabic and Quincy, and a man-engine soon became required equipment at most productive mines. Installing, maintaining, and periodically extending the depth reached by a man-engine was costly (Quincy spent the then sizable sum of $17,500 on its man-engine in 1866). Nevertheless, the man-engine was an unqualified success. In 1866, Quincy's annual report noted, "It is now regularly worked, and its operation is most complete and successful. It saves a great amount of time and labor, and is thus well calculated to benefit both the company and the men." Pewabic's 1868 annual report added its praise for this rather strange contraption:

> By its aid the miners are taken to the surface from the 120-fathom level [720 feet] in ten minutes, thus saving much time and strength that would be lost in climbing by the ordinary method of ladders. For this reason the miners prefer to work here, rather than in other deep mines which are not [yet] supplied with this machinery.[18]

The man-engine was as successful as any new technology could be; it was a boon to employee and employer alike. It was truly an innovation worthy of celebration, and the *Portage Lake Mining Gazette* did celebrate it on 4 February 1865, just a month after the Cliff's man-engine went to work:

Song of the Man Engine

Ho! ho! for my arms of steel!
That are always strong, and never can feel
The weakness that creeps o'er the muscles of man
(Though they fashioned mine by some curious plan).
As he trembling clings to the ladder's round
Two hundred fathoms under the ground!

Weary with toil for one-third of the day
I hear their footfalls coming this way;
Put on thy strength, O arms of mine,
Carry them up where the sun doth shine.
One by one they come without sound
From two hundred fathoms under the ground!

Hammers and drills they've left behind.
Down in the caves their strength hath mined.
And quickly their fears all pass away,
As swiftly they rise to the upper day—
Trusting the friend whose arms they've found
Two hundred fathoms under the ground!

O noble's the task that is mine to perform.
'Tis ever the same in sunshine and storm:
A while I am master, yet ever a slave,
For the lives of men my energies save.
And they bow to me when my arms reach down
Two hundred fathoms under the ground!

Then ho! again for my arms of steel.
And ho! for my rods and strong-toothed wheel.
Ho! ho! O man! for the MAN ENGINE!
I'll save your strength, if you manage mine,
And ne'er shall you reel on the ladder's round
Two hundred fathoms under the ground!

By 1890, many companies began phasing out their man-engines, because they had about reached their limits, in terms of depth. While Calumet and Hecla had afforded multiple man-engines, other mines had only one. While these man-engines spared men much ladder climbing, they did require them to make long walks underground from the man-engine shaft to their work stations. More mines turned to man-cars, a simpler, less capital-intensive way to deliver men closer to their work.

The Phoenix and Central mines had used man-cars by at least the mid-1870s. These long, specialized skips usually had about ten bench seats, each with room for three men. (In very steep shafts, such as those at the Champion mine, men rode standing, rather than sitting, in their man-car.) At the start and end of a shift, shafthouse workers swapped a man-car for an ordinary rock skip, and the men rode up and down in the same shaft in which they worked. The Osceola mine adopted man-cars around 1885, and C&H did so in several shafts in 1889–90. Quincy abandoned its man-engine after nearly 30 years of service in 1895, when its oldest shafts were 4,000 feet deep, measured along the incline.[19] At the Tamarack mine's five vertical shafts, and at Calumet and Hecla's one vertical shaft, neither man-engines nor man-cars were employed. Instead, the shafts were fitted with elevator-like "cages," some of which were double-decked to carry more men per trip.

However delivered underground, men worked there for about nine to 10 hours before starting back to the surface. Over that time, certain biological needs had to be met. Halfway through their shift, underground workers broke for dinner, taking their food and drink from a tin dinner pail. Cornish immigrants provided one of the staples of the mine worker's diet: the pasty. The pasty—a meat and vegetable pie in a folded-over pastry crust—was wrapped in paper to keep it warm, but men often reheated their meal by placing their dinner pail over a mining candle.

When men relieved themselves, they used primitive facilities, or no facilities at all. On the surface, privies stood amongst the surface works; they were often located close to the "dry," or change house, where underground workers changed clothes before and after work. But underground, lacking formal privies, the men improvised. They urinated almost any place where another man wasn't working. They defecated into empty powder or candle boxes. Occasionally, they shoveled a little mine dirt or lime into this expedient latrine to make it a bit less foul. Sometimes they simply left these boxes underground, tucked away in a remote corner of the mine; sometimes they hoisted them to the surface.[20] Overall, sanitary conditions in the mines started out as primitive in the 1840s and stayed that way into the twentieth century.

Perhaps the dominant feature of the underground was its pervasive darkness, and the short range of sight it allowed men to have. From the 1840s through the early 1870s, men lit their way with tallow candles, either spiked to the mine walls or affixed to their stiff felt hats with small balls of clay. Lacking reflectors, these candles provided a broad, dim light that could not be focused on any particular thing, and what little light they provided was swallowed up by cavernous stopes and rough rock walls. The circle of candlelight surrounding workers was small. Men could not see along the mine from team to team. The dark underground never opened up to the worker. Instead, it closed in around him. The poor lighting did have one advantage for laborers—it made it hard for anyone to observe them from afar or sneak up on them. The bobbing of a candle flame in the distance signaled an approaching captain or boss.

Longer-burning stearine candles replaced tallow in the mid-1870s. Paraffin-based fuels burned in small lamps (such as "sunshine" and "moonshine") replaced stearine candles in the 1880s and 1890s, and carbide lamps (burning acetylene gas) became commonly used after 1910.[21] The newer fuels reduced lighting costs, provided more candlepower, or both. (Quincy switched from stearine candles, which it bought by the ton from local manufacturers, to "sunshine" lamp fuel provided by Standard Oil in 1897. After much close figuring, the mine agent calculated that "sunshine" reduced lighting costs by 81 cents per man per month.)[22] But, even if brighter, the successive lamps and fuels still provided only localized lighting, as had the original tallow candles. Although the mines demonstrated their progressiveness by quickly installing arc and incandescent lights in and around surface buildings, they were slow to put electric lights underground. As late as 1912, a writer for the *Engineering and Mining Journal* noted that "the absence of electric lighting" at hoisting stations and along drifts was "one of the striking features" of Lake Superior copper mines.[23]

For the first 50 years of copper mining on the Lake, two environmental conditions in particular benefited both the quality of work life and the companies' cost sheets. The mines, despite their proximity to the greatest freshwater lake in the world, were remarkably dry, and they achieved good ventilation and high air quality through natural, not mechanical, means.

The dryness of the Lake mines surely surprised the Cornish immigrants. In Cornwall, water plagued even shallow mines. The Cornish had wet mines and no local fuel source. This prompted them to develop a special type of fuel-conserving steam engine to unwater their tin and copper mines. Near a Cornish mine shaft, one commonly found two stone engine-houses, one smaller than the other. The small one housed the steam hoisting engine with a bore of 2½ to 4 feet; the large one housed the steam pumping engine, often a behemoth with a cylinder 5 to 8 feet in inside diameter.[24] On Lake Superior the situation was reversed. The companies put their capital into impressively large hoisting engines, while erecting quite modest facilities for unwatering their mines.

Between 1845 and 1910, the companies removed water in several different ways. Those located on a hillside could use an adit as a drain. They simply captured mine water in gutters or "launders" and carried it along a drift, down the adit, and out of the mine. The early mechanical systems for unwatering consisted of a steam engine on the surface, a pump rod (similar to a man-engine rod) descending into the mine, and a pipeline rising back out. The reciprocating wooden rod moved the underground pumps, which piped the water up to the surface. By the 1890s companies generally replaced these rather cumbersome mine pumps and rods with electrically driven pumps or with water-bailing skips.[25] In either case, on one or several levels, launders channeled water to an underground reservoir, or sump, cut out of solid rock by miners. When the sump filled, water was pumped up to the surface, or pulled up in special cask-like skips that could hoist about 1,500 gallons per trip.

The general absence of water added to the personal comfort and safety of most underground workers. They did not have to tromp through streams of water on the mine floor, or work under a constantly dripping mine roof. And the mines' air was not overly humid, which made workers' hard physical labor less exhausting. Also, men drilled and blasted without fear of suddenly breaking into underground streams. In other districts, such events had been known to flood a level and drown the men.

The health, comfort, and productivity of underground workers depended in large measure upon the quality of air they breathed. If too stagnant, it filled with rock dust and noxious gases produced by explosives. Underground air was also heated by the surrounding rock, whose temperature increased with depth. Large quantities of fresh air needed to be circulated to scavenge the underground atmosphere of pollutants and to cool the mine.

Throughout the nineteenth century, the Lake mines were exceptionally well ventilated to the extent that they became something of a scientific curiosity. In other parts of the world, many mines did not fare nearly as well. By the early 1860s, one Cornish mine suffered air temperatures of 102 degrees Farenheit at a depth of 1,200 feet, and in the American West by 1880, mines on the Comstock lode recorded air temperatures of 105 to 110 degrees in some drifts.[26] Under such conditions men required regular and lengthy rest breaks, and companies had to use mechanical systems to force fresh air into the mines. The Michigan mines, meanwhile, never reached such high temperatures, even as they obtained depths of as much as 9,200 feet in the twentieth century.

Several factors abetted the high air quality and low air temperatures found in the copper mines for their first half century. Their rock formations produced no poisonous or explosive gases, such as methane. The mines did not encounter underground hot springs, such as those responsible for the elevated temperatures on the Comstock lode. The rock on the Keweenaw Peninsula indeed grew hotter from the surface down, but the temperature gradient was gradual. Finally, the annual mean air temperature on the surface was low. All these factors allowed the companies to minimize their investments in mechanical ventilation and to circulate air by means of natural ventilation.[27] In short, the air moved itself. Natural drafts were set up because the air stood in columns (the mine shafts) of different height, and the temperature of the air varied from the surface and the cooler upper portions of the mine down to the warmer lower portions. Temperature gradients and varying densities set the air in motion. Fresh air entered the mine in downcast shafts, circulated through drifts, stopes, and winzes, and then exhausted through upcast shafts.

Natural ventilation worked well in these mines, but not perfectly—and not forever. Air currents traveled a path of least resistance, and moved most freely through well-opened parts of a mine. Fresh air, however, tended to bypass other areas where it could not blow straight through. There were always "close" parts of a mine, where men worked in dustier, more polluted air: the very end of a drift, the bottom of a shaft, or the uppermost portions of a stope that had not been holed through to the level above. And by early in the twentieth century, the deepest mines in the region were warm, and becoming more uncomfortable.

Around 1900, the bottom of the C&H mine reached 80 degrees, and by 1910 working levels at the Tamarack mine averaged 82 to 84 degrees. The difference in temperature between shallow and deep mines could be seen within the works of just one company. At Quincy, in 1904 the sixtieth level of the company's deepest shaft, No. 2, recorded air temperatures of 80 degrees. Meanwhile, at its newest and shallowest shaft, No. 8, men worked on the tenth level in 57-degree air. South of Portage Lake, newly opened Champion, Baltic, and Trimountain mines took advantage of the warm air that was beginning to trouble older mines when they set their wage scales. They figured they could attract good workers, even though they paid somewhat less, simply because they were shallower and cooler.[28]

On rare occasions, older companies recognized they had a ventilation problem, which they tried to remedy through mechanical means. In 1902 Quincy's southernmost shaft, No. 7, suddenly suffered a reversal in its air flow. It had been a downcast shaft, taking in fresh air at its end of the mine. Now it became upcast. This change in air flow was "the cause of preventing good ventilation in the bottom levels of the vicinity of No. 7 shaft, and the accumulation of foul air in places had hampered the miners and underground laborers, in general, to quite an extent."[29] Quincy tried to return No. 7 to a downcast shaft by placing a large fan, nine feet in diameter, over a nearby abandoned shaft. This fan extracted 70,000 cubic feet of air per minute from the mine. It aided the flow of air in Quincy's No. 2 and No. 4 shafts, but failed to help matters at No. 7, which quickly gained the reputation as the worst shaft to work in at the Quincy mine.

Few companies before 1920 made much effort to supplement natural ventilation with electrically driven fans or other systems intended to improve underground air quality or to lower temperatures. Generally, they were slow to recognize the heating up of the mines as a problem. Well into the twentieth century, mine managers and outside experts alike still proclaimed the air in the Lake mines to be "clean and comfortable."[30] But by 1910, mine workers, who had a different perspective on the situation, disagreed with this assessment.

Particularly in the winter months, underground workers experienced great temperature changes as they came and went from work. Before leaving his dwelling in February, a man bundled up warmly to protect himself from the bitter cold. He strode through tall banks of snow, four feet deep, with his head tucked down from the wind. At the dryhouse, he put on his hob-nailed boots, his work pants and shirt, and his lighter weight mining coat. Then he dashed over to the shafthouse. On the way down into the mine, initially the increasing warmth felt good. But when he climbed up in the stope to run his drilling machine, the 80-degree air soon proved too much. He shucked off his jacket and shirt and worked bare to the waist. The more he worked, the more he sweated. When he drilled a dry up-hole, all the dust came down on him, caking his face, arms, and upper body with grime. Despite what any bosses said, when you had to strip down to do this work, it was too damn hot. And when you hacked and your spit was "as black as pine tar," it was too damn dusty.[31] The mines needed better air.

Overall, between 1845 and 1910 the workers' underground world had been remarkably transformed in some ways, while remaining very traditional in others. The companies moved from a focus on mass copper deposits to an overwhelming reliance on taking their profit from conglomerate and amygdaloid copper rock. They moved from having many shafts, each producing limited amounts of copper, to fewer shafts, each being far more productive. For about half a century, most mines exploited their lodes using open stopes and the advancing system. They took out good copper rock wherever they found it and neglected to support their hanging walls systematically. Local hanging walls finally protested by becoming less stable. Some mining companies switched over to the retreating system to escape the growing hazards of collapsing mine roofs. Others stayed with the advancing system, but could no longer take out so much ground with impunity. Next to their shafts, and up in their stopes, they had to leave some good copper rock behind as supporting pillars.

In terms of adopting new tools and techniques, two fairly early periods were most revolutionary. In the 1860s, the mines tracked their drifts and replaced wheelbarrows with tramcars; they tracked their shafts and replaced kibbles with skips. The greatest mechanical innovation of the 1860s was the man-engine, which replaced the ladder as a means of getting up and down.

The next revolution came around 1880, with the successful adoption of nitroglycerine dynamite and the machine rock drill, powered by compressed air. Compared with hand drills and black powder, these technologies allowed miners to bring down far more rock per shift. They also opened the door to mining jobs to men having less experience and knowledge. Miners no longer had to have such a "nose" for copper, or an ability to read the mine rock to determine how to attack it. A neophyte miner, armed with a machine drill and dynamite, could bring down more rock than the most skilled hand-drilling miner using black powder.

Once the companies put air drills and dynamite into use, the pace of change slowed. They flirted with mechanized tramming after the mid-1890s, but almost all failed to find a new technology that would pay. The few types of machines used underground generally continued to be powered by compressed air. The companies benefited greatly from the rise of the electrical industry; it created a new and large market for copper. But, ironically, the companies made precious little use of electricity in their own mines prior to 1910, either to drive motors or to light up the underground.

Workers still carried their own lights with them wherever they went. Underground sanitation remained crude. Men still breathed air moved almost exclusively by natural forces, but now the air was getting uncomfortably warm. Many remained engaged in very physical and demanding work, especially trammers. In important ways, this job had distinctly worsened over the years. Trammers grew more weary in the warmer air, and since the companies operated fewer hoisting shafts, they now had to push loaded cars greater distances. And the mines' less stable hanging walls, which forced the companies to rethink their mining methods, also forced greater fear onto many underground workers. The further and further they worked from the surface, the less safe they felt.

By 1910, nobody seemed pleased with underground conditions. Workers felt less comfortable and secure. Mine agents, efficiency experts, and mining engineers had an entirely different complaint. The last major technological breakthroughs to sweep the entire district had occurred thirty years earlier. Surely some new technologies and new ways of organizing men were long overdue. The mines required modernizing; they needed more productivity increases. They were working in a nineteenth-century mode, and to remain competitive they had to be brought up to date.

3

The Surface: A Celebration of Steam Power

> The numberless powerful engines, pumps, compressors, and the ponderous winding drums that one may see at this mine [Calumet and Hecla] are indeed a sight worth beholding. Nowhere else on this continent, if indeed in the world, is there so much powerful and costly machinery employed in mining work.
>
> Annual report, Michigan's commissioner
of mineral statistics, 1885

The United States celebrated its centennial in 1876 by mounting a proudful exposition in Philadelphia's Fairmount Park. Within the large Machinery Hall, filled with mostly American products, manufacturers displayed their state-of-the-art goods: metal and wood-working machines, pumps, printing presses, rock drills, steam traction engines, locomotives, cannon, corset looms, and even popcorn machines.[1] Industrial equipment that normally toiled in functional mines, mills, or factories now performed for an admiring public under colorful bunting and amidst patriotic regalia. A massive Corliss steam engine, embodying 1500 horsepower and weighing 700 tons, served as the centennial's centerpiece. Its vertical cylinders stood high in one of Machinery Hall's transepts. Visitors came to admire it, to look up to it. The power of this great engine drove the other marvels of modern technology throughout the hall.

The American-built Corliss engine was an appropriate symbol of a young nation on the move. The young Republic had started the nineteenth century as a weak sister of Great Britain and other western nations. Technologically, the country was a late starter in the Industrial Revolution. For a time it borrowed, rather than created, new technologies.[2] But by mid-century, Britishers and Europeans took note of American progress in the mechanical arts. The Corliss engine at the Centennial Exposition was doubly meaningful. First, it testified to the fact that America had earned a high rank for its technological acumen among the industrializing nations of the world. Second, the marvelous engine made it clear that, during the process of industrialization, the harnessing of steam power was key to much economic growth and technological change. The nineteenth century was the Age of Steam.

Nowhere in America did the steam engine bring more change than at the Lake Superior copper mines. The industry grew to international importance as a copper producer by applying steam power to numerous tasks—and many of the engines it used came to dwarf the Centennial's famous Corliss in size and horsepower. By late in the nineteenth century and up until 1920, this northernmost tip of Michigan had one of the nation's most impressive assemblages of stationary steam engines.

The mines were never far removed from bodies of water, from Lake Superior, Portage and Torch Lakes, and numerous small streams or rivers. Yet local topography made it virtually impossible for most companies to harness water power. Mines and waterfalls rarely coincided. Only the Copper Falls, Nonesuch, and Victoria mines made much use of water wheels or turbines to power stamps or compress air.[3] When other companies replaced human or animal power with mechanical power, they used water—but only by converting it into pressurized steam within a boiler. When piped into an engine's cylinder, this steam acted first on one side of a piston, and then on the other. The pressurized steam drove the piston back and forth, and a crankshaft converted the reciprocating, linear motion of the piston into rotary motion. Gears or pulleys affixed to the engine's rotating shaft could power a variety of machines, or a drum mounted on the shaft could wind up rock from the depths.

The transformation of the Lake copper industry at the hands of steam power started early. The first steam engine, erected at the Lake Superior mine near Eagle River, arrived in 1845.[4] By 1857, 48 engines worked in the district. The Cliff and North American mines each had six, and Copper Falls ran five. The Minesota mine had four engines and the Toltec and Forest mines operated three a piece. Altogether, 20 separate companies had purchased and erected steam engines having an aggregate 1300 horsepower and an estimated value of $309,000.[5]

These engines embodied various configurations; "British style" beam engines with vertical cylinders, using either high- or low-pressure steam; "American style" horizontal, high-pressure engines; and smaller portables. The mines' early engines were not designed specifically to do only one job. Instead, mechanics could gear up these general-purpose prime movers to perform various functions. In 1857, about 14 engines operated as hoists; 18 drove stamps; and only one served full-time as a mine pump. The remaining engines could hoist, pump water, or power stamps, as needed.

The Keweenaw's native forest supported the diffusion of steam-power technology. No coal deposits existed in the vicinity, and shipping coal up from the lower Great Lakes would have been an expensive proposition, especially before the opening of the Soo Locks in 1855. But woods surrounded the mines, and for a quarter-century companies used this natural resource almost exclusively to fire their boilers. They hired woodchoppers, often French Canadians, to harvest the forest and to split and stack vast amounts of cordwood. When companies exhausted the closest woodlots, they bought up more distant timber tracts and turned them, too, into acres and acres of stumps. The steam engine carried an environmental cost: the companies clear-cut much of the Keweenaw to feed its voracious appetite.[6]

Readily available wood encouraged the early use of steam power, but other factors caused companies to be somewhat conservative when first investing in this technology. Working on the frontier, and shut off by ice from their shipping lanes for nearly five months a year, early companies found it difficult to keep large and complex engines in good repair. George M. Bird & Co. of Boston built many of the mines' first engines. Getting replacement parts from such a distant manufacturer could take half a year or more. By the late 1850s and early 1860s, conditions improved when two Detroit firms, J. B. Wayne & Co. and Hodge and Christie, began providing the mines with engines and foundry service.[7]

Still, before the Civil War era the mines did not have the full range of facilities needed to tend to steam engines and other heavy equipment. If a casting broke or a part wore out, a mine blacksmith was limited to what he could fix at his forge with hammer and anvil, and companies had very modest machine shops. But as the mines grew deeper they became more dependent on steam power and less willing to suffer breakdowns in equipment. They required better-equipped shops, where machinists could both repair and fabricate equipment. Also, sizable foundry and machining companies located on the Keweenaw to offer the mines technical support. The mechanization of mining proceeded more smoothly once the Lake Superior Iron Works and the Portage Lake Foundry and Machine Works entered business.[8]

Under certain circumstances, a mining company—particularly a new, developing one—was wise to invest minimally in steam technology. Expensive machinery sometimes drained a company of capital, causing it to skimp on the work that counted most—opening up the underground. Spending too much on surface improvements was the bane of many young firms. Smart companies used their invested capital to buy labor, sink shafts, and drive drifts.[9] On the surface, they erected only essential facilities, which they often housed in primitive buildings. Instead of buying a separate steam engine to hoist rock from each shaft, they often located an engine where its rope or chain could be run to as many as three different shafts.[10] Companies were wise to defer the erection of an elaborate physical plant until it could be paid for not by investors' capital but by money made from the sale of copper.

Early steam engines on the Keweenaw often saw decades of service and numerous types of duty. Companies put a premium on durability, on ease of repair, and on flexible engines that could be uncoupled from one task and bolted down elsewhere to do another. For three decades, they engaged in much horse-trading and jury-rigging of equipment, and sought short-range, expedient solutions to their power needs. Two things changed all this: the rise of Calumet and Hecla, which spent money on equipment more lavishly than any other mining company in the world; and the sinking of mines like Tamarack and Quincy to greater depths, which caused them to purchase specialized hoisting engines that reached down further, hoisted larger loads faster, and yet ran at less cost.

It was fortunate for Alexander Agassiz that he presided over Calumet and Hecla, and not one of the area's lesser mines. Only the singularly rich C&H could afford Agassiz's technological style and fully satisfy his ample ego. While residing on the Keweenaw in the late 1860s, Agassiz became convinced that the Calumet conglomerate was one of the world's truly great mineral deposits. A great lode merited a great mining company—one that would do things its own way and never settle for less that the best. In terms of his technological style, Agassiz believed that foresight and planning could eliminate many risks involved in mining. He believed it wiser to spend money up front to prevent a problem, rather than to spend it later to fix a problem. That philosophy immediately set C&H apart from other mines on the Lake.

Under Agassiz's leadership, Calumet and Hecla built two surface plants. The first was needed to operate the mine. The second was needed if anything went wrong with the first. Michigan's Commissioner of Mineral Statistics, upon observing C&H in 1882, offered the opinion that "it may be with companies as with individuals, [that] an immense and assured income begets a seeming extravagance."[11] Agassiz never admitted to extravagance, because the word implied wasteful excess. "Extravagance" was the wrong word; Agassiz preferred "insurance":

> Your Directors have deliberately incurred large expenditures in order to insure the stockholders, as far so practicable, against all possible contingencies. The ordinary delays and unlooked for mishaps and accidents incident to so large and varied a business . . . make it imperative that the provisions for insurance should be on an extensive scale. Nothing short of a practical duplicating of the machinery and plant, could guard the stockholders against the disasters incident to a mining company's work.[12]

When Agassiz joined his penchant for duplication with his belief in centralized technological systems, the result was a collection of steam engines second to none in the world. He did not choose to scatter small engines and boilers all across C&H. Instead, he consolidated C&H's steam power into a more finite number of large engines. Agassiz believed it more efficient to operate fewer but larger engines, and to have each one power several different things. Also, Agassiz thought that expensive equipment should meet long-term needs, not immediate ones. When C&H ordered a steam engine, he expected it to serve for several decades.

Erasmus Darwin Leavitt translated Agassiz's philosophies into working engines. Leavitt was born in 1836 in Lowell, Massachusetts, which was then one of America's premier industrial cities. Along with ten major cotton mills, the city boasted the Lowell Machine Shop, a notable builder of textile machinery, water-turbines, and steam locomotives. Leavitt apprenticed at the Lowell Machine Shop for three years before leaving in the mid-1850s to work for a series of engine-building firms. He stayed at Corliss and Nightengale in Providence, Rhode Island, for a year, then went to the City Point Works in South Boston, where he had charge of constructing the engine for the flagship U.S.S. *Hartford*. From 1859 to 1861 he served as chief draftsman for Gardner & Company of Providence, and after serving in the Navy during the Civil War he was detailed to the U.S. Naval Academy for about two years as an instructor in engineering. In 1867 he left the Academy to strike off on his own as a consulting mechanical engineer. Leavitt soon achieved great success in designing water-pumping engines for Lynn and Lawrence, Massachusetts. Upon the recommendation of Lowell's famed hydraulic engineer, James B. Francis, in 1874 Leavitt became a consultant to the Calumet and Hecla Mining Company.[13]

Leavitt was affiliated with C&H for 30 years, and the mine provided him with excellent opportunities to showcase his talent and enhance his reputation. Leavitt presided over the American Society of Mechanical Engineers in 1883, and upon his death in 1916, his eulogy in the society's *Transactions* ranked him at the top as a machinery designer:

> . . . it can be said that he did more than any other engineer in this country to establish sound principles and propriety of design. He appreciated the importance of directness and the absence of ornamentation in strictly utilitarian designs, and he firmly believed that beauty in machine design came from propriety.[14]

On his first job for C&H, Leavitt designed a steam pumping engine for the company's stamp mill. In 1876 he designed a second pump for this purpose, having twice the capacity of the first. In 1877, C&H put its third Leavitt engine into service. This engine, put to work on the Hecla branch of the mine, initiated C&H's reliance on large engines serving multiple purposes. The 1,000-horsepower Hecla engine drove four hoisting drums of 24-foot diameter: a pair of air compressors; plus rock breakers and other mine machinery.

Leavitt's fourth engine for C&H, destined to become one of the company's most recognized symbols of technological prowess, bore the accurate if immodest name of "Superior." This "monster" engine, contracted for in 1879 and started up nearly two years later, was "the largest stationary engine in the world" and cost nearly $100,000. Built by I. P. Morris, the noted engine-constructing firm of Philadelphia, the Superior engine incorporated German steel from the Krupp Works in its main shafts. The engine tipped the scales at 700,000 pounds and presented 4,700 horsepower, of which 2,700 were considered its normal working load. The engine drove four winding drums 20½ feet in diameter, which could hoist from a depth of over 4,000 feet. Running at 50 to 60 rpm, the engine also turned two flywheels, 32 feet in diameter, which served as belt pulleys. These 45-ton wheels carried 30-inch-wide belts that took power off the engine for myriad purposes. In addition to hoisting from four shafts, the Superior engine, through belts, gearing, or wire-rope power transmission, powered the surface tram road, two large air compressors, two mine pumps, and two man-engines. Power from the engine traveled as far as 2,000 feet along the mine site before being applied.[15]

Calumet and Hecla added other large steam engines to its works throughout the nineteenth century, and the company individualized these costly machines by giving them distinctive names. The engines were sources of corporate pride and symbols of prosperity. With their bright iron work and polished brass, they stood for the power that was C&H.

On the Calumet branch, in addition to the Superior, one found the "Baraga," "Rockland," and "Mackinac" engines, the latter a triple-expansion "steel giant" of 7,000 horsepower. On the Hecla branch, Leavitt's 1877 engine was joined by the "Frontenac," "Gratiot," "Houghton," and "Seneca," engines of about 2,000 horsepower each, plus auxiliary Corliss engines such as the "LaSalle" and the "Perrot." On the South Hecla, major engines bore the monikers of "Hancock," "Pewabic," "Detroit," and "Onota."[16] At the Red Jacket shaft, C&H installed ten 1,000-horsepower boilers. An engine house, 220 feet long and 70 feet wide, sheltered four steam engines embodying about 8,000 horsepower. Like the Superior engine built 15 years earlier, Red Jacket's two primary hoists, the "Minong" and "Siskowit" engines, were designed by Leavitt and built by I. P. Morris.[17]

By the late 1890s, C&H had a least 50 stationary engines at the mine, plus more at its mills and smelter, plus a stable-full of steam locomotives serving its railroad. Its stationary engines alone totaled about 50,000 horsepower. C&H produced "as much power as is now being generated by the great electrical plant at Niagara Falls, and about equal to the power used in the average manufacturing city of 200,000 people."[18] At the mine, this power was distributed to some 40 major structures. It drove machinery at eleven three-story tall shaft-rockhouses standing in a line over a two-mile stretch of the Calumet conglomerate lode. It ran a mill for sawing, mortising, and tenoning 13 million board-feet of square-set timbers per year; it served carpenter shops equipped with every woodworking appliance; several large blacksmith shops housing steam hammers and forges; a large shop for forging and sharpening drill steels; and a "mammoth" machine shop "filled with [the] costliest and best planers, lathes, drills and machinery of all kinds."[19]

Calumet and Hecla's surface facilities dwarfed those of all other copper mines on Lake Superior. Still, all the mines, large and small, used steam power to mechanize their surface work. At every productive mine, steam engines drove machinery within a variety of shops that repaired or fabricated equipment. The application of steam power had its greatest effect, however, on two technologies in particular: the hoisting and breaking of rock.

At mines opened before 1870, steam power was often the third technology applied to hoisting. Typically, the earliest companies used men, then animals, then engines to raise copper rock. Men cranked windlasses, whose chains or hemp ropes drew up kibbles from the shallow bottoms of new mines. Two men on a single windlass could hoist about four tons of rock from 100 feet in one shift. Once several drifts branched off from a shaft, allowing more miners to produce more rock, windlassmen could not keep up with the output, so companies turned to horsewhims. A whim carried a winding drum on a vertical axis. Horses, harnessed to booms or sweeps, walked around in circles all day, winding up kibbles from the depths. A single horse could raise about nine tons per day from 150 feet underground. When a shaft's output and depth began exceeding those figures, a company next moved to steam.[20]

By 1870, all the basic pieces of steam hoisting technology were in place, and they stayed in place into the twentieth century. Wheeled skips had replaced kibbles; wire rope had replaced hemp rope or hoisting chain; and steam engines powering large winding drums replaced men or horses. But while hoisting technology remained the same in gross outline after 1870, it changed immeasurably in all details.

Companies started with a single rock skip in each shaft that had a capacity of about three tons. Most moved to skips having six- and then eight-ton capacities, and many firms turned to balanced hoisting and put two skip roads in each shaft.[21] Balanced hoisting saved on energy costs. While a full skip was being hoisted, a just-emptied skip was descending. The considerable weight of the descending skip and its hoisting rope helped raise the loaded skip, reducing the load on the engine. Also, balanced hoisting nearly doubled the rock tonnage raised per day. While one skip was being dumped at the top of the mine, the other was already at the bottom, waiting to be filled.

Companies started with steam hoists that were general-purpose engines connected to winding drums via friction or gear drives. They turned to engines designed specifically for hoisting, where the drum and the engine were integral and direct-connected. They went from engines having a single cylinder to hoists having two, three, or even four cylinders. They saved on fuel costs by purchasing compound engines that used the same steam at least twice. Many later engines were also equipped with condensers that made them more fuel efficient. Finally, companies changed their winding drum designs. Straight, cylindrical drums remained common, but tapered or conical drums made an appearance. As a hoist started to lift a loaded skip, its rope wound from the smaller end of the drum up onto its wider portions. This drum shape gave the engine a mechanical advantage in initiating each lift, and it also smoothly accelerated the skip as it rose, because each revolution of the drum wound in a somewhat greater length of hoisting rope.[22]

In the mid-1850s, 80-horsepower engines were the largest on Lake Superior, and all 48 engines in use represented a total of 1,300 horsepower. Fifty years later, single hoisting engines rated at 2,000 to 2,500 horsepower were common, and the largest generated 7,000 horsepower. Early engines lifted a ton or two of rock at 500 feet per minute from shallow depths; later engines raised eight or even 10 tons of rock at 3,000 to 3,500 feet per minute from as deep as 5,000 feet.[23] Calumet and Hecla showcased its Leavitt engines; Quincy, Tamarack, and other mines, particularly after 1890, showcased engines built exclusively for them by E. P. Allis & Co. and the Nordberg Manufacturing Company, both of Milwaukee.

Successive generations of steam hoists made great leaps in production levels possible at individual shafts and allowed the mines to capitalize on the enlarged market for copper that accompanied the rise of the electrical industry. In 1905, when reflecting on "improvements between now and twenty years ago," agent John Harris at Quincy expressed the view that the greatest and most significant changes had been wrought by "better mechanical appliances," such as steam hoists.[24] The district was indebted to mechanical engineers who lived at considerable distance from the Keweenaw Peninsula and who worked for firms such as I. P. Morris, E. P. Allis, and Nordberg.

It did a mining company no good to install larger and faster hoists unless the rest of its surface plant could handle the greater tonnages of rock delivered to the surface. New hoists spurred companies to eliminate production bottlenecks at other facilities handling or treating rock. Surface technologies for rock receiving, sorting, and crushing changed dramatically, as best illustrated, perhaps, by the history of the Quincy mine. Quincy started out using the labor-intensive, traditional technologies in vogue in the 1850s, but by 1910 had the most efficient rock handling system on the Lake.

In Quincy's early years, rock arrived at the surface within a simple board-and-batten shaft house erected over the shaft collar. This structure contained no machinery, but did have equipment for tipping the kibble or skip. Dumped rock slid down an incline into small cars, which men pushed into an adjacent sorting house. This structure, too, featured no machinery or significant mechanical aids. Laborers picked over and sorted the material by kind and size. They sent poor rock, barren of copper, to nearby "burrow" or waste piles. They packed small pieces of mass copper into barrels, which later went directly to a smelter. Pieces of copper rock, already less than three or four inches in diameter and thus small enough to fit into Quincy's stamping machines, went directly to the mill. All other copper rock, and all mass copper in need of cleaning before smelting, was sent to one of three or four kiln houses to be "calcined."

At each kiln house, 10 to 25 laborers "burned and dressed" copper within a shallow pit about 75 feet long and floored with heavy cast-iron plates. They built a bed of timber, some 20 feet square and four feet high, which they set on fire after covering it with four to six feet of copper rock and mass copper.[25] The heat cracked the rock; laborers sometimes aided this process by pouring water on the rock just as the fire was dying out. When the rock cooled, men descended on it with sledge hammers, breaking up any pieces still too large to go to the stamps. While men in one end of the pit wielded their sledges, men in the other end started up another fire.

The slow hand-sorting of rock, coupled with the technology of calcining, required numerous laborers, consumed considerable fuel, and created a production bottleneck on the surface. Companies searched for appliances and machines that could hasten the flow of rock from shaft collar to mill, and major improvements were not long in coming. Quincy and other mines simplified the job of sorting rock on the surface by being more conscientious about this job underground. They left most broken poor rock underground, thus saving the time and expense of hoisting it. When they did hoist poor rock, they tried not to mix it with copper rock. They could then send poor rock directly to a waste pile without picking over it. Men still hand-sorted the mine product to some extent, but this largely became a matter of picking out the barrel copper that did not need to be milled. To sort copper rock by size, the mines eliminated much hand labor by adopting the simple yet effective "grizzly."

A grizzly was nothing more than a heavy-duty screen set on an incline. Parallel steel rails, set about four inches apart, ran longitudinally from the high end to the low end of the grizzly. Copper rock, destined for the stamp mill, was dumped onto the top of the grizzly. As it slid down the rails, smaller pieces fell through the screen into cars or bins below; this rock then went straight to the mill. Bigger pieces that tailed off the grizzly still needed to be broken to three-or four-inch size.

To do this, Quincy and other companies replaced calcining with mechanical means of rock breaking. They acquired steam hammers and drop hammers. Steam hammers cleaned barrel copper of adhering rock. Heavier drop hammers cleaned larger masses of copper and also broke up the biggest pieces of copper rock coming from the mine, which were sometimes two feet or more in diameter. But the most important machine applied to rock breaking was the Blake jaw crusher, which was belt-driven by a steam engine.

In 1852 Eli Whitney Blake, named for his uncle, the famous small-arms manufacturer and inventor of the cotton gin, served on a committee superintending the macadamizing of a road near Westville, Connecticut. The committee encountered problems finding inexpensive broken stone for the roadway, so Blake turned his hand to designing a rock crusher. Finally successful in this quest, he patented his crusher in 1858. The machine carried two cast-iron jaws, one fixed and one movable. The jaws sat in nearly upright positions, with the space between them narrowing from the top down to the bottom. Rock dropped into the machine wedged between its jaws. The movable jaw repeatedly moved in and out, crushing the rock against the opposite jaw each time until it was fine enough to pass out the bottom of the crusher.

A machine intended to help build a short road in Connecticut began having an important impact on the American mining industry in 1861, when Blake set up a crusher to replace 25 sledge-wielding laborers at a gold mine in California: "When the first block of quartz was dropped between the jaws and disappeared below in a heap of fragments, the Chinamen crowded around in amazement, and, realizing that their occupation had gone, threw their hammers away."[26] Lake Superior kilnhouse workers began throwing their hammers away in the mid-1860s. By 1865, the Evergreen Bluff, Ogima, National, and Pennsylvania mines had already acquired jaw crushers, and other mines followed suit within the next decade.[27]

Once introduced, grizzlies, drop hammers, steam hammers, and jaw crushers remained in use throughout the mines' working lives. The individual pieces of this technology remained largely the same, but over time the companies greatly altered the arrangement of this equipment. Only one rock-breaking "revolution" occurred in the district when, in the 1860s and 1870s, jaw crushers and rockhouses replaced kilnhouses. But the technology evolved over the next half-century, and periodic reorganizations of the hardware whittled away at the cost of breaking rock and the amount of labor involved.

In exploring technological change, it sometimes proves seductive to emphasize the dramatic effects, wrought almost overnight, by the arrival of a new machine, such as the jaw crusher. But "breakthrough" machines are relatively rare, and once in place they may stay there for decades. Much important technological change, then, does not involve new inventions; it involves using tried and time-tested inventions in new ways.

In 1873 Quincy replaced about four kilnhouses with a single rockhouse containing two steam engines, grizzlies, five Blake crushers, a one-ton drop hammer, and a steam hammer. In building a single rockhouse to handle the output of several shafts, Quincy followed common practice on the Lake. It cost less to build a centralized rock house than to put breaking machinery at each shaft. Besides, the new breakers were so efficient they could easily handle all the rock coming out of the entire mine. Even the giant C&H had but one rock house for its Calumet branch, and one for its Hecla branch, until the early 1890s.

Quincy's three-story rockhouse contained a small steam engine that powered an endless rope tramroad running past the shafts. At each shafthouse, mine product fell into tramcars that traveled along a trestle and into the rockhouse's top story. The cars dumped their contents onto a grizzly. The fine rock fell straight through to the bottom of the building, where it was chuted into cars that headed toward the mill. Larger pieces spilled out over the end of the grizzly and onto the floor. Here, men continued to do some sorting by hand. They sent barrel copper to the steam hammer for cleaning, while mass copper went to the bigger drop hammer. They put most material, however, into the crushers, which were powered by the second, larger steam engine within the building. Large crushers gave the copper rock a preliminary break, and this rock then fed into a second set of smaller crushers on the floor below.

Quincy had some early troubles with a balky tramroad and with crushers that broke themselves instead of rock, but after a brief shakedown period it put its new technology into economical operation. Calcined rock cost the company 48 cents per ton in 1872. Machine-broken rock cost only 30 cents per ton by 1875, when Quincy broke a total of 72,000 tons of copper rock and produced about 3 million pounds of copper.[28] Such cost savings could not have been effected at a more opportune time. The price of copper stood at a high 33 cents per pound in 1872, but steeply declined over the next several years, falling to a low 16.5 cents per pound by 1878. Quincy protected its profit margin in part by adopting the new rock-breaking technology. The price received for its copper declined, but so did Quincy's production costs.

Several features of Quincy's early rockhouse were notable. By bringing material in at the top and processing it as it moved downward, the building used gravity to advantage. Still, the building was only three stories tall, and its system of storage bins, gravity feeds, and chutes was rather undeveloped. While the new technology eliminated much hand-sorting of copper rock by size and required no sledgehammer-wielding men, by no means did it eliminate all human labor. Men still picked up tens of thousands of tons of rock that tailed off the grizzly and came to rest on the rockhouse floor. They dropped the rock into crushers, or rolled it over the floor to the drop hammer. In 1883, a decade after inaugurating the use of jaw crushers, Quincy still employed 22 men over two shifts to feed the crushers.[29]

Eventually, with the advent of machine rock drills and larger, faster hoists, each mine shaft became a much bigger producer. Companies had to dispense with their one, centralized rockhouse and put rock-breaking machinery at each shaft. They grafted a rockhouse onto a shafthouse and created a new, hybrid mine structure, the "shaft-rockhouse."

A few shaft-rockhouses had been erected in the 1870s and 1880s, but they did not become ubiquitous fixtures at the mines until the 1890s, when the two largest companies started building them. Between 1891 and 1893, C&H built eight shaft-rockhouses along the Calumet conglomerate lode. Quincy built its first one at No. 6 in 1892, and put others over No. 2 in 1894 and over No. 7 in 1900.[30] These structures all went up just when Quincy was equipping these three shafts with new E. P. Allis hoists rated at about 2,500 horsepower.

Quincy's shaft-rockhouses, standing tall atop the hillside overlooking Portage Lake, became the most dominant and photographed man-made features on the Keweenaw landscape. Framed of heavy timbers, clad with clapboards, and covered over with a step-like series of double-pitched roofs, Quincy's 100-foot-tall "many-gabled" shaft-rockhouses presented a singular architectural feature. Inside, however, little was new in terms of rock-breaking technology. Quincy's No. 6 shaft-rockhouse contained almost the same machinery as the company's first rockhouse, built nearly 20 years earlier.[31] Two skip roads rose on a steep angle to near the top of the structure. Each skip road dumped rock onto its own grizzly. Rock too big to pass through the grizzly still came down to the main crusher floor, where men had to handle it. If no bigger than about a foot and a half, the rock went to one of two small crushers; bigger rock went to a larger crusher, which fed into smaller crushers on the floor below. The No. 6 shaft-rockhouse also held a steam hammer, a drop hammer, and a horizontal steam engine that drove the equipment.

Quincy's first shaft-rockhouses represented improvements over its 1873 rockhouse, but these mostly lay in the flow of rock into and out of the structures. Because they were located right at the shafts, these buildings eliminated the need for a tramroad running the length of the mine to collect and deliver rock. By being so tall, they also provided room for large storage bins for stamp rock. These bins stood over railroad sidings, a part of the company's newly constructed Quincy and Torch Lake Railroad. The bins collected rock from the grizzlies and crushers and periodically a train of rock cars passed under them. They spilled their contents into the cars, and the train headed off to Quincy's new stamp mill on Torch Lake.

While the early shaft-rockhouses expedited the receiving and shipping of rock, they did little to speed up or reduce the labor costs of sorting, crushing, and cleaning operations. In 1904, for example, 28 men worked in Quincy's No. 7 shaft-rockhouse, where they handled no more than 900 tons of rock per day. In search of cost-cutting, labor-saving measures across all parts of its mine, Quincy rethought its traditional arrangement of grizzlies, crushers, drop hammers, and steam hammers. By rearranging and rebuilding part of its No. 7 shaft-rockhouse, by 1907 the company cut rockhouse labor teams at that shaft from 28 to 5 men, while doubling its production capacity.[32] Quincy then incorporated these improvements into an entirely new shaft-rockhouse built at No. 2 in 1908.

Quincy's steel-framed, 150-foot tall No. 2 shaft-rockhouse was one of the best-engineered, most efficient structures ever erected on the Lake. In one neat package it facilitated the handling of skips, tools, poor rock, copper rock, and mass and barrel copper. On the ground floor, overhead cranes stood ready to help men attach rock skips, water-bailing skips, or man-cars to the two hoisting ropes. Nearby, a lander worked a set of levers to dump skips right where he wanted. Pulling a lever caused movable sections of the skip tracks to slide apart, in effect increasing their gauge. Because the front wheels of a skip were narrower than the rear wheels, when they encountered this broader gauge they switched onto dump tracks instead of continuing upward. As the hoist engineer wound in a bit more rope, the rear of the skip stayed on the main track, tilting the skip and spilling its contents.

The lander emptied skips laden with tools (such as bundles of dulled drill steels) at the first dump above the shaft collar. The tools slid down an iron-clad ramp onto a landing served by a rail line passing through the building. Quickly loaded onto railcars, these tools went to the mine's blacksmith, drill sharpening, or machine shops. The lander unloaded poor rock at the second dump. The rock spilled into a small storage bin. If Quincy just wanted to get rid of the poor rock, it chuted it directly into railcars. However, if the company needed some broken rock for concrete aggregate or railroad ballast, the rock passed through a small crusher on its way to the railcars.

The third and uppermost dump handled mass and barrel copper and vast tonnages of copper rock. Skips loaded with these materials spilled their contents onto a novel grizzly. Actually it was two grizzlies, one placed over the other. Large steel bars, 20 inches apart, formed the top grizzly. All material larger than this tailed off the grizzly and fell right next to a 3,000-pound drop hammer. Men no longer had to pick up or roll these large pieces by hand, because the area was served by a traveling air-lift mounted on an overhead track. If the large piece was mass copper, a few strokes of the drop hammer cleaned it of adhering rock, and a crane lowered it down the outside of the building to a flatcar bound for the smelter. If the piece was copper rock, one blow from the drop hammer broke it into fragments, which were carried by the air-lift over to nearby crushers for a final breaking.

All the rock passing through the 20-inch grizzly fell onto a second grizzly with bars only 2¾ inches apart. Any rock fine enough to slip through these bars fell into a large stamp rock bin, elevated over rail lines. The rock tailing off the second grizzly followed a new and improved path. For more than 40 years, such rock had always fallen off the grizzly onto the floor, where men had to pick it up. Quincy finally eliminated this work, by tailing the rock over into an elevated bin. Air-operated bin doors then chuted the rock directly into one of two large crushers, each capable of breaking 40 tons of rock per hour. Men no longer lifted the vast majority of the mine product once it reached the surface.

One man tended each crusher. He watched the rock slide toward it on the chute. Most of it passed right by him, through the crusher, and fell into a large stamp rock bin, which chuted its contents to fill a railcar in 10 seconds. If the crusherman spotted a piece of poor rock, he lifted it out, dropped it through a hole in the floor, and a chute carried it to the poor rock bin. If he spotted a piece of barrel copper, he dropped it on a short incline, only a few feet away, that delivered it at the foot of a steam hammer. A third man on the crusher floor ran the drop hammer and the steam hammer. Barrel copper, once hammered clean, got dropped into a chute of its own that led to a storage bin for this material, again elevated over a rail siding for gravity feeding into railcars. Just three men working on the crusher floor could process 1,000 tons of rock per day.[33]

The rail lines running beneath the mines' shaft-rockhouses presented further evidence of the importance of harnessed steam power. This industry produced vast tonnages of materials that had to be moved, and ultimately steam power, in the form of locomotives, transported stamp rock from the mines to the mills, and then mineral concentrates from the mills to the smelters. Railroads played a major role in the expansion of the copper industry in the late nineteenth and early twentieth century.

In the earliest decades, when production levels were low, men or draft animals shouldered the burden of moving rock along the surface. Men first pushed rock-laden wheelbarrows. Improving upon this, companies laid narrow-gauge rail lines across their surface plants, and they replaced wheelbarrows with rock cars either pushed by men or pulled by horses or mules. Later, by the late 1860s or early 1870s, some firms improved surface transportation around their shafts by building endless-rope tramways (short "cable-car" systems). The real transport bottleneck, however, was not at the mine proper but between the mine and its stamp mill.

Some early, small operations had been able to site their mills on small ponds or lakes located close to their shafts. Later, larger companies required far vaster quantities of water. They also wanted their mills to stand beside deep lakes, so that year after year they could dump hundreds of thousands of tons of mill tailings into them. This meant that the region's major mining companies had only three places to site mills: the shorelines of Portage Lake, Torch Lake, and Lake Superior.

Some companies were fortunate; their mines were close to one of these bodies of water. Particularly advantaged were Pewabic, Franklin, and Quincy. They were less than half a mile from Portage Lake, and their mines were elevated some 550 feet above it. In the 1860s, these three companies connected their mines and mills in a very cost-effective way: they built gravity tramroads, double-tracked, down the steep hillside.[34] At the top of each incline sat a drum house, so equipped that it wound in cable for one set of tracks, while letting out cable for the other. The winding drum did not require any steam engine for power, because gravity did the work. The weight of loaded cars traveling down to the mill drew empty cars back up.

Other companies were not so well situated. Their mines and suitable mill sites were often 5 to 15 miles apart. And even the mines close to Portage Lake found themselves with a transportation problem by the 1880s. By then, the federal government had set boundaries for how far out into this navigable waterway they could extend their stamp sands. Quincy, after using its incline and Portage Lake mill for a quarter-century, had to move its milling operations over to Torch Lake, six miles distant from the mine. Quincy now faced the same transportation problem that many other companies had already solved, and it solved it in the same way: it built the Quincy and Torch Lake Railroad.

The first steam locomotive arrived on the Keweenaw in 1864, when the jointly managed Pewabic and Franklin mines purchased a small engine and put it to the task of delivering rock to their stampmill inclines.[35] The first real railroad company to provide service was the Hecla & Torch Lake, which in 1868 began carrying stamp rock some four miles from the then independent Hecla mine to its mill. This line, expanded, later carried the full mine product of the combined Calumet and Hecla company over to Torch Lake for milling, and on their return trips the trains carried timber, coal, and other supplies taken off company docks.[36]

Following Calumet and Hecla's lead, other mining companies spun-off their own railroads, and independent investors erected railroads as well. By the 1880s, the Keweenaw had a network of narrow-gauge, standard-gauge, and dual-guage railroads running along its spine and radiating off to the lake fronts where the mills and smelters were. By the mid-1880s the first railroad bridge spanned Portage Lake, tying the northern and southern ends of the Keweenaw together, and a rail line heading south from Houghton connected with others that linked the mining district with such major midwestern cities as Milwaukee and Chicago.[37]

In addition to railroads operated for the benefit of Quincy and Calumet and Hecla, the Isle Royale Railroad connected that company's mine and mill, as did the Atlantic and Lake Superior Railroad. The Copper Range line ran from a select group of mines out to their respective mills on Lake Superior, and then back to a shared smelter on Portage Lake. The Delaware, Mohawk, and Allouez mines also operated their own railroads at one time. The greatest concentration of rail service, not surprisingly, was in the "loop" from Hancock past the Quincy and Osceola mines up to Calumet; then over to the myriad mills along Torch Lake; and back to Hancock via the Dollar Bay, Quincy, and Lake Superior copper smelters. The Mineral Range and Hancock & Calumet railroads formed the perimeter of this loop; the mining companies' smaller, individual lines tied into it at convenient points.[38]

The mine product broken by steam-powered jaw crushers, and transported by steam locomotives, fell into receiving bins at the stamp mills. Here, too, the mining companies assembled an impressive array of steam equipment—

particularly the stamps that crushed the rock and the pumps that provided water for transporting, washing, and separating the copper and the rock. Because 95 to 99 percent of the mine product received at the mill was waste rock, the companies had to stamp and wash vast quantities of it to produce much copper. So, in designing their mills, they put a premium on machinery that achieved high capacities in an efficient manner.

Similarities existed among all the stamps ever used on Lake Superior. They all reciprocated a cast-iron stamp shoe, mounted on the bottom end of a vertical stem, that crushed rock confined in a mortar box at the base of the machine. When fed into the mortar box (accompanied by a strong flow of water), the rock was as large as three or four inches in diameter. Each blow of the stamp shoe reduced the rock in size. A perforated plate formed at least one side of the mortar box, and it served as a screen controlling the outflow of material from the stamp. When the rock had been crushed finely enough, down to about three-sixteenths of an inch, the water carried the rock and liberated copper particles out through the screen and down to the machines that used differences in the specific gravities of the two materials to separate them.

From 1845 to 1855, the region's mills used batteries of "drop" or "gravity" stamps like those found in Cornwall's tin and copper mines.[39] Each timber-framed battery carried four or five small stamp heads sharing a common mortar box. A steam engine drove the stamps via shafting, pulleys, and leather belts. The main shaft on each battery carried cams, one for each stamp head. As the shaft rotated, the cams sequentially lifted each of the machine's stems. As the shaft rotated further, each stem in its proper turn fell off its cam. It dropped by gravity, and its stamp shoe struck a blow against the rock. Then the cam came around again to repeat the action.

Cornish drop stamps were not lifted very high; they were not very heavy; and so they did not apply exceptional force to the rock. When mass copper mines gave way to mines exploiting more finely disseminated copper, these later firms, with few exceptions, quickly sought to replace the Cornish stamp with a more powerful technology. Briefly they tried an altogether different technology: instead of stamping the rock, they tried crushing it between two corrugated iron rollers. These promised to be faster acting, because their rotary crushing motion was continuous. Reciprocating stamps broke rock only on their down stroke; they did no useful work while being raised for the next blow. But while theory supported the use of rollers, practice taught the mines that roller crushers broke down as much as they broke rock.[40] Instead of adopting rollers, in the late 1850s the mines turned to steam stamps, which were akin to the Nasmyth steam hammer, invented some 20 years earlier.

These new machines reciprocated the stamp head in a very different way. The vertical stamp stem ran up to an overhead steam cylinder, passed through a steam-tight packing gland, and connected fast to a piston. Steam drove the piston up, lifting the stamp. Steam also drove the piston down. The traditional Cornish stamp merely dropped onto the rock; the new technology started its stamp downward with a strong push of steam. The steam stamp differed in other important ways from its Cornish predecessor. Instead of carrying four or five small heads per mortar box, it worked one stem and one larger, much heavier stamp shoe. Some early mills had housed ten or more batteries of Cornish stamps, giving them 50 or more stamp heads. Later mills crushed far more rock using only two to five steam stamps.

The Lake Superior copper mines made greater use of steam stamps than any other mining district in the United States, and they periodically improved these machines, both to increase their capacities and to make them more fuel-efficient. The Copper Falls mill put the district's first steam stamp into operation in 1855.[41] Edwin P. Ball of Massachusetts had developed this machine, and Ball's steam stamp remained the industry's unchallenged standard until the early 1880s, when Erasmus Leavitt, serving again in his consultant's role, designed powerful steam stamps for Calumet and Hecla's mills. Leavitt's stamps incorporated two pistons per machine and had novel features for conserving steam. E. P. Allis & Co., beginning in the late 1880s, and the Nordberg Manufacturing Company, beginning in 1902, also engineered and built massive steam stamps for this district.[42]

Using Cornish drop stamps, a single head crushed approximately two tons of rock per day. Early Ball stamps, with nine-inch pistons and a working steam pressure of about 80 pounds per square inch, stamped up to 50 tons per day. The machine's reciprocating parts weighed about 2,000 pounds, and the stamp delivered 70 blows per minute. By the 1870s, Ball stamps carried 12- to 15-inch pistons, ran with 90 pounds of steam, reciprocated a weight of 4,500 pounds, and delivered 90 blows per minute. The Ball stamp then achieved production rates of 125 to 150 tons of rock daily. The later Leavitt, Allis, and Nordberg stamps reciprocated up to 7,800 pounds of machinery and crushed from 225 to 700 tons each day.

These mammoth steam stamps eliminated production bottlenecks at the mills and supported the mines' greatly increased production levels. They also demanded immense quantities of water. By the 1890s, a mill required up to 30 tons of water for each ton of rock stamped.[43] Of necessity, near the tall steam stamps sat some very large steam-powered water pumps. As usual, Calumet and Hecla was unsurpassed in this category of equipment. Its mills housed a main pump, the "Michigan," which was the largest pump in the world (with a capacity of 65 million gallons per day), and three auxiliary pumps, which had a combined capacity of 50 million gallons.[44]

Harnessed steam power radically transformed the Lake Superior copper mining industry after the mid-1840s. The nineteenth century was the Age of Steam, and the Lake mines took great advantage of it. On the surface, steam power proved extremely flexible, and the mining companies effectively applied it to myriad tasks. It drove shop machinery. It powered generators that provided the new wonder of electrical lighting. It drove air compressors. It lifted rock up from 5,000 feet below. By looping a broad leather belt around an engine's flywheel and then around a jaw crusher's drive pulley, it could be made to break rock. It transported the rock 5 to 15 miles, from mine to mill. At the mill, steam stamped and pumped and drove washing and separating machinery.

Outside observers often noted that the Lake mines modernized their surface operations more than their underground ones. The observation was correct. "Building" an underground mine was very different from building surface facilities. Underground, workers kept moving from place to place, whereas surface facilities were more permanently sited. Underground, it was dark, and it was hard to carve out the perfect workplace for men or machines. Of key importance, too, was the fact that the mining companies could not apply steam technology directly to underground work. On the surface, companies could cope with the technology's intrinsic quirks: its demand for vast amounts of fuel, its creation of equally vast amounts of smoke, and its exhausting into the atmosphere of hot and moist spent steam. Underground, these features rendered steam power an untenable technology. The power source that proved so important on the surface could not go underground, and it was at the bottom of the mines, not at the top, where technological change lagged behind.

4

Men at Work: Agents, Captains, and Contract Miners

> [W]e have no means as yet devised for penetrating the rocks which contain mineral treasures but those afforded by the patient and unremitting labour of a great number of men. The regulation, therefore, of this force, and its due application is, after all, more important to the success of mines than even the most ingenious mechanical expedients.
>
> John Taylor, "On the Economy of Mining," 1837

The industrialization of American mining occurred in the period between the Civil War and the First World War. Traditional hand technologies gave way to mechanized means of production, and deploying new machines required considerable capital. It took money to make money in the mining business, and the cost of developing new mines escalated. This fostered a reliance upon corporations as America's chief mining promoters. Incorporated mining firms over these years increased a great deal in size. In the mid-nineteenth century, a large mine employed only two to four hundred men; by early in the twentieth century, a large mine employed two to four thousand. The men who directed these operations changed as well. Traditionally, mine enterprises were led by practically trained men who had learned the business from the bottom up. Over time, mine management came to be dominated by a new, modern breed of experts: engineers, who prided themselves on applying science, and not archaic rules-of-thumb, to the winning of metals.

All these transformations could be seen at the Lake Superior copper mines. Yet in the midst of much change, certain elements stubbornly remained constant. Modernity did not instantaneously brush aside all traditional practices. Prior to 1900, the mines effected far more change in their technologies than in their management structure, their leadership, or in their means of organizing men and putting them to work. At the mines, the agent and the head mining captain continued to hold the most pivotal roles. For leadership, the companies still relied on practical men, not on university-educated engineers. And in organizing their miners and promoting diligent work, the companies still retained the contract system of labor, imported, with some modifications, from Cornish tin and copper mines.

The retention of select traditions worked well for the mines in the nineteenth century, but the first years of the twentieth caused them to initiate changes in parts of their operations previously left alone. Local economic problems and the emerging gospel of efficiency engineering both pressured and seduced the firms into elevating new men into leadership posts, into reducing the stature and importance of others, and into treating contract miners more like common, wage-earning laborers. As a result, by 1910 to 1912 the mines were indeed more modern, but they were also more troubled.

The mines' business bureaucracies remained small throughout the nineteenth century. Atop the hierarchy stood the agent (after 1900, often called the superintendent or general manager). He regularly corresponded with eastern officers, passing along newsworthy information, such as the yields taken from different parts of the mine, any accidents that caused interruptions, the state of labor relations, and recent changes in the company's set routine. The agent passed information up and waited for directions to come back down. He supervised several key personnel serving under him. A chief clerk handled the business end of mining. He kept the books, made payrolls, ordered supplies, and paid bills. Other subordinates managed technologies and workers: a head mining captain oversaw all underground operations; a surface captain or boss managed the mine's shops and machinery above ground; and superintendents headed other discrete parts of the operation, such as its stamp mill or railroad.

The expense of drilling and blasting rock accounted for approximately half the total cost of producing copper, and this fact made the head mining captain's position very important. He decided how to follow and exploit the lode and assayed copper values throughout the mine. He determined which portions of the lode should be stoped out, and which should be skipped over. A truly good head captain proved a tremendous asset to his company. In May 1873, Captain John Cliff wanted to resign from Quincy to take the agent's job with the Albany & Boston company. At this particular time, the copper market was deteriorating and prices were falling. Quincy's agent, A. J. Corey, advised his company not to accept the Captain's resignation:

> I regard it of the highest importance for us to retain his services as long as possible, as the loss of his long experience and practical knowledge of our vein . . . would be seriously felt, and to put a stranger in charge of the underground work, at the present time, could hardly fail to affect considerably our production for the season.[1]

The head captain embodied diverse expertise. He was part manager, part geologist, and part mining engineer. Professional geologists had encouraged the mine rush to the Keweenaw, but once started, the industry had little use for geologists for the next half-century. The copper companies, instead of hiring scientifically trained geologists, relied upon their experienced head captains, using their "nose" for copper, to tell them where and what to mine. Similarly, prior to 1900 the companies employed few university-educated mining or civil engineers. The head captain, always lacking a college degree, "engineered" his company's operations.

Larger companies usually employed a head captain over all the underground, a shaft captain over each principal hoisting shaft, and two shift captains or bosses per shaft, one for day work, one for night. While the head captain orchestrated the entire mine's production, each shaft captain did his own hiring and firing and looked after the work done within his part of the mine.[2] Shift captains or bosses functioned like general foremen. They spent long hours underground, putting crews to work and monitoring their progress.

Compared with a head mining captain, a surface captain enjoyed lesser status. The underground produced a mine's wealth, and the man directing operations there merited greater esteem. Nevertheless, surface captains were bearers of essential knowledge and skill. Usually, they had started working in the mining industry when young, and years of practical experience had made them expert in erecting, operating, maintaining, and sometimes designing mining machinery and buildings. A surface captain had overall charge of his company's hoists, boilers, compressors, and rockhouses, as well as its machine, carpenter and blacksmith shops. A boss within each shop or major facility tended to daily rountines, handled the men, and reported to the surface captain.[3]

Ethnic diversity characterized the mining communities as a whole, but the men filling the companies' top posts presented a picture of ethnic exclusivity. To be sure, a few Germans and Irishmen filled top slots, along with even fewer Scots or Swedes. Native-born Americans or Cornishmen usually held the esteemed positions of agent, chief clerk, mine captain, and surface captain. Americans and Cornishmen followed dissimilar paths into leadership roles. The Americans, who tended to come from New England or New York, did not rise to the top on the basis of their practical knowledge of underground mining techniques. Instead, companies tapped them because of their business acumen, their experience with surface technologies, or prior to 1860, their familiarity with the Keweenaw's geology.

Some participants in the early geological surveys of the Lake district were not content just to discover and study its minerals. They wanted to profit from them. Two notable examples of the "geologist-entrepreneur" type were Columbus Christopher Douglas and Samuel Worth Hill.[4] Both men, when in their late twenties or early thirties, assisted Douglass Houghton in his surveys of the Keweenaw, and both stayed in the region for 20 years after the mining boom began. Over that time, Douglas and Hill actively formed, directed, or supervised fledgling mines in Houghton, Ontonagon, and Keweenaw counties, and on Isle Royale.

The geological route to a mine superintendency worked only for some of the earliest pioneers. Other Americans, such as S. S. Robinson, Daniel

Brockway, and James North Wright, climbed the business side of a company's ladder to serve as agents.[5] Of this type, Wright was clearly the most important. Born in 1839, he received a strong academic education before migrating to Lake Superior at age 20. Wright clerked at the Minesota mine, then became Quincy's chief clerk. In 1872 Quincy elevated him to mine agent, and a few years later, Calumet and Hecla hired him away. Wright superintended the C&H mine until 1888, when Stephen Betts Whiting replaced him.

Whiting and John H. Forster were two Americans selected as agents because of their technical expertise.[6] Forster studied engineering in the East and began working as a civil engineer in 1844. In the first half of the 1860s, he superintended the Pewabic and Franklin mines, which were then important, new producers on Portage Lake. S. B. Whiting, born in 1834, did not arrive in the Copper Country until he was in his mid-fifties. After attending common schools in New Haven, Whiting skipped going to Yale because he could not afford it, and instead attended the "shop culture" school of mechanical engineering. He started as an apprentice and journeyman machinist, and then moved into machinery design and management while working for a succession of engine-building firms and iron works. It was fitting that a man of his background became superintendent of C&H, a mine having one of the most elaborate surface plants in the world.

Americans served as chief clerks or agents, but not as captains. Cornishmen, on the other hand, served as captains, later as agents, but seldom as chief clerks. At first the mining companies employed many skilled Cornishmen as captains, but rarely made them agents. That changed, however, as the Cornish proved their mettle in both technical and managerial matters, and as the companies recognized the wisdom of hiring agents fully familiar with underground mining techniques. At many companies, it became common to have three Cornishmen filling the top posts of agent, head mine captain, and surface captain. The Cornish became the immigrant elite, seldom relegated to unskilled work, but often tapped as skilled laborers, bosses, captains, or agents.

At the midpoint of the nineteenth century, the Cornish mining economy was depressed. Rather than support the unemployed at home, many parishes paid the ship fares that allowed Cornish families to relocate to Canada, Cuba, Australia, South America, and the United States. Lake Superior received numerous Cornishmen, born between 1810 and 1840, who rose to prominent positions at the copper mines: Thomas Wills and William Daniell, first captains of the Hecla and Calumet branches, respectively, of C&H; John Cliff, head captain at Quincy; John Gundry, captain at Cliff and Pewabic; Richard Uren, captain at Copper Falls and agent at Pewabic and Franklin; Johnson Vivian, head captain at the Phoenix, then agent for the Phoenix, Hancock, Schoolcraft, Pewabic, Franklin, and Huron mines; Samuel Harris, captain at Pontiac, Mesnard, and Franklin, and agent at Quincy; and John Daniell, captain at Osceola, later superintendent of the Osceola and Tamarack mines.[7]

The Cornishmen employed in high positions put an indelible stamp on the Lake mines. These were men of a decidedly practical bent, who rose to the top after starting out, quite literally, at the bottom. Beginning as youthful miners, they climbed the occupational ladder a rung at a time. When they became managers in Michigan, they retained their past experiences as workers. These Cornishmen also retained the contract system of mining, brought from their homeland. This contract system defined key economic relationships between managers and miners, and drew important distinctions between miners and less skilled underground workers.

The Cornish contract system evolved after 1700, when Cornwall's tin mining industry started to adopt new technologies and organizational forms. In the 1700s, Cornish miners largely abandoned their traditional tools for picking out the ore, and they adopted the more expeditious technique of blasting it out with black powder. Explosives accelerated the rate at which ore was won and the rate of a mine's descent. Increased production, greater depths, and the problem of water percolating into their mines led the Cornish to put a premium on developing new hoisting and pumping technologies. They found a mechanical ally in the steam engine—a sophisticated and expensive device. The introduction of explosives and of steam power heralded a new era of Cornish mining. The successful taking of tin, or of newly discovered copper ores, now required larger-scale companies which had the capital needed to invest in machinery.[8]

As Cornish copper and tin mines took on a more industrial cast, they adopted organizational forms that allowed for change and expansion, but which still accommodated important local traditions. Inhabiting a spur of land on the extreme southwest corner of England, the Cornish had the reputation of being the most independent of Englishmen—so independent, in fact, that they only begrudgingly acknowledged any ties with the rest of the country. The new, larger mining enterprises needed to attract an independent breed of men, who as miners were long accustomed to working smaller-scale operations with family and friends.

Companies brought these miners into the fold by protecting their sense of self-esteem and self-reliance. Cornish miners did not just sell their time and work as mere wage earners. Instead, they became mine entrepreneurs themselves, albeit on a small scale, by banding together in self-selected mining teams. They chose their coworkers on the basis of friendship, kinship, skill, knowledge, and diligence. These mining teams then contracted for work with their company. Their contracts, often up to six months in duration, took two distinct forms. "Tutwork" contracts covered the "dead" work of shaft-sinking and drifting, while "tribute" contracts covered the stoping of valuable tin or copper ores.[9]

Prior to "setting day," when the mining teams in Cornwall congregated at public auction to bid against one another for contracts, a mine captain toured the underground to decide what work should be done. He divided the stoping ground into "pitches" and assessed any peculiarities that made one pitch more difficult to work than another. Also, he cursorily assayed the mineral value of each pitch. With this information in hand, the head captain determined what he deemed acceptable contract rates for diverse work throughout the mine.

Similarly the mining teams decided what work they wanted and how much they would bid for it. On setting day, each contract went to the lowest bidder, unless the captain found even the the lowest bid too high. If so, he announced his own terms for the contract, and the miners got to take it or leave it.

Mining teams with tutwork contracts divided themselves into two smaller crews to work the day and night shifts. The companies (in Cornwall called the "adventurers") provided tutworkers with tools of a known value and sold them supplies, such as mining candles and blasting powder. The adventurers also charged miners for select services, such as the provision of a mine doctor, the sharpening of drills, and the hoisting out of spoil. At the end of the contract period, the mine captain measured the work the tutworkers had done. He credited them their set contract amount for every fathom (six feet) they had advanced. He then subtracted the costs of any tools, supplies, or services provided by the adventurers. The tutworkers, if they had all worked the same number of shifts, equally divided their final contract settlement.[10]

Tutworkers assumed a degree of financial risk in accepting their contracts, because unforeseen circumstances could impede their progress. A poor contract could tie them into months of hard work for very little money. On the other hand, tutworkers occasionally encountered soft ground, easily worked. When this happened, the miners themselves decided whether to push the work diligently and earn a much higher settlement than usual, or to slack off and still earn an average amount.

Tutworkers' earnings fluctuated, but these fluctuations were less dramatic than those encountered by stoping miners working under tribute contracts. Tributors racked up the same sorts of debits as tutworkers, but the contract system accounted their credits in an entirely different manner. Instead of being based on a simple measurement of work performed, tributor's earnings were set by the amount and market value of the ore they had broken in the stopes, raised to the surface, and dressed for sale to a smelter. After the adventurers sold the dressed ore, they credited the tributors with a set percentage of the proceeds.

Wise tributors did some close figuring before bidding on a contract, because each one posed considerable financial uncertainty. Besides assessing the rock's hardness, tributors assayed its richness. The more copper or tin it carried, the higher its projected market value. Tributors also estimated the costs of raising ore to the surface and of dressing it for sale. Finally, tributors faced the hazard of a downturn in the market over the course of their contract. If the price paid for dressed ore fell, they would be the losers for it. Under such a system, tributors needed a rigorous, practical education grounded in mining and ore-dressing techniques and in the economics of mineral production. This education made Cornish miners much sought after in such far-flung locations as the Lake Superior copper district.[11]

Initially, Cornish miners and adventurers alike favored the contract system, because it benefited them both.[12] For miners, it protected their independence. It rewarded skill and effort, and allowed them to make many important decisions for themselves. For adventurers, contract mining protected their capital by relieving them of some of mining's financial risks. If the market price for dressed ore declined, the adventurers did not suffer the loss alone, because tributors bore a part of it. The contracts tied labor costs to productivity. Adventurers appreciated the fact that contract mining was an incentive system with performance-based rewards. Since the men received little if they worked little or worked unintelligently, they policed their own level of effort and skill, and adventurers escaped having to employ large numbers of underground bosses. And, also important, as long as miners actively bid against one another for work in a competitive environment, managers had little to fear from collective labor actions.

The contract system produced some excellent miners, but it also produced some impoverished ones, men who failed to work the system to advantage. Detractors began faulting contract mining for its inequities in the mid-nineteenth century, when the Cornish mines declined in the face of growing international competition in the metal markets. One critic, mining engineer James Sims, wrote "On the Economy of Mining in Cornwall" in 1849. Sims heaped particular scorn on mining by tribute, which he found "so pregnant with evils . . . that it should be avoided as much as possible."[13] The problem of fluctuating earnings, Sims argued, had a debilitating effect on labor. Men would be healthier, happier, and more productive if they received regular, predictable earnings. Also, deceitful practices had crept into the system that hurt capital and labor alike.

Miners sometimes cheated one another. They stole ore from others and added it to their own pile. Miners sometimes cheated the adventurers. They tried to hide the mineral richness of a pitch, hoping that if a mining captain did not discover just how much copper it really contained, he would allow them exceptionally favorable contract terms. Mining teams occasionally joined forces to dupe the adventurers. Two contract teams working adjacent pitches were sometimes slated to receive quite dissimilar proceeds upon the sale of their dressed ores. One team might be promised ten shillings for each pound Sterling received; the other, only five. Men working on the five-shilling contract did better, then, if they added some of their mine product to the yield of the ten-shilling contract team, and later received a kickback.

Mine captains, in turn, looked after the adventurers' interests, and did not always deal fairly with the men. They knew about how much they wanted miners to average in earnings per contract, and they never wanted them routinely to earn too much. If honest, hard work brought the miners a generous settlement under one contract, the mine captain was likely to try to force less favorable terms on them the next time around, so that their earnings declined.

John Sims faulted the dishonest practices inserted into the Cornish contract system by workers and managers alike.[14] The economic gamesmanship played by contractors and captains interfered with the straightforward conduct of business. Productivity suffered because of time lost in the setting day ritual, and time lost in getting men started on new contracts. Productivity suffered because the contract system no longer worked well as an incentive to work efficiently. Miners with good contracts were not encouraged to work their hardest, because if they did and earned too much, the captain would only trim them back on subsequent contracts. Other men found themselves with unfavorable contracts that required them to work exceptionally hard, just to make their regular earnings of about 50 shillings per month. Instead of doing that, they often just laid off, hoping to do better next time.

Wanting to infuse more productivity and regularity into the system, Sims recommended a radical change in the Cornish mining economy: a switch to straight wages. Barring that, at least the mines should eliminate tribute contracts. Sims believed that stoping miners should work under a simpler, piece-rate system, and receive a set amount of money per ton of ore produced, regardless of its copper content or market value. In short, stopers should be paid more like tutworkers.

Sim's call for a wage system went unheeded, but the Cornish mines did experiment with contracts that paid stoping miners by the ton. Such contracts were rare prior to 1850, but after that date the new practice spread. By the late 1880s tributing "was not by any means defunct," yet it could "strictly be described as being within measurable distance of death."[15] Besides limiting their use of tribute contracts, the mines shortened their contract periods, cutting them down to one or two months. Miners received their settlements more frequently, and in more consistent amounts.

The Cornishmen who emigrated to Lake Superior in the late 1840s and 1850s left their homeland when its traditional employment practices were starting to undergo major change. When they settled in Michigan, they did not reestablish the older Cornish mining economy. Instead, they adopted many of the reforms that in Cornwall were just taking root. Instead of replicating the Cornish contract system, the Lake mines jumped ahead in the direction of where it was going. They abandoned the public auctioning of contracts. Rather than openly bidding against one another for work, individual mining teams negotiated their contracts with a mining captain in his office. More significant, the mining companies rarely let tribute contracts, unless they were in dire financial straits. In Michigan, tributors usually were buzzards, miners picking scraps off crippled or dying companies.

Many notable mines in the district, including Cliff, Minesota, Pewabic, and Franklin, contracted with tributors at one time or another. These companies cut their losses by letting go their regular workers and turning their mines over to tributors, who assumed all the risks, did all the work, and kept most of the product.

Tributors could not afford to follow usual mining practices. Tributing paid, but only if the miners neglected shaft-sinking, drifting, and timbering. They just stoped out already opened parts of the mine and blasted out remnants of copper left behind by earlier miners. One knowledgeable observer called tributing "a form of grand larceny, at the expense of the mine's future." Tributors let a mine go to rack and ruin while robbing it of "its last particle of value."[16]

Mines not facing economic calamity employed tutwork-like contracts.

Shaft-sinkers and drifters, who worked in openings having a regular height, and width, received a set amount of compensation per foot of advance. Stopers, working in irregular spaces (which could open up or close down as the richness and width of the lode varied), were paid on the basis of the number of cubic fathoms of rock extracted. At the start of a contract period, a mining captain visited each work site and recorded benchmarks that he referred to later, when measuring the amount of work completed by each mining team.[17]

As in Cornwall, mining captains on Lake Superior charged certain costs to the mining teams. The companies sold miners explosives by the keg or box, fuse by hundred-foot lengths, and candles by the pound. They also charged for the consumption of drill steel by the pound. At the start of a contract, each mining team received bundles of drill steels of various lengths; the company recorded the numbers forged on the drills, and their aggregate weight. As the miner used and dulled the drills, the steels went to a blacksmith shop for resharpening. The companies bore the labor and machinery costs of forging and regrinding drills, but by reweighing them at the end of the contract, they could charge miners for all steel worn away in the process.[18]

Companies also sold miners sledgehammers, picks, shovels, tamping clay, rope, powder magazines, locks, and ladders. They regularly deducted the dues owed in support of keeping a company doctor, and at least one mine charged its men 50 cents per month for use of the dry or change house.[19] Miners, however, did not pay the costs of tramming or hoisting out their broken rock, and their settlements were not affected by the rock's copper content or market value. Nor did miners have any "subsist" payment deducted from their earnings. In Cornwall, adventurers commonly advanced miners some money, giving them something to live on over their four-to six-month contracts.[20] In Michigan, companies avoided this practice. Instead of carrying the men over long contracts, they shortened the contract periods. Some miners on the Lake worked under contracts as long as three months, but most received monthly settlements in cash at the mine office.

Not all miners worked under contracts. Some "company account" men worked for daily wages. The mines always had small jobs that required but a few days' work. Instead of contracting for these tasks, the companies turned them over to a contingent of miners carried on the payroll. Similarly, the companies carried "stemmers," or underground utility men. When a regular member of a contract team was ill, injured, or otherwise absent, the company assigned a stemmer to fill in for him. Instead of having the contract team share its settlement with the substitute worker, the company simply paid him a daily wage. Also, contract miners occasionally doubled as company account miners. They broke off from their contract work for a few days to block-hole some rock too big to be trammed, or to complete a sump near a pumping shaft. On settlement day, the company credited them with a few days of "extra work" at a daily wage.[21]

The contract system entailed somewhat complicated accounting and bookkeeping procedures, but the Lake mines figured its assets outweighed this liability. Their Cornish captains knew and supported the system. For them it was simply the normal way of running a mine. Beneath the captains, the companies sought to lure and keep as many skilled Cornish miners as possible. For them, too, the contract system represented a valued tradition that made the Lake mines more hospitable.

Some features of contract mining served the interests of the companies. The system allowed them to pass economic burdens down to their men. When neophyte miners secured employment, the companies did not have to pay them a standard wage while waiting for them to become skilled. These men received normal contracts, and as long as they remained poor miners they received little income. The receipt of low monthly settlements pressured the men to increase their skills and break more rock, or forced them to quit.[22] In either case, mine managers figured they came out ahead in the deal. And like the adventurers in Cornwall, the Lake companies saw contract mining as an incentive system. It supposedly kept everyone hard at work, because it tied their earnings to their level of effort.

The most experienced and skilled miners tended to be most supportive of contract mining. These men willingly matched wits with mine captains, believing they could work the system to their benefit. Instead of taking whatever was offered, they sought contracts promising higher than usual settlements. In 1863, the Phoenix Mining Company found itself short of such miners. In part, the Civil War caused the company's problem. The increased demand for copper had encouraged many marginal mines to go back into production, and they siphoned off workers from larger mines. But the Phoenix mine coped with a special problem of its own. Its "Ash Bed" lode was "of very uniform character," making the mine a less desirable place to work for the best miners.

At Phoenix, month after month miners drilled and blasted rock of equal hardness, always completing about the same amount of work and earning the same settlements. The miners made "good average wages," but not much more than that. Consequently, the best miners preferred to work in other mines, "in veins or lodes where the ground is more variable," and where "men make occasionally quite large sums . . . from favorable changes in the ground." Phoenix, in short, did not attract "a large class of miners . . . eager to take the risk of the chances, as are speculators elsewhere."[23]

The notion of the miner-entrepreneur lived on in Michigan, and men sometimes did earn lucrative settlements. But on Lake Superior, as earlier in Cornwall, if miners broke an inordinate amount of ground at a high contract rate one month, they could not expect the bonanza to carry over to the next month, at least not in the same place. The mine captain protected his company's interests by adjusting contract rates. Where the ground proved soft, the contract rate dipped lower.

In a sense, the contract system employed on Lake Superior was an elaborate sham, a ritual played out by managers and workers alike. Neither side openly talked about miners working for a monthly wage. Nevertheless, mine managers definitely had an average earnings figure in mind; mine captains juggled contract rates to achieve this average; and all the miners knew it. So, too, did an outsider who observed the workings of the contract system in 1853: "The miners are employed nominally by contract. . . . These contracts are so arranged by the companies that the men shall receive about $35 per month, and are of short duration, so that in case it turns out that they are for a short time making more, the difficulty may be remedied speedily."[24]

Miners seemingly forgave companies their economic manipulation of contract mining, in return for receiving its social benefits. It bestowed them with higher status. They were skilled, "independent" contractors, who bought their own supplies and paid for their tools. They formed their own mining teams. They chose whether to shaft-sink, drift, or stope. At work, they were little harassed by petty bosses. True, at the end of each contract a mine captain carefully measured the work they had done, but on a day-to-day basis the men largely controlled their own work pace and routine.

At the Quincy mine, Samuel Ley and John Roberts worked side by side as miners. By following them from contract to contract for three years, some inner workings of the system become more apparent.[25] From the start of 1870 though the end of 1872, Ley and Roberts always stayed together, but to field a full mining team they joined forces with two or four other men each month. Altogether, they paired up with 18 different individuals and worked on 35 contracts: 17 for drifting; six for stoping; two for drifting and stoping; four for rising and stoping; one for shaft-sinking; and five for sinking a winze. In addition to their contract mining, twice they received credit for extra days of labor charged to the company account. Besides teaming with numerous men and performing many different kinds of work, Ley and Roberts also moved from place to place underground. Over the three years, they worked on seven mine levels and in four different shafts.

Ley and Roberts each paid out $5 to $10 per month for supplies and medical service. In 1870, their contract rates for drifting varied from $10.50 to $15 per foot, and for rising and stoping from $23 to $26 per cubic fathom. Their individual monthly settlements peaked at $62.64 in April, after a low of $41.09 in March. Over the year, Ley and Roberts averaged $46.14 per month: for all Quincy contract miners, the average settlement was a nearly identical $46.09.

In 1871, the two men each earned a monthly high of $53.85 in June and a low of $32.33 in October. For the year they averaged $47.57 per month, while all Quincy contract miners averaged $47.08.[26] Indeed, in normal times the mining captains could manipulate contracts to see that men earned about what the company wanted to pay them, just as if they were on straight wages.

There were, however, some unusual times, when the miners forced their will on the companies. As the Atlantic mine's agent once noted, "When men are scarce they will do as they please, but when plenty, as they are told."[27] Men had been plentiful in 1870 and 1871, and the companies held down their earnings. This benefited the companies in the short run, but ultimately cost them: low earnings encouraged men to leave the district. By 1872, the Lake mines were labor short while copper prices were high, a combination that provided the men with an opportunity to press for better terms. In May 1872, miners successfully struck the Portage Lake mines and Calumet and Hecla.[28]

In the first four months of 1872, Ley and Roberts averaged monthly settlements of $45.00. In the strike month of May, they earned $51.43, even though off the job for eight days. They benefited most from the strike in June, when Quincy substantially raised its contract rates. In 1870 and 1871, Ley and Roberts never received more than $15 per foot for drifting. In June 1872, they received $22 per foot of advance, worked very hard, and earned $102.40 each—more than twice their usual amount. Contracts for the rest of the year remained high, and for all of 1872 Quincy's miners averaged $60.62 per month, while Ley and Roberts did slightly better and earned $62.66. Monthly settlements at Quincy stayed in $60 range for both 1872 and 1873, when the company sold its copper for 32⅓ and then 26½ cents per pound. When copper dipped to 22 cents in 1874, Quincy's captains reinstituted lower rates and forced miners' monthly earnings back down to their pre-strike levels of about $45.[29]

Initially, the Lake mines did not limit the contract system to miners. Through the 1860s, Quincy's captains wrote contracts for all kinds of work.[30] After miners broke the rock, "wheeling contractors" pushed it to the shaft. Other men had contracts to fill kibbles underground, raise them to the surface with a windlass, and then land and dump them at the shaft-house. Still others entered contracts to pick copper; to burn, break, and dress copper at the kilnhouses; and to run rock from the mine to the stamp mill. So at first, contractors handled the copper rock all the way from the stopes to the stamps. Under most of these agreements, men received a set amount for working a specified time, which was the same as working for wages. Some agreements, however, rewarded the men on the basis of work performed. Men who burned and dressed copper received so much per ton of rock readied for the stamps, or per ton of barrel copper cleaned, picked, and readied for the smelter.

Quincy and other companies used contract labor extensively in their early years, when the production from any one shaft was limited or even intermittent, and prior to investing heavily in machinery. Under such conditions, it made sense to secure workers on a month-to-month basis, putting them to work only where really needed. Later, as production expanded and mechanization progressed, the number of contractors declined. Steam-powered hoists replaced windlass contractors, and jaw crushers superseded kilnhouse contractors. The companies eliminated some occupations altogether, and as they settled into larger, more predictable levels of production, they switched many remaining laborers off of contracts and onto straight wages.

This change entrenched the miners—who remained as contractors—as the working-class elite. They enjoyed superior status; they remained the skilled workers who kept considerable control over their jobs. Trammers, timbermen, and other underground laborers, by contrast, became inferior employees. They sold the companies their time and sweat, not their skill. Besides having greater status, miners earned more money. In the second half of the nineteenth century at Quincy, the contract miner typically earned 10 to 20 percent more than the trammer. At Calumet and Hecla, in May 1913, the earnings differential remained similar: the average miner took home 23 percent more than the trammer.[31]

Trammers had the harder, lower-paying job. They also worked more hours per week. The mines worked two shifts of about 10 hours each, and the day and night shifts rotated weekly. Men on the night crew worked five full shifts during the week: Monday night through Friday night. After getting off work early on Saturday morning, they did not return to the mine until Monday morning, when they began a week of day work, consisting of about five and a half shifts. Day workers did not have the entire weekend off. They reported about seven in the morning on Saturday to work an abbreviated shift. But the companies made this extra shift shorter for miners than for others. They let miners go home at noon on Saturday, but kept trammers and other underground laborers for another three hours.[32]

This practice reinforced the fact that the companies favored contract miners over "common" laborers. This favoritism, expressed in many ways, had important effects on the underground work force. It drove a wedge between miners and trammers and set them off as two distinct groups. Underground workers, in the nineteenth century at least, did not share a common working-class consciousness, because miners and trammers had separate interests. The contract system worked to the advantage of the one group, and to the disadvantage of the other. Mine managers liked the system, believing that it cemented the loyalties of their skilled miners, while at the same time teaching the trammers a lesson. If they worked diligently, avoided fractious behavior, and concentrated on learning the mining trade, then perhaps some day they might be tapped to step up into the privileged ranks of the contract miner.

The first two decades of the twentieth century brought considerable change to the Lake mines' administration of contract mining. They had inaugurated this system to attract and keep Cornish miners. But Cornish migration had not kept pace with the mines' expansion, and by 1900 the mines had handed machine rock drills over to a large number of Finns and Italians. As skilled Cornishmen made up less and less of their force of miners, companies began treating miners with less deference. Changing economic conditions within the industry also caused companies to treat contract miners in less favorable ways.

The mines rode into the twentieth century on a wave of unprecedented profitability. From the mid-1880s to the mid-1890s, copper prices had held at only 10 to 12 cents per pound, and the Lake mines altogether paid out dividends totalling $2.5 to $3.5 million per year. Then, in the late 1890s, copper prices jumped to 16.5 to 17.5 cents, and dividends rose impressively. From 1898 to 1901, the six to eight most successful mines averaged total annual dividends of $9.1 million.[33] When there was a seller's market for copper, these old producers still performed very well.

Nevertheless, the turn-of-the-century burst of profitability only masked, temporarily, some fundamental problems within the industry that became more apparent over the next two decades. In confronting these problems, the companies turned to new managers and experts, to new means of increasing productivity—and in the process they added more strident labor unrest to their growing list of ailments.

The Lake mines greatly increased their annual production of copper between 1880 and 1910, and over these years American copper producers as a whole impressively enlarged their share of the world market. The United States accounted for about one-fifth of world production in 1880, and for 43 percent in 1890. Between 1900 and 1910, mines in the United States annually produced 53 to 57 percent of the world's new copper. This increase, however, was not due to the larger production coming out of Lake Superior. Between 1880 and 1910 the Michigan mines accounted for a relatively constant 12 to 16 percent of world production.[34] The increase was due to the opening up of other copper districts in the United States, especially those in Montana and Arizona.

The development of Anaconda and other mines near Butte, Montana, ended Michigan's domination of the American copper mining industry. Starting in 1887, Montana's mines outproduced Michigan's, year after year. The Arizona mines grew more slowly than their Montana counterparts, but they expanded rapidly in the first decade of the twentieth century and surpassed Michigan's production for the first time in 1905. By 1910, Montana and Arizona each accounted for about 30 percent of American copper, while the Lake mines' share stood at 20 percent, a substantial drop from its 40 and 25 percent shares in 1890 and 1900, respectively.[35]

South of Portage Lake, several new mines operated in the Lake Superior district, most notably the Champion, Baltic, and Trimountain mines (all controlled by the Copper Range company) and the Winona mine. Overall, however, Michigan was an old, deep, and fully mature mining district. Arizona, by contrast, which recently had made "such heavy gains in production," was a still-developing, young district. The Lake mines had a glorious past, but Arizona had "brighter prospects for the future."[36]

Newer districts, mining richer deposits at shallower depths, now dominated the market. Importantly, these districts produced copper at less cost. By 1912, it cost the Michigan mines, on average, 10.7 cents to produce a pound of copper. For the entire American industry, the cost per pound stood at 9.3 cents, and the Arizona mines made theirs for only 8.7 cents.[37] The profit margin per pound was shrinking in Michigan, but was increasing for the industry as a whole. This meant that the continued success of the Lake mines, more so than for western producers, depended on high copper prices.

Checking production costs became of paramount concern to the Lake mines, but achieving cost reductions proved difficult. As the mines went a level or two deeper each year, their production costs increased, unless they found new technologies, such as rock drills that could be run with fewer men, or faster, more fuel-efficient hoists. The companies pinned other hopes on obtaining a more productive labor forces. But just when they needed more skillful, efficient men, they received the bulk of their new laborers from Finland, Italy, and eastern Europe—and these immigrants usually had no mining experience whatsoever.

The companies also grappled with structural problems within their deep mines. Atlantic and Quincy were particularly plagued by hanging walls suddenly grown far less stable, but other companies, too, paid the price for operating at depths not attained by other American mines. At great depths, an especially threatening problem confronted some of the region's best mines: their copper content was diminishing.

Rock from the Tamarack mine yielded 30 pounds of copper per ton in 1900, but only 20 pounds per ton in 1910. Over the same time, Quincy's yield fell from 25 to only 15 or 16 pounds per ton. Even the best of the newest, shallowest mines, the Champion, saw its yield fall from 30 pounds during 1902–4 to 24 pounds in 1908–10.[38] But the most rapid and dramatic falling off occurred at Calumet and Hecla. The Calumet conglomerate lode suddenly ran much poorer, and it threatened to transform this world-class company into merely an average mine within its own district. In 1900, C&H still garnered 50 pounds of copper from every ton of stamp rock. By 1910, C&H's yield had plummeted to only 25.8 pounds.[39]

Confronted by problems, the mines could not afford to conduct "business as usual." Calumet and Hecla scrambled to stay on top. Just before and after the turn of the century, it struggled to make up for the yield suffered by its works on the conglomerate lode. C&H opened new works on the Osceola and Kearsarge amygdaloid lodes crossing its property.[40] Calumet and Hecla expanded its operations in yet another way: in 1905, it began buying up shares in profitable and unprofitable mines alike.[41] To prolong its own life, the company acquired a fifth to a half of the stock of the Osceola, Tamarack, Isle Royale, and Ahmeek mines. All were significant producers, and by 1909–10, all fell under the general management and control of C&H. As for increasing its stock holdings in lesser operations, C&H saw them as a source of copper reserves, and perhaps a hidden bonanza would be found amongst them somewhere.

In somewhat similar fashion, the Copper Range company exerted control over the Baltic, Trimountain, and Champion mines, and by 1910 Quincy worked not only its original short stretch of the Pewabic lode, but additional portions of the lode once exploited by the independent Pewabic, Franklin, Mesnard, Pontiac, and Arcadian mines. The major producers expanded their holdings to sustain the high production levels needed to achieve high profitability.

During this time, companies also turned to new managers and experts, who differed considerably from the men who guided operations in the nineteenth century. Cornishmen and other practically trained men gave way to college-educated men serving as superintendents, geologists, and mining engineers.

In 1902, John Luther Harris followed his father, Samuel, into the superintendent's position at Quincy. The father, born in Cornwall, had worked himself up the ranks. The son, in contrast, had been in the first graduating class (1888) of the Michigan College of Mines in Houghton, and he had also studied at the Massachusetts Institute of Technology. John Harris headed Quincy for only three years, but another college graduate followed him, Charles Lawton, who remained the mine's superintendent until 1946. Lawton, son of a former commissioner of mineral statistics for the state of Michigan, had received a degree from the Mechanical Department at Michigan Agricultural College (now Michigan State University) in 1888, and he had then studied mining engineering at the Michigan College of Mines.[42]

Beginning in 1905, Frederick W. Denton took charge of Copper Range's three mines. Previously, Denton had graduated from the Columbia School of Mines (again, in 1888), and he came to Lake Superior in 1890 to teach at the Michigan College of Mines.[43] At Calumet and Hecla, the same pattern held. In 1901, S. B. Whiting, an engineer from the "shop culture" school of practical education, resigned. James MacNaughton replaced him as C&H's general manager and stayed with the company until 1941. MacNaughton, after a year's study at Oberlin College, had completed a degree in civil engineering at the University of Michigan in 1886.[44]

The nearly simultaneous ascendency of college-educated engineers into the top posts at the region's three most dominant companies had important ramifications throughout these organizations. Lawton, Denton, and MacNaughton deemed themselves modern practitioners. They believed that certain traditional practices had to be overturned and new methods installed. These men advocated scientific problem-solving. Comfortable with numbers, they encouraged a close cost-accounting of all parts of a mine's operation. Numbers told them where their companies stood, what areas needed improvement, and the economic effects of any changes introduced. Besides trusting in numbers, these men trusted in other men holding college degrees.

For nearly 50 years, the companies had given geologists virtually no role in helping to guide their development. Then, at the turn of the century, Copper Range employed a former Michigan state geologist, Dr. Lucius L. Hubbard, to direct explorations. Hubbard proved instrumental in finding the southern extension of the Baltic lode.[45] This gave rise to the new Champion mine, and it also boosted the stock of geologists in the district. Companies began seeing these professionals as useful. As the tenor of the rock diminished on Calumet and Hecla's original works, the company showed a much keener interest in geology, and by 1919 C&H employed five or six professionals in its geological department. C&H geologists launched a thorough exploration of the entire Keweenaw that was directed by Professor L. C. Graton of Harvard University, serving as a consultant.[46] Besides conducting extensive field investigations, Calumet and Hecla assembled copious historical files on virtually all mines that ever operated on Lake Superior, emphasizing where they had looked for copper and how much they had found. This company-sponsored research later formed the basis of the United States Geological Survey's Professional Paper No. 144, *The Copper Deposits of Michigan,* published in 1929.

Early in the twentieth century, geologists made some gains in establishing their profession's importance to an ongoing industry. Their gains were modest, however, when compared with those made by engineers. In an era when the fine-tuning of production methods and the close monitoring of costs became essential, mining and surface captains looked like antiquarians when posed next to professional engineers. When the companies saw the need to change their technologies and management styles, they by no means eliminated their traditional captains. They did, however, strip them of some responsibilities, which they then vested in engineers.

American colleges and universities had produced mining engineers since the Civil War era.[47] The Columbia school of Mines opened in 1864 and the Colorado School of Mines in 1874. In the heart of the Lake Superior copper district, the Michigan College of Mines admitted its first students in 1886. Although operated on a shoestring budget and housed in Houghton's fire hall, the Michigan College of Mines had high expectations for itself. Here was a practical school destined to enhance the economic performance of its state by supplying Michigan's copper and iron mines with technical experts. The college defined a special and narrow mission for itself. It realized at the start that it "must not, as is so frequently and universally done, waste its energies in the vain attempt to teach branches foreign to the object of the institution." One of the college's early presidents stated this object succinctly: "this school should confine its attention wholly to *mining and the subjects relating thereto,*" because its purpose was "to train men to be of real use in any line of work connected with the winning and reduction of mineral products."[48]

In truth, the local copper mines found little "real use" underground for graduated mining engineers until 1905 to 1910. While engineers and scientists played an earlier role in the stamp milling end of the industry, they remained noticeably absent from the mines themselves. The dearth of professionals could be measured in several ways. The American Institute of Mining Engineers, organized in 1871, counted as members "professional mining engineers, geologists, metallurgists or chemists." By the early to mid-1890s, the entire Lake Superior copper district boasted only 10 to 12 AIME members, and about of third of them were faculty members at Houghton's mining school.[49]

In the 1870s, the Quincy Mining Company had no full-time employees who were graduated mining engineers or technical men. When it occasionally needed an engineer, it hired a consultant, Luther G. Emerson, who was described as the "civil and mining engineer of the Quincy, Franklin, Pewabic, Huron, Allouez, Phoenix, Copper Falls and other mines."[50] In 1876, Quincy retained Emerson's services (at $10 per day) for a total of 58 days of surveying and mapping work, both above and below ground.

In the 1880s Quincy employed one technical man; in the 1890s it employed two; and by 1904 it employed four. Still, up to this time Quincy's engineers played a limited role. In 1901, Quincy's agent, in assessing the performance of an assistant mining engineer, noted that he was "good at surveying, drafting and assaying"—and that was apparently all he did. In 1905, the agent wrote that "in our case we do not . . . owe to the Mining Engineer the improvements made in [underground] development work, but are under obligations to the Mine Superintendent and the Chief Mining Captain. The principal duties of the Mining Engineer here has been in connection with the mine surveying, mapping, etc. only."[51]

The engineer's role changed significantly once Charles Lawton headed

Quincy, and once the American mining industry came alive with talk of reducing production costs, of improving efficiency, and of practicing "scientific management." "Efficiency engineering" became a buzzword. At one time, "efficiency" had been an industrial term applied to machinery. Machines were efficient if they did a great deal of work while requiring small energy inputs. Now, in large measure due to Frederick W. Taylor's elucidation of *The Principles of Scientific Management* in 1911, the concept of efficiency would be applied to men as well. Men became more efficient when they performed more work per dollar spent in labor costs.

By practicing efficiency engineering, mine managers sought to transform their traditional industry into a modern one on a par with the most advanced manufacturing industries. Of course, the mining industry depended on some very coarse, seemingly unsophisticated work—such as shoveling mine rock into cars. But the modern gospel of efficiency taught mine managers that no jobs, no matter how crude, were beneath study and improvement. Surely, there was a scientifically correct way of shoveling rock.[52] To find it, hire experts and send them underground with stop watches and charts. Run time studies. Experiment with different shovels. Change the length and shape of the handle. Raise and lower the heights of the tramcar's sides. See how the best men work. How do they hold the shovel? How big a load do they pick up each time? How do they place their feet? How far do they lift and toss the rock? Once the experts determined the best, most efficient shoveling technique, they could train men to use it.

At Quincy, near 1910 the company's engineers became more important, because they became the efficiency experts. Instead of merely surveying and mapping what work had been done underground, they began redefining the work itself. By 1912, Quincy employed six men in its engineering department—two each from the Michigan College of Mines, the Colorado School of Mines, and the Massachusetts Institute of Technology.[53] Quincy's eastern officers did not yet grasp the breadth and importance of the engineers' work within the mine, but out in Michigan the engineers had clearly made their presence known. Charles Lawton wrote Quincy's president:

> You want "Scientific Management"? And here I have been priding myself all this time that that is just what the Quincy is receiving, with all the technical, scientific college-bred men that we have about us, all of them studying and working upon the various subjects to that end, and all of them deeply interested in scientific and efficiency management—all to such an extent that the foremen on the Hill and the mining captains to the uttermost part of the mines think that we have gone clean "Daffy" on the subject.[54]

Quincy was not alone in its quest for greater efficiency. Calumet and Hecla also hunted for it. In 1909–10, it subjected its miners to time and motion studies, usually using "college boys from the [Michigan] College of Mines." The "boys" observed men at work, "with the idea of finding out where the loss of time occurred," so that C&H could eliminate it. In 1912, Calumet and

Hecla contracted with the consulting firm of Playfair & Hurd, "Standard Practice and Efficiency Engineers," to evaluate the performance of its large machine shop.[55] All this sudden emphasis on efficiency seemed to auger well for the mines' economic performance, as reported by the *Engineering and Mining Journal* in 1912:

> There can be no doubt that many economies are going to be effected in Lake Superior mining in the near future, as only lately have the companies commenced in earnest to experiment in regard to supplies, types of rock drills, and different methods of doing operations, using trained men to keep track of the experiments so that the latter will show something definite and their outcome will not be a matter of individual opinion, as so often was the case in the past, when the mine captain had the most to say in regard to underground equipment and practice.[56]

Between 1900 and 1910, worker productivity had declined in the Lake mines. The pounds of refined copper produced per year per employee had dropped from 14,000 to 12,000.[57] The mine managers felt squeezed on all sides: by decreasing productivity, by increasing competition from Arizona and Montana, by diminishing copper values in their lodes, and by their reliance on a labor force growing less and less skilled. In response, managers embraced efficiency engineering and a more rigorous cost accounting, and, in effect, put the squeeze on men up and down their companies' ranks. They told mining captains to start taking new marching orders from engineers. The captains, in turn, told their sub-captains and foremen to be more diligent and get out more work. The sub-captains and foremen then berated their miners and trammers. What started out at the top of the mine as a progressive, intellectual commitment to modern management techniques, needed to protect a company's economic well-being, ended up being seen at the bottom of the mine as a threatening incursion into the workers' domain, as an enforced speed-up of work that made hard jobs even worse than before.

Underground workers, for the first time, lined up in the shaft-rockhouses to have their entering and leaving times recorded daily on cards. Although the mining companies and the men continued to cling to the old vocabulary of using "contract miners," in fact most of them were now just daily wage earners, pure and simple. When Quincy wanted its miners to earn $2.50 per day, that was exactly what they earned. The company still kept track of the supplies drawn by each team, but this was a mere masquerade, a playing-out of an old ritual now without meaning. When it came time to settle up, Quincy deducted not the real cost of the drawn supplies but whatever amount was needed to get the men down to the $2.50 per day wage.[58] Calumet and Hecla, by 1910, had had enough of this sham: the company altogether stopped charging "contract" teams for their explosives, blasting caps, fuses, and drill steel.[59] The miners' status eroded as these changes progressed. In local myth only did miners remain independent contractors. They were machine-drill runners, and their bosses knew how much work their machines could do. Bosses told the men, miners and trammers alike, just how much work was expected of them in return for their daily wages. If men did not meet the standard, they received reprimands or were fired.

By 1910 to 1912, the companies' economic problems, coupled with their shucking off of traditional practices and their embracing of efficiency engineering, had encouraged a widening breech in labor-management relations. Managers looked down with suspicion on a labor force they faulted for lack of skill and diligence. The workers looked up at managers and wondered what was to be the next speed-up or job change. Change itself became a contentious issue, just when the managers believed that more widesweeping changes were necessary. Yet while the breech between management and labor opened wider, one other gap narrowed—the traditional gap between miners and trammers. These two groups had once been separate and distinct, because the companies had treated them so differently. But as the companies chipped away at the last vestiges of the Cornish contract system, and as miners too felt the sting of change, they began to see themselves as less of a working-class elite, and as having more in common with the rest of the laboring class. As efficiency earned more champions on the surface, it spawned more detractors underground among miners and trammers alike.

5

Change and Consensus: Miners, Managers, and Two-Man Drills

Correspondence from agent William Tonkin to company treasurer, John Stanton, regarding the change from hand-drilling to machine-drilling at the Atlantic mine:

"I have been thinking of late of using power drills. I have visited the Calumet and Osceola mines in order to get all the information possible, as to the saving of labor in breaking ground." (March 1880)

"The drill machines work very nicely. . . . This month will be a trial month, and if the miners can make wages with the price I have set them, the drills will be a great saving to the company. . . ." (September 1880)

" . . . [T]he more work we do with them the better I like them." (January 1881)

"When I get enough of the power drills, the sound of a hammer will not be heard in the Atlantic mine." (March 1881)

In the late nineteenth and early twentieth century, the pace of scientific discovery and technological change quickened. Industrial research laboratories sprang up. Mechanical engineers, electrical engineers, and industrial chemists developed innovative ideas into new materials and machines. As the mining companies explored the engines, compressors, drills, motors, generators, and explosives made available to them, they were often seduced into buying by promises of increased production and decreased costs. Some of these modern machines and explosives ultimately had a profound impact on the underground operations of the Lake Superior copper mines. The companies discovered, however, that implementing a technological change was no simple swap of something new for something old.

Besides fitting a new technology into a particular underground setting, a mining company had to fit it into established patterns of work. Change, by definition, upset normal operations and created novel conditions. Many employees opposed novelty and preferred the tried and true. They fretted over technological change, because of its possible adverse effects upon their earnings, their continued employment, their occupational mobility, or the pace of their work. Bosses, captains, and even mine superintendents also resisted change at times, unless they recognized a compelling need for it. Managers, too, found comfort in familiar routines and sometimes did not want to trade old, predictable technologies for new, uncertain ones—particularly when the sponsors of change were not the local mine managers themselves but eastern officers or boards of directors.

Technological changes could spawn divisiveness on at least two levels. They could create tension between eastern officers and local mine officials, or between mine officials and workers. Mine agents could not afford to be overly enthusiastic or overly cautious when implementing change, because either extreme might bring them under fire from above or below. Mine superintendents lived in the middle. They felt the pressures applied from above by company presidents and treasurers, and the pressures applied from below by miners and trammers.

Company leaders in Boston or New York saw themselves as responsible for protecting stockholders' interests by monitoring profits and costs. When they promoted a technological change, they usually expected it to effect considerable savings for the corporation. Local mine officers recognized the critical importance of checking costs and maximizing profits, but they resented being dictated to. The men in Michigan believed that they, not remote officers, directors, or stockholders, knew best how to run a mine.

Samuel B. Harris was a practically trained man who served as agent for the Quincy mine from 1884 until 1902. Over this time, Harris supervised the rebuilding of the mine's surface plant, plus the building of two new stamp mills and a new smelter. He implemented more change than any agent before him, and yet his appreciation for the modern was always tempered by a conservative appreciation for the technological status quo. Harris was guided by the dictum, "If it ain't broke, don't fix it!" This attitude sometimes alienated him from Quincy's eastern officers.

In 1884–85, when Harris was new to his agent's position, he received letters from Nathan Daniels lecturing him on modern steam engines. Daniels, who headed Quincy's stock transfer office in Boston, was disturbed by the "dilapidated condition of the hoisting apparatus" at the No. 4 shaft. The No.4 hoist needed a new, larger drum and Daniels saw this as an opportunity to begin modernizing Quincy's steam-powered equipment. Whenever the Bostonian wrote to Harris, his eastern elitism shown through: "The power required at the shafts and at the stamp mill is large, and the Engines at each place, would not in our Eastern mills be allowed to remain in place longer than the time required to replace them with others of modern construction."[1]

The old engine at No. 4 had a 26-inch bore and 60-inch stroke. Harris argued that the engine was large enough to drive a bigger drum, so there was no need to get rid of it. Daniels, however, faulted the engine as wasteful. He wanted the old slide-valve engine replaced with one having Corliss valves.

Such valves used the expansive nature of steam to greater advantage and required less fuel consumption at the boiler. Daniels made clear his criterion for change:

> The real question is not whether the old slide valve Engine now there is big enough or can be built over and made big enough to do the work at that shaft for years, but rather can the corporation afford to literally throw away the extra fuel required to furnish steam, and hold on to an old fashioned engine, entirely behind the requirements or economics of the present age. It is a simple question of dollars and cents and so becomes directly of interest to stockholders that their servants shall be fully alive and up to the present date in all matters.[2]

In the face of such condescension, "servant" Samuel Harris dug in his heels. The Bostonian wanted sophistication, but Harris wanted "a hoisting arrangement [that] will be plain, strong, easily operated and durable." In 1885, Harris won out: he put a new drum on the old engine. A decade later, he faced the same issue again and made the same decision. He kept the old engine and outfitted it with an even larger hositing drum. To power that larger drum, the old engine needed help from a second engine connected to the drum's shaft. Harris eschewed purchasing a new engine for that purpose, and pulled a used engine out of Quincy's stamp mill. Harris figured the two old engines in combination would "do good service for many years."[3]

Harris remained conservative when expending capital to equip the No. 4 shaft, because he recognized that it was Quincy's least important hoisting shaft. But he demonstrated he could be both conservative and "fully alive" to progress when, in the 1890s, he selected new hoists for Quincy's Nos. 2, 6, and 7 shafts. Although "ancient" engines sufficed at No. 4, Harris ordered state-of-the-art E. P. Allis hoists (with Corliss valves) for the other shafts. With each successive purchase, Quincy acquired the largest, most modern hoist that the Milwaukee manufacturer had ever built.[4]

Still, Quincy's eastern officers were never fully satisfied with Harris's resolve to hold down the company's energy costs. While Harris was installing the finest, most efficient engines on the market, Quincy's treasurer, William Rogers Todd, scolded him for firing their boilers with expensive Mansfield coal. By 1892, Quincy consumed 40,000 tons of coal per year, so a small cost reduction per ton could add up to considerable savings. Todd ordered cheaper coal for trial and had rival coal companies send experts to Michigan to teach Quincy's boiler tenders how to stoke a fire. Throughout the 1890s, Todd believed the boiler tenders were too unskilled or lazy to make cheap coal do good work.[5]

Harris, meanwhile, trusted his own boilerhouse men to know their business. He argued that Mansfield coal was the best ever used at the mine and continued to order thousands of tons of it. The agent refused to defer to the company treasurer, because he believed he knew infinitely more about boilers and coal than Todd. For Harris, coal was a fuel that men worked and burned to create steam. The more easily it burned and the more heat it produced, the better it was. For Todd, coal was an expensive collection of large invoice entries on the books in New York. The two men knew coal in different terms. They never reached a showdown over this particular issue, or over the company's selection of explosives, machinery, or other supplies. Nevertheless, they clearly grew tired of trying to foil one another.

The eastern men and the Michigan men who led the mining companies often concurred on the merits of change. But when new technologies had less than unanimous support, they could initiate brief bouts of corporate politics. The Michigan men held that they were best situated to make assessments, because they were the experts who lived and worked where the change would take place. Men sitting at mahogany desks over 1,000 miles removed from the Keweenaw Peninsula had a mental template of what a mine was, but they didn't experience the daily fray of operations. As a consequence, they had a very different perspective on technological change.

Eastern officers afforded themselves the luxury of focusing on the benefits to be derived once a change was implemented. They did not have to deal with the often messy and uncertain process of change. That was the mine agent's job. He had to solve technical problems and make the new technology work. He had to bring mining captains and bosses into line in support of the change and then monitor how well they did in putting the new technology into ordinary workers' hands. If unrest or rebellion erupted anywhere down the line, the mine agent had to quell it.

Implementing technological change, then, entailed far more than simply acquiring the latest inventions and plugging them into the production process. Before a new technology was ever tried, it could spark dissention between the companies' eastern and Michigan leaders. While being tried, it might raise the ire of the men who were told to use it. Finally, there was the possibility that the new technology simply would not work. All these problems were particularly acute when trying to modernize underground operations. Within the mines, the drilling, blasting, and tramming of rock were the most fundamental tasks. In no instances did the companies succeed in significantly altering these tasks without encountering great difficulty.

The vanguard of modern mining arrived at the mines in 1868 in the form of a power rock drill driven by compressed air. This machine was intended to replace the traditional hand drilling of shot holes. When the companies purchased their first machines from the Burleigh Rock Drill Company of Fitchburg, Massachusetts, they were on the receiving end of innovation. The manufacturer put "into the miner's hands a machine that will drill 2 in. or 3 in. holes in diameter, from 40 to 60 ft. in the shift," but the mining companies had to "have brains enough to handle that power to the best advantage."[6] They had to test the machines and then, if they passed muster, introduce them into general use.

As it turned out, the introduction and diffusion of power rock drills did not come easily. The companies tried Burleigh drills for several years, but then gave up on them and reverted almost exclusively to hand drilling. Only after nearly a decade more, and after trying about 10 different drilling machines, did the mines find one that met their expectations: the "Little Giant" drill manufactured by the Rand Drill Company.

Identifying the right machine, however, was only the first step toward mechanized drilling. The technical features of the Rand drill did not guarantee its success. The companies still had to deploy the new technology in an acceptable and profitable manner. As the mine managers inserted Rand drills into their industry, they were never in absolute control of the innovation. They evidenced skill in manipulating men and machines, yet many complex, subtle, and unpredicted factors aided in the diffusion of the Rand drill. Not the least of these was the irony that the machine worked well—but not too well. Certain limitations kept the new technology from destroying important social and economic patterns associated with the old technology. As a result, miners welcomed the Rand drill and worked to make it a success.

In the mid-nineteenth century, a number of individuals sought to develop rock-drilling machines powered by steam or compressed air for tunneling, mining, and open excavation work.[7] Producing a commercially successful drill proved difficult, because many machines failed to survive the rigor of reciprocating a drill steel through solid rock. In the United States, Charles Burleigh finally brought out the first successful power drill. Starting in 1866–67, the Burleigh drill won international renown for its work in driving the Hoosac Tunnel in Massachusetts.[8]

The 25,000-foot-long Hoosac Tunnel was driven through the Green Mountain range to facilitate a short railroad link between Boston and the Hudson River. Begun in 1851 and not completed until 1876, the tunnel was plagued by a host of economic and technical problems. It became a showcase for machine-drill trials, and Burleigh's new drill proved itself worthy of intensive use. In driving the tunnel, Burleigh drills were gang-mounted on large carriages rolled up to the working faces. Shortly after the drill's successful use on the Hoosac, the Lake copper mines tried to transfer the technology directly from tunneling to mining.

Richard Uren, agent for both the Pewabic and the Franklin mines, first brought power drills to the Lake. In 1867, Franklin's annual report noted that "the introduction of new and powerful explosive compounds, [and] the use of power drills in mining, will tend to such a reduction in cost that a reasonable profit may be expected, even with the low price of copper." The next year Pewabic purchased a Burleigh drill, while the Ogima mine tested one of its rivals, a Gardner drill. Between 1868 and 1873, at least ten mines invested in Burleigh rock drills. The users included small, even intermittent, copper producers (such as the Aztec and Clark mines), as well as the region's four largest producers: Calumet and Hecla, Quincy, Central, and Copper Falls.[9]

The *Portage Lake Mining Gazette* chronicled the initial enthusiasm for the Burleighs. In 1870 the paper noted that the drill "promises to be a great success in our mines, and that, for certain work, it is a mere matter of time as to its general introduction." In 1871, it reported that the machines had "expedited the work very materially" at the Central mine, where they broke 56 percent more ground than shaft-sinking hand drillers. The paper also recorded that at Copper Falls, where three machines operated, they broke rock at a cost of 83.7 cents per ton, while the cost by hand was $1.03. In 1872 the *Mining Gazette* continued its favorable reports: "The movement that has been made towards introducing drilling machinery into the copper mines of Lake Superior, even in its inception, foreshadows magnificent results to the industry in the future."[10]

Enthusiastic companies invested rather heavily in machine rock drilling. This technology required substantial additions to a mine's physical plant. To put the machine to use underground (where boilers and live steam could not go), a company had to purchase a steam-powered air compressor, erect a plant for it on the surface, and lay iron pipes and flexible hoses throughout the underground.

The Quincy Mining Company entertained no doubts whether the Burleigh would prove a success. At a meeting in April 1872, directors appropriated $100,000 for improvements to the mine's plant, reserving a considerable portion of this for rock drills. By the end of the year, Quincy's air compressor and drill account amounted to $25,093, making it one of the largest investments in a new technology that the company had ever made. Quincy purchased an air compressor, installed pipelines, and ordered three "tunnel" and two "mining" drills from Burleigh, as well as an assortment of spare parts, mounting clamps, and two "mining carriages," which were low, wheeled trucks having vertical screw-posts to carry the drills.[11] By October 1872 the company had two machines driving drifts 1,200 feet underground.

The first two Burleighs were large piston drills. A piston drill held the drill steel firmly in a chuck which in turn connected fast to the forward end of the machine's piston. As compressed air reciprocated the double-acting piston, the entire drill steel was alternately driven forward and then pulled back. The machine, using a rifle bar and a ratchet and pawl mechanism, rotated the drill steel a partial turn prior to each forward stoke. It also had an automatic feed. The tunnel drill could advance a hole 36 inches before miners had to back the machine off and insert a longer drill. The machine was 67 inches long and weighed 550 pounds without its clamps or mounts. Burleigh's mining drill, a foot shorter and 75 pounds lighter, could feed a drill forward up to 26 inches.

Sheer size was the dominant feature of the Burleigh drills. This feature had not impeded progress on the 24-foot-wide Hoosac Tunnel, but it became a major liability in the mines. In drifting with Burleigh drills, Quincy increased the cross-sectional area of these openings more than threefold just to give the drills room to be worked. While men advanced the larger, machine-drilled drifts a little faster than smaller, hand-drilled ones, it cost the company "considerably more than it would have [if] done by hand-power." The net cost of the large drift depended partly on the nature of the rock. When a machine drilled through a pocket of rich ground, the large drifts recovered copper that otherwise would have required stoping, and this copper could make the method pay. But Quincy's agent, A. J. Corey, noted that "we cannot afford to carry a 10 x 10 drift through poor ground."[12] So when the first two Burleighs ran into long stretches of barren ground, Quincy stopped using them for a time.

Quincy did not fault the machine's power. The Burleighs worked well once they were set up, but they were too cumbersome. Miners lugged the heavy machines and their mounting posts or carriages into position at the start of each shift; as each hole was finished, they repositioned the drill on its mounts to start a new one; and, as the shift's end neared, the miners, before blasting, hauled the equipment back again. Quincy never did a time study of its miners using a Burleigh drill, but Copper Falls did two such studies in 1871. Over shifts of 660 minutes, one team ran its drill for 278 minutes, the other for 252. The teams spent just as much time moving and adjusting the machinery and changing the drill steels: 252 minutes in the first case, 253 in the second.[13]

A major part of the handling problem lay with the long and heavy drilling machines themselves, and with their first-generation clamps, posts, carriages, and mounts, which did little to facilitate fast setups. Another difficulty lay with the miners, who initially failed to appreciate an important difference between hand and machine drilling. They spent too much time struggling in confined quarters to get the Burleigh drills into exactly the right position. The miners needed to learn not to be so careful. To maximize their productivity and earnings, sledge-wielding miners traditionally located and angled their holes precisely, to assure that just a few, shallow holes brought down a good amount of rock. But with drilling machines, it was not so important to worry about the mine rock's lines of least resistance. Instead, miners were well advised just to point the Burleigh in the right general direction and to drill as many deep holes per shift as they could.[14] The increased number and depth of the holes would more than compensate for any errors in alignment.

After it stopped drifting with Burleighs, Quincy unsuccessfully tried them in shaft sinking and stoping. In the middle of 1873 the company finally received two smaller Burleigh "stoping drills." These machines, only 38 inches long and weighing 206 pounds, were closer to what Quincy was hoping to find: a compact yet powerful drill of about 175 pounds. But while the first Burleighs were too big and overpowered, the smaller stoping drills were apparently underpowered. They did not break "as much ground as had [been] hoped."[15] After expending $26,557 on rock-drill equipment in 1872–73, Quincy acknowledged that "our success with them had not been such as to warrant their general introduction into the mine."[16] Near the end of 1873 the company shelved its power drills and deactivated its compressor.

The Pewabic mine shelved its Burleigh drills, too, and the machine proved a "disgraceful failure" at the Aztec. The Central mine used Burleighs for shaft sinking from 1869 to 1872, but then reverted to hand drilling. The Franklin mine ran only two Burleighs in 1877, "from which a little profit over hand labor is made monthly." Calumet and Hecla ran one or two Burleighs early in 1872 and reportedly placed an order for 20 more. This company's conglomerate lode contained the hardest rock in the district, and C&H made a concerted effort to succeed with the Burleigh. Yet here, too, the machine could not be made to outperform hand drills.[17]

In the hands of select workmen willing to put out extraordinary effort, the Burleighs had sometimes effected cost savings. At Copper Falls, a team of 12 men and three boys, running two drills over both shifts, had broken rock at a cost of 20 cents less per ton than hand-drilling miners. But this team rigged special blocks and winches to move around 3,000 pounds of power drill and mining carriage, and its members were hardly ordinary workers. Four of the 12 were "foremen," four were "engineers," and only the remaining four were "assistant miners."[18] On top of its other faults, the Burleigh failed an important test of an effective mining drill: that it be manageable in the hands of men having no special skills. Instead of revolutionizing mining in the Lake Superior copper district, Burleigh drills became mechanical curiousities gathering dust in spare corners of the mines.

The copper companies were among the first mines to try Burleigh drills. In 1872, the *Mining and Scientific Press* of San Francisco was still "anticipating" the introduction of the machine into "some of our large mines on this coast," and in a trade catalog published about 1873, the Burleigh company boasted nine Michigan copper mines as users of its machine, but only one western mine.[19] The Michigan mines suffered for having been among the first to invest heavily in Burleigh technology. While that machine still earned praise for tunneling, its record in the mining industry quickly proved spotty, at best. An 1874 article in the *Transactions* of the American Institute of Mining Engineers listed only four "practical mining drills." The Burleigh was one of the four, but it was said to be so superseded by others "that it is placed almost out of competition."[20]

The copper mines were not so disenchanted with the Burleighs that they shunned rock drills altogether, and the *Mining Gazette* kept them abreast of change by running reports of drill trials and descriptions of new machines. Between 1866 and 1880, numerous manufacturers entered the rock-drill market, and in the United States alone at least 160 patents were issued for rock drills and their constituent parts, including mounting apparatus.[21] The copper companies, more cautious this time around, experimented with rock drills manufactured by Wood, Winchester, Ingersoll, Horton, Brown, Bryer, and Duncan. In other regions some of these machines proved successful, but in the copper district their trials ended in failure.

Other machines tried included the No. 2 and especially the No. 3 "Little Giant" drills manufactured by the Rand Drill Company of New York, which had started production in 1872.[22] The Rand piston drill was similar to the Burleigh, but embodied important improvements. It had a protected, internal tappet valve, moved directly by the piston, that controlled intake and exhaust to the cylinder. The first Burleigh had carried a heavier valve gear outside the cylinder, where it was more prone to damage. Unlike the first Burleigh, the Rand had no automatic feed, a feature that practical experience had shown to be an advantage. (A hand-cranked feed reduced the machine's weight and complexity and eliminated many maintenance problems.) Instead of being mounted on a threaded column attached to a cumbersome mining carriage, the lighter Rand clamped to a smooth post, and miners could shift its position more rapidly. The Rand drills were simple, strong, relatively easy to repair, and they finally transformed the Copper Country into machine-drilling territory.

This transformation began in 1879 and was completed by 1883. Quincy acquired 22 Rand drills over this period, and bought no other manufacturer's machines. Such an exclusive reliance upon the Rand drill was typical. The Atlantic mine ran 17 Rand drills in 1881 (when it reported "there is no more hand drilling"), and 21 machines by 1885. Calumet and Hecla, by far the largest purchaser of Rand drills, operated 65 by 1882. The machine proved a phenomenon in this mining district. In the early 1880s, 17 to 20 mines operated, and at least 15 used Rand drills by 1883, including all major producers.[23]

The "Yankee miners"—as the Cornishmen dubbed the new rock drills—pushed mining work faster, at less cost, and with fewer men. In 1880, a direct test of hand drilling versus machine drilling at the Northwest mine showed that the Rand could drive a drift 2.2 times faster and break four times more ground in stoping. In that same year, the Osceola mine claimed cost savings of 40 percent for stoping and 20 percent for drifting; the company also announced that with ten machine drills it could dispense with 80 to 100 miners. Meanwhile, the Atlantic mine reported that "the cost per fathom with the [power] drill is about $10, and by hand labor it is $17." In 1881 hand-drilled stopes at the Pewabic mine cost $22.53 per fathom, while machine-drilled stopes were $14.08. The Franklin mine reported similar cost reductions: $16.70 per fathom by hand, $12.40 by machine. By 1882, with Rand drills Calumet and Hecla produced over 20 percent more copper, using 20 percent fewer miners.[24]

The machine also brought sharp increases in output, profits, and worker productivity to the Quincy mine. The introduction of Rand drills coincided with Quincy's jump in production between 1880 and 1881 of nearly 50 percent—from 3.7 million pounds of ingot copper to 5.5 million pounds. At the same time, the mine's running expenses increased by only 3.5 percent, and the number of contract miners declined from 192 to 167. In 1882, production reached 5.7 million pounds, a figure attained by only 90 contract miners.[25] About 60 men worked Rand drills, and 30 remained as hand drillers.

In another important way, the Rands were highly successful. They altered some traditional work patterns, but the companies introduced them without incurring any recorded labor unrest. The possibility of labor protest was thwarted both by good planning and by circumstance. The companies used discretion and did not brutishly force the new technology into place. Managers encouraged miners to accept the machine by offering incentives. And once they gained some experience with the machine, miners acknowledged that it offered them several benefits.

Quincy received its first Rand drill in the summer of 1879. Earlier, in November 1878, the company had reactivated its compressor and an old Burleigh drill for the first time in five years. By running the Burleigh, Quincy reacquainted a small number of miners with machine drilling. Miners entered contracts to drift and stope with a Burleigh from November 1878 through February 1879. Quincy idled the Burleigh from March through July, then put it to work again in August, just when it gave its first Rand drill to contract miners. The head-to-head test of the Burleigh and Rand lasted but two months before Quincy permanently retired the older drill.[26]

Late in October, Quincy passed its Rand to the all-Irish contract team of Daniel Crowley, John Crowley, Cornelius O'Neil, and Michael Sheehan. In August and September, these men (along with two others) had used the Burleigh drill in stoping north of the No. 2 shaft on the twenty-fourth level. For 19 days in October, the Crowley team stoped with hand drills at the same location, and then from 26 October to 5 November it remained there but used the new Rand drill. Thus Quincy saw the results of hand drills, Burleigh drills, and Rand drills, all operated in the same place by the same men.[27]

The Crowley team participated in another comparison of hand drills and Rand drills in July and August 1880. North of the No. 2 shaft, on the twenty-second level, the team executed successive contracts (covering 26 working days) both to drift and stope, first by hand and then by machine. By hand, the miners drifted 19.2 feet and stoped 7 cubic fathoms. By machine, the numbers rose to 27 feet and 16.7 fathoms.[28]

The Crowley team, after learning to run the Rand drill themselves, taught other miners to run the machine. When Quincy added additional machines, John Crowley and Michael Sheehan temporarily split off from their usual team to work with miners new to the drill. And from October through December 1880 the Crowley team entered contracts for shaft sinking that appear to have been training classes in Rand drill operation, because nine additional miners joined the team during each of those months.

The first Rand drills went to volunteer contract miners, and those men instructed other contract miners. Quincy did not import new men to run the drills and did not violate the established contract system by putting drills in the hands of men earning wages on the company account. It did, however, encourage miners to use the drills by temporarily suspending the usual manner of settling contracts.

Men running unfamiliar machines would worry about making slow progress, which could cost them a loss of monthly earnings. To alleviate such worries, Quincy awarded its men contracts that guaranteed a monthly payment higher than the average earnings of contract miners. For example, in July 1881 Quincy ran four Rand drills. It paid one team $57 per man for 26 days worked; one team, $55; and the two remaining teams, $52. During this same month, hand-drilling contract miners averaged $47.92.[29]

As miners became more proficient with the machines, they were paid once again on the basis of the number of feet driven or the number of fathoms stoped. Rand drill contractors received less per foot or fathom than did remaining hand drillers, but the speed of the machine allowed them to break more ground and earn more money. In 1878 and 1879, Quincy contract miners' monthly earnings averaged $40.80 and $38.34, respectively. In 1880 and then 1881, the figures stood at $49.14 and $47.63. In 1882 and 1883, when machine drillers outnumbered hand drillers, the average monthly earnings rose to $53.15 and $50.10.[30] Thus a 25 percent increase in earnings attended the introduction of the machine drill, which certainly made the Rands more acceptable to the labor force.

Quincy did change the contract system in one important way to make the rock drill a more effective labor-saving device. A traditional stoping team consisted of six hand drillers, three working each shift, and Quincy used the same number of men to stope with Burleigh drills. It never ran a Burleigh with fewer than three men per shift, even in drifting, where just two hand drillers were the norm. An important part of the Burleigh's failure had been that it was not a machine that two men could "carry anywhere" and operate.[31] But from the start Quincy ran the Rand drill with four-man teams, or only two men per shift, and it did not wait until it had a full complement of Rand drills before making this change. In February 1881, when running only four Rand drills, Quincy switched to four-man stoping teams throughout the mine, even for hand drillers.[32] The timing of this move suggests that perhaps Quincy wanted to go to smaller stoping teams early, while most miners were still hand drilling, so that the new machines were less obviously the cause of the change.

Quincy put two men on each drill per shift in 1880 and 1881. In 1882, however, Quincy and Calumet and Hecla started a new practice indicative of problems with the two-man-per-machine rule. Both companies put extra hands underground, but they were not adult contract miners. Quincy's 20 to 25 extra hands were "drill boys."[33]

Prior to the Rand drill, few boys under eighteen worked underground; the 1880 U.S. Census reported only 20 throughout the entire copper district.[34] Quincy and C&H did not plan from the start to use more child labor in conjunction with the machines. The idea arose as they experimented with means of maximizing the machine's economic benefits by keeping them constantly in operation.

A fully outfitted Rand drill weighed about 600 pounds. That weight was distributed in three separate pieces—the drill itself, its clamp ring, and mounting post. Because the machinery could be broken down, two adult miners could move, set up, and operate it. Yet two miners could not run the drill full-time while attending to other chores, such as obtaining supplies. Compared with hand drilling, one liability of machine drilling was that it dulled far more drill steels. Bundles of dull steels had to be hoisted to the surface, where blacksmiths applied new edges. Then the resharpened steels, lowered underground, had to be returned to the contract teams. To meet this need the companies turned to child labor. The boys fetched drill steels, water, tamping clay, and other supplies, while the men stayed at their machines. The drill boys provided miners with extra help and companies with cheap labor. In 1882, if Quincy added an adult to a Rand drill team, he would have expected $50 to $55 per month. Drill boys received about half of that, usually $1 per day.[35]

The increasing use of older boys underground did not result in protests from mining families. On the contrary, many parents happily received the extra income that their sons provided. One dissenting voice came from the *Mining Gazette*, which faulted parents for pulling their "young lads" from school and putting them to work. It urged that the drill boys' educational needs be met by evening schools or private classes.[36]

By running errands for miners, boys took away one excuse the men might have used to stop drilling. The machine itself took away another excuse: unlike tired men wielding sledgehammers, it did not require frequent breaks. The machine altered the pace and rhythm of the miners' workday, and the extent of the change depended on company policy. In 1881, the Atlantic mine told its men that a mining team was not supposed to break for lunch as a group. While part of the team ate, the rest were to keep the drill going. Similarly Osceola required its miners to "push things." The company knew how much work Rand drill teams could do, and "they must do it, or they fail to get contracts."[37]

At least one company temporarily pushed its miners by putting their work under daily scrutiny. The Conglomerate mine, instead of paying stoping teams per cubic fathom (and measuring the work once a month), paid them on the basis of the cumulative footage of holes drilled per shift.[38] The miners quit work after drilling, without doing their own firing. A "charger" toured each work site, measured and recorded the hole depths, and then charged them. Quincy, however, took no extraordinary measures to drive its men. It never gave a shift team of two men, drifting or stoping, two drills to run at once. Shaft-sinking teams occasionally ran two drills under a single contract, but the team was always increased in size. Since Quincy did not add any shift bosses, foremen, or assistant captains to its ranks, the Rand drill miners worked under no more scrutiny or control than before. The company still relied on its traditional contract system and on lower prices per foot or per fathom for Rand drill teams to encourage the men to work diligently.

Rand drill miners worked more continuously than hand drillers, but their work was less taxing. Historians sometimes write nostalgically about the dignity and pride inherent in hand labor, even when the labor is coarse work and the hand tool is an eight-pound sledge. Most copper miners willingly surrendered these hammers for machines powered by compressed air. The machines substituted short periods of hard labor (mostly heavy lifting) for the longer periods of hard labor that attended hand drilling. For hand drillers, the arduous task of sledging the drill steel consumed five or six hours of a 10-hour shift. With machine drilling, the hardest jobs were moving the drill to and from the working face and erecting it. While the machine worked, one man at the rear fed it forward with a hand crank—not an extremely hard job, but one that called for prolonged attentiveness. The second man stood at the front of the machine, ready to rap the side of the drill steel with a sledge if it stuck in a hole. He also squirted water in the holes drilled downward, to clear chips for faster cutting and to keep the dust down, and he chucked longer drills into the machine when necessary.

In hard-rock mining districts in the American West, men called early drills such as the Burleigh, Rand, and Ingersoll "widow makers," because the rock dust they produced, when inhaled, could cause silicosis (also known as phthisis or "miners' consumption").[39] The reciprocating drill steels pumped rock particles out of shot holes, and unlike later drills the machines did not automatically inject water into the holes to allay this dust. But silicosis was not the problem in Michigan that it was in the West, perhaps because there was less quartz or silica in the rock. In the copper mines, the introduction of the machine rock drill did not raise any issue of occupational health and safety. On the contrary, the drills were credited with improving the miners' work environment.[40] The technology adopted to achieve more economical shot-hole drilling had the beneficial side effect of carrying more air into the mines. The companies and their miners used this feature to good advantage, particularly in long drifts or high stopes. Miners opened compressed air valves as soon as they reached their work stations, which helped blow out any lingering dust or noxious gases. Once the drill was under way, the compressed air powered the machine and then exhausted into the mine, helping freshen the atmosphere. And finally, companies ran compressed air underground between shifts to augment natural air currents.

Other factors, not anticipated or planned, caused miners to accept the machine rock drill. Although the Rand machines rose to dominance in a matter of only two to four years, their diffusion was somewhat gradual, and this made them appear less threatening. After waiting 10 years for the proper machine, companies could not switch from hand drilling to machine drilling instantly. Typically, a mine purchased one or two drills for a trial. Then it acquired more, up to the capacity of its first-generation air compressor. Only later could more drills be added, as more and larger compressors were added.

Another aspect of the machine's introduction seemingly made it less threatening to miners: it did not completely eliminate hand drilling. If miners opposed the power drill, they could still find employment at Quincy and, in all probability, at several of the other mines. Although the machine drill dominated at Quincy by 1882, some 30 hand drillers regularly contracted for work in 1890, and indeed hand drillers worked at least through 1906.[41] The old technology survived because in one sense it was more flexible than the new technology—because some areas of the mine did not carry copper in sufficient quantity to justify the trouble and expense of running in air lines for the machines. The company's hand drillers worked marginal deposits, and perhaps "scrammed" in areas already stoped out, extracting copper rock missed the first time around.

Quincy used its remaining hand drillers only for limited stoping; in the development work of shaft sinking and drifting, where speed was important, it used machine drills only. Besides being limited in the kind of work they could do, hand drillers paid a price for adhering to the old technology, in terms of earnings that ultimately ran some 10 percent less than those of machine drillers. In April 1892, there were 163 Rand drill miners who averaged earnings of $54.04, while 33 hand drillers averaged $48.30.[42]

In 1882, after receiving a new compressor and doubling its force of machine drills, Quincy dispensed with the services of 75 miners and laborers, and at the same time reduced the size of surviving hand-drilling teams from four men to two.[43] By April 1882, a full 12 of 13 hand-drilling teams had but two men, while the thirteenth had four. Interestingly, of the 13 teams, only four contained men who did not share the same last name. The other nine teams were filled by Cornelius and Patrick Regan, Michael and Daniel Casey, James and John Harrold, John and James Coombs, John and Peter Stoppart, and so on.

The extremely high incidence of kinship cannot be explained with great precision, but it may reflect two seperate responses to the Rand drill, one from the men and the other from the company. For those at Quincy who remained as hand drillers, kinship ties may have been an inducement to adhere to the old way of work. In the copper mines, it would seem that men did not accept or reject the Rand drill as individuals. They had worked together as teams and probably made collective decisions regarding the drills. As for the company's choices, it may have used the machine drill as an opportunity to cull a certain type of miner from its ranks. The surviving hand drillers were not lone, transient miners, but men with fathers, sons, brothers, or an extended family on Qunicy Hill. Quincy preferred family men to single men, believing them more reliable and dependable. It seems that those were the men it chose to keep.

The Rand drill definitely displaced many hand-drilling miners, but it seems unlikely that they were unemployed for long, or that the drills had only a deleterious effect on employment in the region. Quincy more than halved its number of contract miners between 1880 and 1882, dropping from 192 men to 90. But its total employment declined by only 53 men, or 11 percent. For the copper district as a whole, employment actually rose with the introduction of the machine drill: from 5,000 men in 1880 to 6,300 in 1884.[44] While the drills meant fewer jobs for miners at some companies, at others they protected jobs. The drills helped sustain the Atlantic mine, which worked, at a profit, a lode charged with less than 1 percent copper. And in at least one important case, the drills created jobs by making possible the opening of a mine that became an important producer. The Tamarack owned mineral rights to a portion of the rich Calumet conglomerate lode, but that lode crossed onto Tamarack's land at a depth of 2,270 feet. The company sank a vertical shaft that deep through poor rock to reach the lode and exploit its copper. Using Rand drills exclusively, Tamarack accomplished the feat between February 1881 and June 1885. It was "the most rapid sinking in hard rock, that had anywhere been done," and without the machine drill it would not have been tried.[45]

Finally, miners accepted the Rand drill because certain limitations blunted its impact on the labor force. The machine ran on 60 pounds of air pressure per square inch, meaning that it required large pistons to make it powerful enough to drill rock. With large pistons, the overall machine remained large, because the manufacturer could not pare down metal and weight while still preserving durability. As a result, two men were needed to transport, set up, and operate the machine. This limited its effect on miners' employment and preserved the social nature of contract work and the tradition of men working together on teams. Every miner still had his buddy.

The mechanization of shot-hole drilling did not come easily to the mines. Transferring a technology from tunneling to mining proved more difficult than expected. From their first purchases of machine rock drills, the companies expected the innovation to lead to lower mining costs through higher productivity, but they had no precise blueprint to follow in implementing the new technology. First the Burleigh drill failed, and then, when the Rand drill arrived, it proved no simple substitute for the hand drill. Companies had to fit the new machine into their mines and into their labor forces. In doing so, they made several changes in their mining and employment systems, not all of which were obvious or anticipated.

That miners accepted these changes was a key element in the Rand drill's success. Their acceptance resulted in part from the sagacity the companies had shown in introducing the machines. The companies gave miners incentives to use them, and because the companies shared the machine's economic benefits, they brought miners higher earnings. The companies introduced the drills in a manner that did not destroy the established contract system of mining, even though they did alter that system. They did not use the drills as an excuse to shift contract miners over to daily wages, which preserved the miners' high status in the hierarchy of underground laborers.

But miners' acceptance of the Rand drill was also due to some unplanned circumstances. Because the local industry was in a period of expansion, men displaced by the machine at one mine often found work at another. The machine also led to the employment of drill boys, miners' sons who could now contribute more to their families' incomes. The machine did not totally eliminate hand drilling, because some copper was still best taken out by the old technology. Rand drills were driven by compressed air, and that same air also improved underground ventilation. And the machine's limited technology made it too heavy for one man to handle, thus ensuring that miners were not required to run it alone.

6

Change and Conflict: Nitro, Trolleys, and One-Man Drills

> Question: "How many kinds of drills are there—one-man drills?"
> Answer: "I believe there are many different kinds."
> Question: "They are trying out a lot of different kinds to see which one is best, aren't they."
> Answer: "The old machine is the best of all."
>
> Worker's testimony before a
congressional hearing on the Michigan
copper strike of 1913–14

Machine drills arrived at the mines in the late 1860s, met with initial praise, fell out of use altogether, and finally achieved widespread acceptance in the early 1880s. Another product of the industrial revolution—high explosives—followed an almost identical path. The fortunes of nitroglycerine-based explosives and of rock drills rose, fell, and then rose again in remarkably sychronized fashion. Their delayed acceptance, however, was due to two entirely different causes. Early drills failed because they were too cumbersome. Early high explosives failed because they were too frightening. From the start, high explosives were as capable of detonating labor unrest as they were of blasting rock.

Even after coming into general use, high explosives sometimes triggered dissent on the part of miners, as well as disputes between Michigan mine managers and eastern officers. The explosives, as manufactured by several companies, differed in their chemistry, reliability, force, and in the noxious gases they produced. They also differed in their cost. Miners wanted predictable explosives that detonated only when supposed to. They wanted explosives that did not foul the air unduly, whose gases did not burn their eyes or lungs or cause headaches. Because miners always remained sensitive to the issue of powder selection, mine managers deemed themselves fortunate whenever they found an effective explosive that miners liked. They eschewed changing powders frequently, because each switch might result in miners' protests. Eastern officers, on the other hand, tended to see explosives as expensive consumables, and they hoped to find new ones that cost less per pound. Consequently, debates over high explosives were not confined to the early 1867 to 1880 era, but continued into the twentieth century.

The black powder used by the copper mines prior to 1880 was a mixture of saltpeter, sulphur, and charcoal. When a burning fuse ignited it, the explosive detonated at a rate of about 1,500 feet per second. This relatively slow rate of burn produced gases (principally nitrogen and carbonic acid) that exerted a pushing or heaving effect on the rock.[1] High explosives, in contrast, burned much faster. Their gases expanded more instantaneously and had a shattering effect. They brought down more rock per charge, so mining companies saw high explosives as a key to increasing productivity.

The use of high explosives followed from the discovery of nitroglycerine by the Italian Ascanio Sobrero in 1846. Sobrero produced this oily, clear liquid by reacting nitric acid, sulphuric acid, and glycerine, but did nothing to commercialize the explosive. It remained just a scientific curiosity until 1859, when two Swedes—the father and son team of Emmanuel and Alfred Nobel—began working on the problems of manufacturing and detonating nitroglycerine "blasting oil."[2] By 1865, their novel explosive had made its way to New York City, and two years later the first shipment of nitro reached Lake Superior.

In the summer of 1867, the Phoenix mine was the only copper company to use nitroglycerine oil. In putting it into service, Phoenix miners probably poured it over black powder in tin cartridges, which they then corked. They put the nitro cartridges into shot holes and packed additional black powder around them. Miners detonated the black powder with normal fuses, and the black powder fired the nitroglycerine.

At Phoenix, J. G. Jackson praised the blasting oil to the directors of his company. His plaudits sounded just like early praises sung on behalf of the Burleigh drill. The blasting oil had "effected a wonderful change." Miners using nitro drove drifts over 50 feet per week, whereas with black powder they progressed only 25 or 30 feet per month. In stoping, nitro broke four times more ground than black powder. The "general use" of nitroglycerine, Jackson wrote, "will revolutionize the business of mining."[3]

Jackson's predicted revolution came to pass, but far more slowly than expected. Nitroglycerine oil worked wonders but also created mayhem. Simple shocks could detonate this touchy explosive. Even before its trial at the Phoenix mine, blasting oil had earned a deadly reputation. In 1865, ten pounds of it blew up in front of the Wyoming Hotel in New York City. In 1866, the steamer *European,* carrying 70 cases of nitroglycerine, exploded while being unloaded at Panama, and another shipment of nitro blew up in the Wells Fargo Express office in San Francisco, killing 14 persons.[4]

Following the Phoenix mine's experiment with nitro in 1867, the next known shipment of blasting oil reached Lake Superior in August 1869. It did not receive a welcome reception. The powder vessel's captain proposed docking at Houghton, but this made "quite a number of people nervous," so he put in at the Isle Royale mine's dock. The Aztec mine took fourteen 60-pound cans of the nitro, and the Mabbs brothers took the other 47. All this nitro was not rapidly consumed. In April, 1870, the Mabbs brothers were still trying some of it at the Huron mine.

When nitro arrived at the Huron mine, it prompted a strike. The Mabbs brothers' nitro spooked the men because it was over a year old. They feared it had decomposed and become less stable. Also, the plant that had manufactured this nitro had just blown up. The workers declared they would not return underground until the Mabbs brothers took away the offensive explosive, but Huron's agent insisted that the nitro would have a fair trial. Since the Mabbs brothers' own crew would test the explosive underground, ordinary miners had nothing to fear. But the men disagreed. If the explosive proved successful, they would surely be told to use it.

Miners candidly admitted "they were afraid of the oil," and in their strike they had "the sympathies of almost the entire community, including all our leading mining men." Fear of nitro, it seems, was felt by many mine agents, as well as by the Huron miners. Those miners finally ended their standoff with Huron's management by blowing up the despised nitro. In the middle of the night, some miners apparently "got into the shanty, heaped the powder from the keg around the glycerine, attached a long fuse, and the shock from the powder exploded the oil." Residents heard two explosions: one light, when the black powder detonated, and one very heavy, when the nitro went up. They saw the explosion too: "The country was illuminated, as by an electric flash."[5]

This sabotage ended the mines' brief flirtation with blasting oil, but nitroglycerine soon came in other forms. Alfred Nobel found a way to package the explosive that made it safer to transport and handle. In 1867 he combined liquid nitro with an inert absorbent called "kieselguhr earth" that held the explosive in a more stable form.[6] Nobel called his product "dynamite," and it came in different grades having more or less power, depending on its percentage of nitroglycerine. Users could custom-tailor their explosives. They could buy 60 percent nitroglycerine dynamite, or 50 or 40 percent grades, depending on their needs. To safely and surely detonate this new explosive, Nobel turned to the use of blasting caps—small charges that set off the larger explosion.

Nobel's dynamite quickly became a much-sought-after alternative to black powder, and its successful marketing encouraged many competitors to produce other high explosives. "Dynamite" soon applied not only to Nobel's explosive but to a broad class of nitroglycerine-based powders. Some, like the original kieselguhr dynamite, had an inert base or absorbent. Others had an active, explosive base that contributed to the power of the blast. This latter category of high explosives—those with active bases—soon displaced Nobel's original dynamite.[7]

Between 1870 and 1880, high explosives proved to be both more powerful and more predictable than black powder, and numerous mining districts across the country adopted them. Only one hundred miles east of the Copper Country, mines on the Marquette iron range adopted nitroglycerine-based explosives in the early 1870s, and an explosives manufacturing plant located there, the Lake Superior Powder Company, to enter the dynamite business.[8] On the Keweenaw Peninsula, however, the copper mines largely maintained a "hands-off" policy regarding dynamite until the late 1870s.

In 1874, for instance, Quincy's secretary-treasurer, William R. Todd, and the company's agent, A. J. Corey, squabbled over the mine's choice of explosives—but dynamite was not the issue. The men disputed the merits of two black powders: "saltpeter" powder, made with potassium nitrate, and "soda" powder, made with sodium nitrate. Todd ordered 50 kegs of soda powder for trial, because it was cheaper than DuPont's saltpeter powder. Corey agreed to try it, but without enthusiasm. He noted that Quincy had tried and "condemned" soda powder years before. The mine agent preferred the DuPont powder because it "breaks from 25 to 33 percent more ground than any *soda* powder," and because it "gives universal satisfaction among the men."[9]

While Quincy's leaders debated the merits of various black powders, in 1874 the Phoenix mine experimented with Dualin, an active-base dynamite. The trial ended quickly and tragically when 100 pounds of it exploded in the mining captain's office. The men in and about the office were "instantly hurled into eternity," their bodies cut in half and blown 250 yards away. The *Northwestern Mining Journal* reported, "It was one of the most frightful scenes that [the] human eye ever witnessed." Quincy's agent assessed the accident as "Six more victims to the little understood, possibly unknown, properties of Dualin."[10] Corey's remark typified the sense of fear and suspicion that most local mine agents still felt toward high explosives. In 1877, Thomas Egleston, professor of mining engineering at Columbia, reported that at the copper mines, "[Black] powder is used exclusively in blasting. The miners do not like dynamite, nor any of the modern explosives, as they say the air after a shot gives them a headache."[11]

Two factors delayed for a decade the mining companies' adoption of high explosives. The first was fear. The second was an unwillingness on the part of mine agents to force these new explosives onto their miners. Agents acted conservatively when selecting powders; they bent to their skilled workmen's protests. Not until high explosives had proved themselves safer than black powder in numerous other mining districts did the Lake Superior mine managers, or the miners themselves, willingly convert to dynamite.

The conversion finally happened in 1878 through the early 1880s—just as the new Rand drills were introduced. For mine managers, this proved a fortuitous time to experiment with dynamite, because the coupling of high explosives and machine drills produced impressive results. The one drilled more and deeper holes. The other brought down more rock per charge. Together, the piggy-backed innovations promised a tremendous increase in productivity.

The Altantic mine began switching to high explosives in 1878, and by 1880

C&H's chief explosive was Hercules powder, nitroglycerine in a mixture of magnesium carbonate, potassium nitrate, and sugar.[12] Quincy also began using high explosives in 1878, when it made some rather small purchases of Hercules powder. In 1881, Quincy first purchased a different high explosive, Excelsior powder, and this became the company's standard explosive in 1882. Excelsior powder proved a happy coincidence for the company in the early 1880s. Mine officials preferred it to Hercules powder because it cost less per pound. And the company changed over to Excelsior without objections from the miners, because "they like this powder here better than any other they have used."[13]

Once the companies and miners finally accepted high explosives, they were in the mines to stay. But the permanent adoption of high explosives did not end all controversies regarding their use. Whenever company officers considered selecting a new high explosive having a different chemistry, the change raised the possibility of labor unrest. As a consequence, local mine managers often preferred to stick with their current explosive, as long as the men liked it. But in siding with the status quo and the men, these superintendents invited criticism from eastern officers.

William Rogers Todd, who served first as Quincy's secretary-treasurer and later as its president, challenged at least three mine agents on their powder selections. Todd, the man who searched for cheaper coal, also searched for cheaper explosives. For him, the price was the chief criterion, while for Quincy's superintendents, effectiveness and labor contentment ranked more highly. In the early 1870s, Todd had made A. J. Corey defend his choice of DuPont saltpeter powder over Oriental soda powder. Todd next put Samuel B. Harris in a defensive position.

Harris believed that the men at the mine were the best judges of explosives. He trusted the reports from miners that filtered up to him through the captains. Based on these reports, Harris strongly communicated his evaluations to explosives manufacturers and to Quincy's eastern officers. In 1900, Harris expressed a willingness to try different explosives, but demonstrated a conservatism rooted in his deference to the opinions of miners:

> We are using the Lake Superior Powder Co's. powder and have done so for several years. We have tried almost every other kind of powder in the market—and on the whole like this a little better than any other. . . . We have no complaints from the men in regard to ill effects of the gases. . . . Of course, *any* powder in a close place will make more or less nasty fumes—but, in fact, we have had less trouble or complaints, in this respect, than from any other high explosive we have used.[14]

By 1901, Quincy used about 960,000 pounds of high explosives annually. Early in that year, William Todd wanted to switch from the Lake Superior Powder Company's nitroglycerine dynamite to another high explosive whose chief ingredient was ammonium nitrate. The ammonia powder could be had for half a cent less per pound. For Harris, Todd's advocacy of ammonia powder was a misguided affront to the agent's technical expertise. Harris made his sentiments clear in a letter to Todd:

> We have tried *lower* and *cheaper* grades of powder. . . . such as is used in some of the neighboring mines—but *were losers* by it. We can get that same kind of powder today for 10¼ cents per lb.—*but don't want it.* We at the mines are certainly the best judges.

In this instance, Harris lost the battle but won the war. Under pressure from Todd, he reluctantly contracted for a six-month supply of ammonia powder. But after that trial, straight dynamite remained Quincy's primary explosive as long as Harris remained agent.[15]

Charles Lawton became the next manager to run afoul of Todd in matters of powder selection. Lawton, a university-educated mining engineer who had never been a miner himself, put less stock than Harris in the opinions of ordinary workers. He saw himself as a progressive advocate of modern mining. He liked being in the vanguard of change and liked beating other mine managers to something new. But in 1912, even Charles Lawton incurred the wrath of president Todd, because of a reluctance to change explosives. Lawton learned to read the signals coming up from the miners' ranks, and those signals told him that 1912 was not a good time to change powders.

The high explosives that proved contentious in 1912 were ammonium nitrate (again) and gelatine powder. Unlike ammonia powder, gelatine dynamite was another nitroglycerine-based explosive, but the nitro was not put into the explosive as a liquid. Instead, it was first transformed into a gelatinized mass by reacting it with collodian cotton. To produce the final explosive, the gelatine was mixed with sodium or potassium nitrate, plus meal and soda. Ammonia powder and gelatine dynamite were both deemed more powerful than the straight nitro dynamite that Samuel Harris had preferred.

In 1907 Quincy had finally switched from 50 percent nitroglycerine dynamite to 50 percent ammonia powder. According to Lawton, "We did it purely and simply for the benefit of our miners underground, two of whom had been asphyxiated from the nauseous gases of the old nitro glycerine dynamite." But by 1909, Quincy was purchasing more than $150,000 worth of ammonia powder per year, and Todd criticized the mines' rising powder consumption and costs. He faulted the general manager's choice of powders, and he faulted the miners for being too wasteful when charging shot holes.

Lawton countered with the argument that powder usage increased as the mine got deeper, because its rock became tighter and more difficult to blast out. But to appease Todd, in 1909 he switched to a less expensive, less powerful 40 percent ammonia powder. The mine's powder consumption rose dramatically, but the overall cost of expended explosives remained about the same. In 1910, Lawton tried again. He substituted 40 percent gelatine dynamite, which he had used earlier in a Utah mine, for ammonia powder. He believed the gelatine powder was more powerful, "just as cheap," and "far better for the ventilation of the mine" because of its "freedom from forming deleterious gases when detonated."

By 1912, however, Todd wanted the company to return to ammonium nitrate. As usual, his main concern was cost, and he made this clear to Lawton:

> It was you who, I believe innocently, but ignorantly tumbled into the blunder of introducing Gelatine Powder in place of Ammonia. . . . I believe your introduction of Gelatin Powder underground has caused a loss to the company of at least $50,000 without any corresponding benefit.[16]

Todd and Lawton read the same account books, but arrived at opposite findings regarding the economics of gelatine powder. Lawton believed that gelatine had *saved* the company $50,000. The general manager was willing to return to ammonia powder, but in letters to Todd he expressed "great fears" about the effect of the change on labor relations:

> I have been planning for some time to make the change to low-freezing ammonia powder, as per your instructions, but . . . I would be fearful of some labor disturbances underground. As you know, we never make a change in powder without having more or less labor disturbance.[17]

Lawton characterized 1912 as "chaotic labor times," and the last thing he wanted to do was to exacerbate labor-management problems by switching back to an explosive he deemed more expensive and less effective than gelatine dynamite. But while Lawton became particularly careful during an era of labor uneasiness, his exercise of due caution in changing explosives was typical of Quincy superintendents. From the days of black powder right through the days of advanced dynamites, they believed that an extra penny a pound was a small price to pay if an explosive kept the miners content. Eastern officers saw explosives as expensive commodities consumed by the hundreds of thousands of pounds. Miners saw explosives in a more personal way. They risked their lives handling them. Their respiratory systems inhaled the by-products of blasting. When the miners complained, the local agents tended to listen, even if the eastern officers did not.

By the early 1880s, machines drilled the shot holes for blasting rock and powerful hoists raised it to the surface. Jaw crushers reduced the rock in size, and then it passed through stamps and washing machinery at the mills. Every major step in processing and transporting copper rock from the mine through the mill had been mechanized—with one notable exception. Underground, men still mucked out broken rock by hand, hand-shoveled it into tramcars, and then pushed the cars to the shaft. These labor-intensive tasks required virtually no skill. Nevertheless, the mining companies were long frustrated in their attempts to get machines to do this work.

Geological conditions impeded the mechanization of mucking and tramming. Narrow copper-bearing lodes extended for hundreds or even thousands of feet across a mining company's property. To get out large-scale production, the companies had to spread out their miners, trammers, and unskilled laborers. Men worked not on just one level but on several. And on each level, small teams of men worked in different stoping areas. This decentralization of copper rock production made it difficult to put mucking and tramming machinery into economic operation. There was no one place to set up such machines, where they could operate almost continuously on large stockpiles of rock.[18]

Mucking was a two-step process. Broken rock had to be gotten from the stopes down to the drift, and then it had to be loaded into tramcars. Throughout the nineteenth century, the first step proved relatively simple. Most lodes dipped at a sufficiently steep angle to cause the rock, when blasted, to roll down to the drift. The copper lodes, however, often flattened out at depth. By 1900, as their stopes became less steep, companies sent "rock busters" up into them to give balky rock a bit of a shove.

Men proved better than machines in doing this work. Unlike machines, laborers required no setup time, and they were versatile. If there was no rock to pull down at the moment, they could perform other tasks, while mucking machinery would only sit idle. Not until 1915 to 1920 did Quincy and a few other mines succeed in using mechanical stope scrapers to advantage. At the foot of a stope, a compressed-air-driven "pony engine" wound and unwound wire ropes that ran up to the top of the stope. The ropes connected to a scraper blade. They lifted the scraper up into the stope, and then dragged it back along the mine floor, pulling down the broken rock in its path.[19]

Trammers finished the mucking operation by loading rock into cars. This job proved particularly labor intensive when the copper rock rolled from the stope right onto the floor of the drift. Trammers then had to lift it up four feet to get it over the sides of the cars. They scooped up smaller rock with short, D-handled shovels, but more than half the broken rock was four or more inches in diameter. Trammers constantly stooped to pick up larger rock and then stood to toss it into the cars. When they encountered rock too heavy to lift, they rolled it up a plank into a car. When rock was too big to be rolled, then they called in a block-holer, who put a drill hole in it and blasted it into finer fragments.[20]

Prior to 1920 the mines made virtually no progress in mechanizing carloading. They could not get enough rock together at one place to justify the installation of machinery. Compounding this problem, most drifts were small, and many were filled with twists and turns. They were too confined and irregular to permit the use of power shovels (which looked like scaled-down versions of steam shovels). Lacking workable loading machinery, the companies improved the productivity of the trammers by other means. They stopped letting copper rock spill out of the stope and onto the drift floor. Timbermen built a heavy wall of stulls and planks at the base of a stope, somewhat elevated above the drift and its tram tracks. Through openings in the wall, men chuted rock directly into tramcars. Alternatively, timbermen erected a "high sollar," a heavy wooden platform built over the top of a drift. Rock broken in the stope rolled onto the platform. Trammers then shoveled or dropped it through holes in the sollar into cars below.[21]

Starting in the early 1890s, companies began searching for a way to mechanize the hauling of loaded tramcars from stopes to shafts. Over the next twenty years, only one major company—Quincy—made substantial headway in this effort. The industry as a whole failed to mechanize haulage, but in the process companies did "succeed" in alienating their trammers. They pushed their attempts to increase trammers' productivity in a harsh, even punitive way. When they failed to find a new technology that eliminated many trammers, the mine managers found themselves stuck with a group of hostile workers.

When the companies introduced rock drills and high explosives, they showed concern over how their skilled miners would react to the changes, and they shared the economic benefits derived from the innovations. But when the same firms tried to mechanize haulage, they exhibited no such sensitivity toward trammers. They did not care to appease or please these unskilled laborers. They wanted to rid themselves of as many trammers as possible, and then pay lower wages to those who were still left.

Calumet and Hecla initiated the search for mechanized haulage in 1893. By then, trammers had already become somewhat fractious. The work bred discontent, as only the strongest men could face up to its physical challenges without risking exhaustion. Some men took pride in their ability to stand up to the grind. Others simply rebelled at their exploitation. President Alexander Agassiz made clear his motives for wanting to mechanize tramming in a letter to C&H's superintendent:

> We must be prepared to do our tramming in some other way than man power very soon, for the men, judging from scraps in the paper, are beginning to talk about the use of men as beasts of burden. I shall be very glad not to have any of them.

Agassiz knew that no new technology could eliminate trammers altogether. Still, he held out high hopes for mechanization: "We ought to get rid of many trammers, perhaps half or anyway 2/5ths, by introduction of machinery."[22] Agassiz's estimate proved overly optimistic for more than two decades.

The mines (with only a few exceptions, such as the Isle Royale and the Ahmeek) had never figured out how to replace trammers with mules or horses, and replacing them with machines was no easier. The crux of the problem was not the moving of loaded tramcars along the drift. Obviously, animals or machines could perform that task. The problems lay at the beginning and end of a tramming run. At the loading end, in the absence of loading machinery the companies still had to hire many men to put rock into cars. At the unloading end, tramcars waited for their contents to be dumped into skips, and this operation also required trammers. In short, the mines had to employ numerous trammers just to fill and dump the cars, and that fact would not change by putting animals or machines underground to transport the cars.

Animals or machines would not pay—in fact, they would represent an additional expense—unless the new tramming system somehow reduced the labor involved in car filling and dumping.[23] If it did not, then the companies found it cheaper to retain trammers to fill the cars, push them to the shaft, and then dump them.

In 1894–95, Calumet and Hecla tried a cable-car system on some of its longest tramming runs. Compressed-air engines, using wire ropes, drew two-and-a-half-ton tramcars to the shafts. Alexander Agassiz and his mine managers also considered using electric mining locomotives. General Electric had exhibited haulage locomotives at the Columbian Exposition, and these had gone into service at the Cleveland-Cliffs iron mine, only 100 miles from the Copper Country. C&H, however, prior to 1913–14, did not implement cable or electric haulage on a large scale. It still relied predominately on hand-tramming. In 1914, when quizzed about why the great C&H had not modernized tramming, general superintendent James MacNaughton replied, "It had been sort of an impossibility."[24]

In 1900, William Rogers Todd had pressured Quincy's agent, Samuel Harris, to mechanize tramming. Harris did not object to the change in principal, but he admitted to Todd that he did not know how to effect such a change successfully:

> Your remarks concerning underground haulage are suggestive, but the subject is not at all new to us. You say "difficulties don't seem to us insurmountable," etc. Mr. Mason [formerly Quincy's president] used to say—"You can do anything with men, and money." Perhaps, but in this case with us . . . , we fail to discover or invent any method that will *pay*—here is a chance for you to immortalize [yourself].[25]

Although unsure of the entire process, Harris tackled the mechanization of tramming. Early in 1901 he was set to copy C&H's use of cable-drawn cars, but by the end of the year he opted instead for a trial of General Electric trolley-type locomotives. Quincy was the first of the mines to try electric haulage and, prior to 1913–14, the only one in the district to use locomotives to tram a majority of its rock.

Quincy successfully addressed the issue of how best to load and unload tramcars, while keeping its locomotives more constantly in use. It initially equipped each locomotive with nine cars. While three were being loaded, three were being pulled, and three dumped. Quincy worked four men on loading; used a boy to drive the locomotive; and put only two men to the task of unloading (a job made easier through the use of some special appliances). By 1902, this force of six men, a boy, and a locomotive could tram 132 tons of rock along Quincy's longest level in a 10-hour shift. Earlier, without the locomotive, reaching this output had required 12 men.[26]

Six months after introducing electric haulage, Quincy's cost accounting showed a savings of three cents per ton of rock trammed. After another six or eight months, the company reported a savings of seven or eight cents. Later,

Quincy achieved further cost reductions by getting rid of unloaders altogether. Instead of dumping the rock from tramcars directly into skips (which required men to uncouple, turn, and tip the cars), Quincy started chuting rock into skips from underground storage bins of 500-ton capacity. Locomotives pulled tramcars slowly by the mouths of these bins, and new "automatic side-dumping" cars discharged their rock without ever stopping.[27]

From the 1890s onward, attempts to increase trammers' productivity did not sit well with labor. Quincy mechanized tramming, but did not share any economic benefits of the new technology with workers. To the contrary, it penalized men for the innovation's success. In 1903–04, Quincy lowered trammers' wages from $60 to $55 per month, arguing that they deserved less money because electric locomotives made their work easier. The trammers had to call a brief strike to restore their higher wage.[28]

Other companies that failed to make a new tramming technology work still strove to obtain significant productivity increases. When they could not get them out of machines, they tried to get them out of men. They raised the height of tramcars with boards, so that each carried more rock. They reduced hand-tramming crews from three men per car to only two. They required men to deliver a set number of loaded cars to the shaft per shift (such as 15), and then had foremen ride the men to see that it was done.[29]

In 1893, Alexander Agassiz had noted that trammers were "beginning to talk about the use of men as beasts of burden." By 1910, such talk was loud and ubiquitous. Tramming, always a hard job, now was even worse:

> Question: How do the men push them [tramcars], as a rule when they get into a hard place?
> Answer: You have to put your shoulder under.
> Question: What would be the condition of your shoulder after pushing the car?
> Answer: Sometimes takes the skin right off where you put your shoulder to the car.
> Question: Now, would it ever become necessary for you to turn around and push the cars with your back when your shoulder got sore?
> Answer: Oh, yes. One night when I was pushing the car my shoulder got sore and the skin came off, and when it gets sore you have to take your back and . . . push it with your back.[30]

Trammers, in more ways than one, had their backs against it early in this century. They worked at low-paying, exhausting jobs in a hazardous environment. Some, at least, held a hope of escape. If they hung on and stayed out of trouble, they might be freed from tramming and offered a miner's job. But after 1910, ongoing quests for productivity increases jeopardized even that hope. The mines brought a new drill into the district that threatened miners with unemployment and promised to kill trammers' prospects for upward mobility.

After adopting the two-man Rand piston drill in the early 1880s, the Lake mines made precious little change in their drilling technology for three decades. By the turn of the century, they clearly lagged behind western mines in deploying a new generation of smaller drilling machines developed by men such as J. George Leyner. Leyner had only an incomplete public school education, but he parlayed his mechanical skills into a career as an engineer and manufacturer. In 1891 he became proprietor of a small machine shop in Denver, and repairing mining machinery formed much of his business. By 1893, Leyner manufactured his own piston drill, and by 1897 he marketed his first "hammer-type" drilling machine.[31]

The hammer drill represented a distinct break from the older piston drill. The machine still had an internal piston reciprocated by compressed air, but the drill steel itself (perhaps eight feet long) was not firmly bolted to the end of the machine in a chuck. Instead, the drill was carried loosely in a sleeve or collar. Thus the entire drill steel did not move in and out of the shot hole as the driving piston reciprocated; the piston more simply hammered it in by delivering a blow on each forward stroke. The older piston drills delivered fewer, but more powerful stokes per minute; the hammer drill's piston reciprocated much faster, but delivered lighter blows.[32]

Because the drill steels on Leyner's machines did not reciprocate, they did not help pump rock chips out of the shot hole. To accomplish this in another manner, Leyner used hollow drill steels. On his first machines, compressed air passed down the drill shank to help clear the shot hole for faster cutting. The air, however, blew back so much dust in miners' faces that they refused to operate the machines. So Leyner injected water into the drill steel and shot hole. The water allayed the dust, and the "Water Leyner" drilling machine quickly won favor in many western mining regions. In 1902 Leyner formed the J. George Leyner Engineering Works Company to manufacture and sell Water Leyners and other equipment, and in 1912 he agreed to have the larger Ingersoll-Rand Company manufacture and market his drills.[33]

A few small rock drills arrived in the Copper Country about 1906, but they were not intended for general mining use. These jackhammers or pneumatic chisels were used for cutting up mass copper, for block-holing, and for a few other special purposes. Not until about 1910 did Quincy and other major producers seriously investigate the use of Water Leyners and other light machines.

In 1904 the J. George Leyner Engineering Works Company sent two Water Leyner drills to Quincy for trial. The mine's agent at the time, John Harris, returned them unused. He sent along a letter saying he could not foresee how their use would help him cut mining expenses.[34] But late in 1905, Charles Lawton replaced Harris, and Quincy became far more interested in new drills. Before arriving at Quincy, Lawton had put smaller drills into a Utah mine. He strongly advocated the new technology, and on a personal level, he dearly wanted to see his mine, and not Calumet and Hecla, be the first to adopt it on Lake Superior.

For several years, Lawton put his plans for new drills on hold. When he first arrived, it "got noised about the location" that he was "going to institute many objectionable western customs in the mine."[35] So Lawton let the matter rest until 1910 and 1912, when he agressively sought out a lighter drilling machine suited to Quincy's underground works. At Calumet and Hecla, general manager James MacNaughton entered the same hunt.

Lawton, MacNaughton, and other mine managers believed a change to smaller drills was absolutely necessary. Their mines, compared with those of Montana and Arizona, were high-cost producers exploiting low-grade copper deposits. To remain profitable, they dearly needed to reduce the costs of drilling and blasting mine rock that became denser and harder as the mines went deeper. Also, some lodes, including Quincy's, were "pinching down," or becoming narrower, and a smaller machine was needed to work in more confined spaces.[36] The ideal technological fix seemed to be a powerful drilling machine small enough to be run by one miner. If they could obtain the right hardware, the mines could maintain current production levels while halving their force of miners. They could swiftly double productivity.

The two-man Rand piston drill that Quincy used in 1910 weighed 293 pounds. It required a 145-pound mounting post and a 164-pound clamp ring. When Lawton started searching for a one-man drill, he hoped to find a machine weighing about 125 pounds that would save about 50 pounds each on its clamp and mounting post.[37] Ideally, the smaller machine, operated by one miner, would drill the same footage of holes per shift as the bigger machine worked by two.

Lawton believed this could be accomplished through a series of technical improvements, starting with the mine's air compressors. They had to provide air at pressures raised from 60 to 100 pounds per square inch. With higher pressure air moving the drill's piston, that piston and its cylinder could be made smaller without loss of power. Correspondingly the entire machine could be scaled down, and the use of stronger, lighter materials, coupled with good engineering designs, would allow these lighter machines to be as durable as their predecessors.[38] Finally, their drilling speed would be increased if Quincy switched to shot holes reduced from $1\frac{1}{8}$ to $\frac{7}{8}$ inch in diameter.

In 1911 Quincy readied itself for switching to one-man drills. It installed its first 100-psi air compressor. It ran larger pipelines underground to handle the increased air pressure and to cut down on power losses due to friction. Lawton corresponded with all major drill manufacturers, and between 1911 and 1913 Quincy fully tested at least five small machines and took a cursory look at another six to 10.[39] The essential test for each was the same: Could it drill as fast as the older two-man machine?

Most machines Quincy tested were standard models put out by the manufacturers, which were of several different types. Lawton ordered trials of a scaled-down piston drill weighing about 140 pounds unmounted. The company tried various hammer drills, especially the Water Leyners that Lawton had used in Utah. He felt they were reliable, fast-drilling machines, and he favored their water injection system for keeping dust down. The Water Leyners first used at Quincy weighed 154 pounds unmounted.[40] The machine worked well, but its weight bothered Lawton. He thought a drill had to be lighter still to be a true one-man machine.

Quincy had made limited use of small, hand-held jackhammers since 1906.

These machines had no water feed and generated a great deal of dust. Also, since they were not column-mounted, they were suited only for drilling down holes. If trying to drill a horizontal or upward hole, "then even the devil could not hold it."[41] Nevertheless, the jackhammers had proved themselves useful, and they captured Lawton's attention. What the mine needed, he reckoned, was a jackhammer-sized drill that was column-mounted and had water injection. Since no drill already on the market carried these attributes, in the spring of 1913 Lawton took his case directly to W. L. Saunders, the president of Ingersoll-Rand.

Lawton pushed Ingersoll-Rand to build a hammer drill to his specifications. As he wrote to Saunders, "we wish to fit the drill to our requirements, rather than fit the mine to the drill."[42] The machine would drive steels only ⅞ inch in diameter. It would have a small 1½" or 2" piston, be column-mounted, and have water injection. If Ingersoll-Rand built such a machine, Lawton predicted that Quincy and other Lake copper mines would buy it.

In October 1913, Lawton wrote Quincy's president that, "after fifteen years of hard-fighting, I have at last been able to get a one-man drill that now possesses all the qualities I have demanded of such a machine."[43] Quincy had just received from Ingersoll-Rand its first "Baby Water Leyner" drill, weighing only 90 pounds. As an engineer, Lawton had persevered. He had had a clear vision of the machine he wanted, and he finally got it. But while Lawton the engineer had continued to press for the proper machine, while faulting all those already on the market, Lawton the mine manager had plunged ahead with those imperfect one-man drills. He passed them off to miners as one-man machines, even while telling drill manufacturers that they were too heavy for that. Other mine agents had done the same thing.

After experimenting with one-man machines in 1910–11, in 1912 the mines demonstrated just how eager they were to cut production costs. They rushed to adopt an imperfect technology, rather than wait any longer for better machines. Calumet and Hecla made its impatience known to Ingersoll-Rand in April 1912, when it sent in an order: "We want you to ship us as soon as they are made three hundred number eighteen Leyner-Ingersoll drills; that is, we want the first three hundred you make as soon as you can possibly ship them."[44] Early trials with the drills had not gone well; the miners did not like them. In 1912, while the companies determined to make one-man drills the new standard in their industry, workers grew just as resolute to frustrate that attempt. As a result, the one-man drill became the most contentious technological change ever made at the mines.

Even Charles Lawton, a leading champion of the new technology, was afraid of the strife it promised to bring. The technical details and economic potential of the new drills clearly excited him, and he believed their adoption was essential for the survival of these old mines. Yet Lawton's enthusiasm flagged at times as he struggled to effect change:

> One must have a little encouragement occasionally, in order to keep his grit and determination. If there was ever a position in God's world where a man needs

grit, all he has and all he can get, he needs it right here where he is placed between two fires—the employees of the mine versus the cost sheet.[45]

Miners objected to one-man drills for many reasons, not the least of which was their size. Managers blundered tactically when they told miners that a 140- or 150-pound drill was easily operable by one man. While two men had worked drills weighing almost twice that, miners protested that the Water Leyners presented too much weight for one man to handle:

> It is too heavy for one man to rig up; too heavy for him to lift up and down on the post; and too much work; because you have got to be at the front and back of the machine at the same time.[46]

At Quincy, Charles Lawton knew full well that 140- or 150-pound machines were "a little heavy for one-man drills," so he had pushed Ingersoll-Rand to develop the 90-pound Baby Leyner drill.[47] But by the time that "more perfect machine" arrived, late in 1913, the larger drills had long since antagonized the labor force. The damage had already been done.

Some miners objected to the water injection system.[48] It kept the dust down, but made a damp mess of a man's work station. Workers also saw the drills as dismantling the long-established tradition of contract miners working on teams. As Charles Lawton noted, "the old habits, or customs, of men working as partners together, is a hard one to break up."[49] Miners in the district had never worked alone before. Since the 1840s, friends and relatives had organized their own small mining teams. This gave the worker a compatriot to talk to, to share the work with. It also gave the miner a friend to look after him if he got in trouble. For the sake of socializing and safety, the men rejected the idea of working alone.

In addition to losing a partner, miners strongly sensed that they were losing independence and control over their work. The new drills came in when the companies were consciously applying the principles of scientific management and were superintending their workers more closely. Between 1901 and 1918, for example, Calumet and Hecla doubled its percentage of underground workers serving as captains or bosses.[50] Also, by about 1910, mining engineers looked for cost-cutting measures and made technical and work-related decisions once made by mining captains or the miners themselves.

At this same time, miners still entered monthly contracts, but the companies treated them for all practical purposes like common laborers working for a daily wage. They set standards of performance, checked on workers' output, and reprimanded them if they fell short. Miners felt that managers were engaged in a speed-up of work:

> The main complaint is about . . . work, it is too hard. Ever since I have been in the mine it has continually become harder. . . . Generally they want more work of a man. When I first started if a man drilled five holes they never wanted more. . . . Several months before I left there [the Quincy mine] the Captain

came into the office and said: "Six holes won't do anymore—8, 9, and ten." It was six holes up to that time.[51]

Miners perceived the one-man drill as a burden they did not want to assume in exchange for an increase from $2.80 to $3.13 per day.[52] The men using the machine would have to work harder, and they would be the fortunate ones. The machine threatened others with unemployment or with a reduction in occupational mobility. Because of the impact it promised to have on employment, the new technology spawned resistance from a broad spectrum of underground workers.

In 1912 and the beginning of 1913, workers demonstrated their displeasure with the one-man machines in several ways. They became sullen, mute, and uncooperative. At Quincy, many simply refused to use the drills. Others accepted them, but then practiced "soldiering," or the act of cutting back on productivity. Given one-man machines to run alone, miners saw to it that they did only half the work they were capable of, only half the work of a two-man drill. They also caused the machines to suffer severe reliability problems:

> The men in their opposition to the one-man drills are breaking them beyond all precedents; the cost for repairs is the very maximum that it has ever been and is beyond reason. . . . We think that there is a united action in this regard.[53]

Despite workers' sometimes subtle and sometimes overt protests, mine managers purchased hundreds of the new one-man drills. Buying them was one thing; developing a strategy to soften resistance to them was quite another. The companies were prepared to force the drills into use if necessary, but they preferred that workers accept the machines willingly. This hope caused managers to vacillate, to attack and then retreat. At times labor protests "bluffed" mine managers out and they proceeded more slowly with the change. They hedged and were "a little more moderate and conciliatory." They stopped adding new machines underground, or they allowed two men to operate them, instead of just one. On other occasions, the companies behaved more forcefully; if men refused to use the new machines properly, a captain told them to get out of the mine.[54]

At Quincy, Charles Lawton looked for any advantage he could while deploying the new technology. Late in 1912, he thought he had found a helpful ally to the cause: winter weather. For some time, Quincy had 40 new, small piston drills sitting in storage at the mine. These were "to be used to take the place of the large drills as we merge into the one-man drills," but Lawton had held off putting them underground. Throughout the summer and fall the men had posed too much opposition. Now Lawton awaited regular blasts of cold air out of Canada and a deep blanket of lake-effect snow, courtesy of Superior: "I am waiting for good hard winter weather to make the change, as our men are then more amenable to changes."[55]

Lawton was correct. On the Keweenaw, labor troubles always subsided with the onset of winter. The men found life on the Keweenaw hard enough in that season, even when work was going well. Winter was no time to instigate a strike, leave for work in another place, or get fired for breaking a new machine. But Lawton chose to look only at the short-term, positive effects of seasonal changes on labor relations. Winter would send the dispute over the one-man drill into hibernation, but would not kill it.

Even on the Keweenaw, winter did not last forever. In April the tall banks of snow would slowly recede and then disappear, and in May the brown, beaten-down fields would begin to regenerate themselves. The men, too, would start to stir. There was a corollary to Lawton's rule that labor troubles quieted in winter; labor unrest always peaked in the summer. The winter that began in November 1912 was one of festering discontent; the summer of 1913, largely due to the one-man drill, would be one of open rebellion, the likes of which the Copper Country had never seen.

7

The Cost of Copper: One Man Per Week

"CASUALTY AT THE SUPERIOR MINE—On Thursday the 12th . . . , a German boy named George Berger was sent into the adit at the Superior mine, and as he was not seen for several hours afterwards, and did not come home to dinner, Capt. Richards sent a man to find him, who examined the adit and drift for its entire length, without result, but on coming to the No. 2 shaft a hat was observed floating on the water. He immediately procured assistance, and the body of the boy was drawn out of the sump. He had probably lost his light, and becoming bewildered had wandered the wrong course, and walked into the shaft in the dark. He was met when on his way to the shaft, by another lad, who enquired his age, to whom he replied, "It does not matter what age we are when we come to die.""

Lake Superior Miner,
21 November 1857

Miners opposed the one-man drill in part because if a man was hurt, he wanted somebody to be there. He did not want the nearest co-worker to be several hundred feet away, running his own drill in isolation. In the dark and noisy underground, a man that far away might never hear or see his neighbor go down. An injured miner could die by the time anybody found him. So the safety issue raised by workers in opposition to the one-man drill was not an expression of mass paranoia or a trumped-up charge aimed at the new technology. The machine arrived at a time when workers believed that the copper mines posed greater threats to life and limb than in the past. Certainly, far more men were dying each year than ever before.

In the 1850s, 12 men died underground; in the 1860s, 54. The death toll nearly doubled to 106 underground fatalities in the 1870s, and it almost doubled again in the 1880s, when 195 men died. In the 1890s, 284 fatal accidents occurred. Then, early in the twentieth century, the death toll rapidly escalated. Between 1900 and 1909, the mines claimed 511 lives. Starting in 1905 and running through 1911, the region averaged nearly 61 underground deaths per year, or more than one per week.¹ It was the worst era in terms of the yearly death toll. By 1910–11, about one-tenth of all annual fatalities in the United States metal-mining industry occured in the Copper Country.²

Life-threatening hazards came in various forms and from different directions. Danger stared the miner in the face when he returned to check on a charge that had not exploded when it was supposed to. For the trammer, hazards resided both above and below. Rock breaking lose from the hanging wall could crush him, or he could stumble and fall to his death in a steeply pitched shaft or stope. Anybody walking along a drift could be struck by a tramcar and squeezed against unyielding timbers or rock. No mine in the district was perfectly safe, and no amount of skill or experience shielded a man from hazard. Accidents killed men of all ranks, and ethnic groups. The weekly human sacrifice could be any one of the men or boys working underground.

From the opening of the mines in the 1840s, until their closing in the 1960s, at least 1,900 men died in underground accidents. The deaths were not evenly distributed across time. The mines suffered few fatalities in early decades, when the industry was starting, and few deaths in later decades, when it was in decline. Fatal accidents proved a macabre measure of economic growth: the more men employed, the more men who died. While 1,900 men died over 120 years, 27 percent were killed between 1900 and 1909, and 23 percent died in the following decade. Half of all fatalities occured in just 20 years.

Mining deaths were also distributed unevenly across the 60 companies recording fatalities. Again, economic growth chiefly determined a company's fatal accident record. The largest, longest-lived companies invariably suffered the most accidents. The Central, Copper Falls, Victoria, Ahmeek, and Mohawk mines each killed 20 to 25 men. Thirty-four men died underground at the Atlantic mine, 41 at the Franklin, 45 at Trimountain, and 46 at Kearsarge. Three mines suffered between 50 and 100 deaths: Baltic (59), Isle Royale (71), and Osceola (92). The Champion mine did not open until the turn of the century, but soon became an important producer and a major contributor to post-1900 accidents: 116 men died there. The Tamarack mine, opened some 20 years before Champion, killed 168 men. Not surprisingly, Quincy and Calumet and Hecla recorded the most fatalities. Quincy accounted for at least 253 deaths, and 470 men died from injuries in Calumet and Hecla's shafts, drifts, and stopes.

In terms of the personal characteristics of the accident victims, 60 percent were married. This percentage remained remarkably constant from the 1870s onward, so the industry produced many widows and fatherless children. The mines were carried on the backs of young men: the average age at death was 34, and the single most frequent age of death was just 21. Men aged 18 to 29 accounted for 41 percent of all deaths; men in their thirties for 27 percent; men in their forties for 16 percent; and those aged 50 to 59 represented 9.3 percent of all underground deaths. Boys under 18 and men over 60 worked and died in the mines, but they were distinct minorities. Each group represented about 3 percent of all fatalities.

The industry depended on a constant influx of immigrants to fill underground positions. Through 1880, 92 percent of all men killed in the mines were foreign-born. Over the next several decades, local society grew and matured. Nevertheless, foreign-born workers still accounted for 80 percent of all fatalities after 1900.

The rooted Copper Country population never provided a self-sustaining labor pool adequate to staff the industry's underground jobs. Many sons born to local mine workers did not join their fathers underground. They either left the region to find employment, or opted for surface work. As for Copper Country sons reared in mining communities by mining families, if they did work underground their background by no means advantaged them when coping with hazards. Compared with immigrants, they went into the mines earlier and consequently died younger. The average age of death for the foreign-born accident vicitim was 35; for the American-born, only 30.

The underground dead included men from 25 ethnic groups. About half of these groups (including Greeks, Syrians, Armenians, Danes, and Mexicans) suffered five or fewer deaths—testimony to the fact that they were never more than a trace element in the work force. Six ethnic groups accounted for 15 to 25 deaths each: the Scots, Norwegians, French Canadians, Russians, Hungarians, and Croatians. Three groups lost 55 to 65 men: the Swedes, Poles and Germans. Approximately 85 Irishmen died, while the Italians lost at least 230 men. Local society loosely identified a diverse group of immigrants from the Austro-Hungarian Empire as "Austrians," and 250 of these men died underground, along with no fewer than 380 Cornishmen and 475 Finns.

Ethnic deaths occurred in roughly the same rank order as ethnic underground employment. By 1910, for example, at Calumet and Hecla the Austrians formed the largest ethnic group in the mine.[3] They were joined, in descending order, by the Italians, the Cornish, and the Finns. From 1900 to 1910, the same order held for fatalities: the dead included 37 Austrians, 33 Italians, 27 Cornishmen, and 19 Finns.

The ethnic mix of accident victims varied significantly from mine to mine. Quincy employed a far higher percentage of Finns than Calumet and Hecla, and a much lower percentage of Austrians. From 1900 through 1910, Finns represented only 16 percent of the deaths at C&H, while they comprised 45 percent of all deaths at Quincy. The distribution of fatalities across ethnic groups also changed tremendously over time. The Cornish, together with the Irish, accounted for about three-fourths of all mining deaths through the 1870s; for half the deaths in the 1880s; a third of all fatalities in the 1890s; and for only one-fifth of the deaths between 1900 and 1909. Decade after decade, as skilled Cornish miners became a smaller percentage of the underground work force, the burden of fatalities shifted away from them and onto Finns, Austrians, and Italians. The shift ultimately had social repercussions. Finns, Austrians, and Italians did not hold mine work in high esteem and did not readily accept its risks. Compared with Cornishmen, they were more likely to complain about hazardous working conditions.

Underground accidents struck down men or boys of all occupations. Of

1,700 victims whose occupation is known, 800 died while working as miners, 316 were trammers, 200 were timbermen, 153 served as common underground laborers, and 45 were crew bosses or foremen. Death claimed, in almost equal numbers, both the sagest veterans (28 mine captains) and the rankest newcomers (30 boys under age 18).

The various occupations entailed different levels of skill, different types of work, and somewhat different hazards. On the whole, however, these factors balanced out in a way that placed men in all underground occupations at nearly equal risk. The best evidence of the even distribution of risk across occupations comes from Calumet and Hecla. The years 1895 through 1918 represented the mine's deadliest era, in terms of the absolute numbers of men who were killed. Over this span, miners comprised 29 percent of all underground employment and 29 percent of all deaths. Trammers made up 24 percent of the work force, and 25 percent of all fatalities. Timbermen and laborers combined (C&H used the terms interchangeably in this era) represented 39 percent of the workers, and 40 percent of the accident victims. At the top of the occupational hierarchy, mining captains comprised 2 percent of the work force and 4 percent of the deaths. At the bottom end of the labor force, the single greatest discrepancy between the two percentage figures is found. Boys made up 6 percent of C&H's underground workers, but accounted for only 2 percent of all fatalities.[4]

All underground workers faced the risk of accidental death, but they were not equally compensated for assuming their risk. Supervisory personnel and miners earned more money for putting their lives on the line than did trammers or unskilled laborers. What was true for occupations was also true for ethnic groups—some received better compensation than others for accepting hazardous work. Of Cornish accident victims, a full 80 percent died while serving as miners, bosses, or captains, and only 18 percent died while toiling as trammers or laborers. Amongst Finns, virtually none died while filling the highest paid supervisory posts, because they were not employed for such work. Half died as miners, and 35 percent died as trammers or unskilled laborers. Austrians and Italians fared even worse. Besides being locked out of supervisory positions, only one-fifth of the Austrians and Italians killed underground had worked their way up to miners' jobs. More than half of them had risked their lives—and lost them—while still on the very bottom rungs of the wage-earning ladder.[5]

Underground, falling materials proved the biggest killer, especially rock dropping down from the hanging wall, or rock rolling down a stope or shaft. Such accidents killed at least 720 men between 1860 and 1929, accounting for 45 percent of all deaths of known cause. The second most common category of accidental death involved machinery, principally the equipment for transporting rock or men. Machinery-related accidents claimed about 20 percent of all fatalities and killed a minimum of 316 men. Some 252 men (16 percent of all fatalities) died due to explosives; in the local parlance, they were "blasted." A nearly equal number, 250 workers, fell to their deaths from ladders or while walking in a stope, drift, or shaft. Since most of the mines carried little timbering and produced no combustible dust or explosive gases, fire was not an omnipresent danger. Through 1929, fire or smoke inhalation claimed only 3 percent of all fatalities. Nevertheless, the worst mine disaster on the Keweenaw was the Osceola mine fire of 1895, which killed 30. Altogether, mine fires claimed 54 victims.

For every known workplace hazard, known precautions existed. But men did not always follow safe practices and sometimes, even when they did, hazards conspired to kill them. This was particularly true in the case of falling materials. Workers never held a guarantee that the mine roof above their heads would hold. In 1873, Michael Sullivan walked into a stope at the Quincy mine. Without warning, two tons of rock came down on him from the hanging wall: "The bones of the head were all mashed, the brain was forced out through his eyes. His right arm was torn from his body; the right thigh and knees were literally crushed; in short the remains were one mass of jelly."[6]

Sullivan's accident was typical in the sense that the rockfall that killed him was localized and involved only a small portion of ground. The vast majority of the victims of falling materials died in similar collapses that killed just one or two men at a time. One tragic exception occured at the Copper Falls mine just a year after Sullivan's accident. There, a large shelf of ground came down, catching seven men beneath it. Before fellow workers could find and recover the buried bodies, "they were so badly eaten by the rats as to be almost unrecognizable."[7]

Underground workers took several measures to protect themselves from falling materials. Men working at the very bottom of a shaft were particularly vulnerable to this hazard, because they had no place to run to safety if rock, tools, timbers, or run-away skips crashed down upon them. To protect shaft-sinkers, it was standard practice to deepen shafts with the aid of a solid rock "pentice."[8] From the lowest level of the mine, men sank a small passageway paralleling the route of the shaft, but off to the side. After sinking this passageway some 10 to 20 feet, miners then enlarged the opening to the full width of the shaft, which they proceeded to sink further. This left a pentice—a ledge of rock ten to twenty feet thick—as a safety barrier above them. After sinking the shaft one or two more levels, miners blasted out the pentice so that the new shaft section could be timbered, tracked, and put into service.

Miners working in drifts and stopes also protected themselves from rockfalls. When entering a work station, safety-conscious miners visually checked the hanging wall. They sounded out suspicious ground by rapping it with a long steel bar. If it returned a crisp ring, they deemed the ground solid and secure. If it gave out a dull thud, the hanging wall merited closer attention. Miners looked for cracks. If they discovered one, they might drive a wedge into it. Later checks told them if the rock had dropped. As long as the wedge remained tight, the ground was not moving. Larger cracks gave men a place to insert a pinch bar to bring the rock down. Standing off to the side, where they hoped they were safe, miners pried the rock free, letting it crash to the mine floor. If a troublesome piece of ground refused to come down safely, then the miners could call for a timber crew to prop it up with a stull.

As a last defense against rockfalls, miners staked their lives on keen hearing and a quick step. In theory, at least, loose rock usually did not fall all at once. Many falls started with some preliminary movement, accompanied by a crackling sound that signaled miners to beat a quick retreat.[9] Of course, rock did not always oblige and send an audible warning. And even when it did, busy men performing noisy work did not always hear it.

In a very literal sense, miners could not always afford to tend to their own safety. An appreciation for self-preservation motivated contract miners to look after overhead rock, but a desire to maximize earnings encouraged them to not worry about every last piece of the hanging, and to take risks. Time spent securing the hanging wall was time taken away from the contract work that determined the miners' monthly earnings. As one miner noted, "at the place where we worked up in the Quincy mine, in some places if you begin barring in the hanging where it is not timbered sufficiently you can bar all day."[10] No miners wanted to stay off the drill that long, unless they received extra compensation for tending to the hanging. If contract miners had to spend an entire day working with the hanging, they might negotiate with a mining captain to receive a daily wage for that effort, to be added to their monthly contract settlement. But rather than go to the bother, many miners quickly fixed the hanging wall's worst spots and ignored the rest. Such skimping jeopardized not only the miners but all other underground workers.

Miners, aided by timbermen, were chiefly responsible for the safety of the hanging wall. After securing or neglecting it, they moved ahead, and less experienced workers came into the area. Trammers and common laborers were "not supposed to know an unsafe place from a safe one."[11] If miners left an unstable hanging wall behind, or if the hanging loosened after miners had left a drift or stope, the hazard might go undetected until it fell on some unskilled workman. Such accidents were not uncommon. While falling materials accounted for slightly more than 40 percent of all miners' deaths, they killed nearly 60 percent of all trammers who died in the mines.

About one-fifth of all underground fatalities involved machinery. While working in stopes, men were relatively free of life-threatening mechanical hazards. Rock drills often toppled over, causing numerous injuries to extremities, but they almost never killed anyone. While working in drifts, too, men had little fear from machines until after 1900 and the introduction of electric tramming. Heavy tramcars, pulled at six to eight miles an hour, could strike an unwary worker, causing him to fall or crushing him against nearby rock or timbers. Also, trolley locomotives picked up their current from a bare wire strung overhead in a low drift, and this wire could electrocute a man. These new hazards made drifts more dangerous places. Still, men were always more likely to run afoul of machinery in the well-traveled mine shafts.

Several factors made the shafts particularly dangerous. Steeply pitched and dark, they were hard on pedestrians. Their floors, taken up with crossties, stringers, and rails, were uneven. As uninterrupted openings extending from a mine's top to bottom, they allowed men or materials to fall great distances, accelerating all the way. Skips moved up and down these passageways at considerable speed, and little clearance existed between a skip and the mine's hanging wall or the shaft's timbering. Little clearance meant little margin for error, should a man have to scramble out of a skip's way.

The hoist engineer controlled the major mechanical systems in a shaft. Sitting at the controls to start, stop, throttle, and reverse the hoist, he watched a "miniature" (a gauge marked off with numbers representing different mine levels) that told him where the skips were. He received bell signals from fillers underground, and from a lander in the shafthouse, to tell him what to do. But he could not react instantly to the signals, and he could not count on the engine to give him a telltale pause when something went wrong. The powerful hoists in use by the 1890s did not miss a beat if a skip encountered a minor obstacle, such as a man. The men underground who could see danger had no way of stopping it. The engineers who could stop it had no way of seeing it.

The steam-powered systems designed to transport men up and down—man-engines, man-cars, and cages—presented several hazards. Men riding a man-engine repeatedly had to step back and forth while the up-and-down rods briefly paused, and the rods themselves could break free.[12] A worker riding in an uncovered man-car could be killed if he rose up too high and caught his head against the mine roof, or if a sudden start or stop pitched him off his seat. The man-car's hoisting rope posed the greatest threat by far. If it broke, 30 to 40 men might die in a rapid fall. Elevator-like cages posed a range of hazards similar to those of man-cars. Nevertheless, in the 70 years following 1860, only 41 men are known to have been killed while riding on or in man-engines, man-cars, or cages.

After 1890, given the general abandonment of man-engines and the rarity of vertical shafts with cages, the vast majority of workers rode man-cars to and from the underground. For new immigrants, the man-car was certainly one of the stranger and more frightening (or exhilarating) features encountered at the mines. Having left a village in Finland, Italy, or some other far-flung place, they had crossed the Atlantic, milled about in some large eastern or midwestern cities, and then headed to the remote Keweenaw on a train or steamer. They traveled thousands of miles, with each step of the way presenting new challenges and experiences. No part of this long journey, however, fully prepared the new mine worker for the last three to five thousand feet of his trip, the ride into the mines.

At the shafthouse, he watched men, using overhead cranes and long, clanking chains, pull the rock skip off the hoist rope and attach the man-car. When everything was ready, he moved with a gang of trammers to the loading platform, then stepped into the man-car. The men climbed up its step-like seats, then sat three abreast, filling the seats from the top of the car down. His co-workers warned the newcomer to sit still, keep his head down, and his arms at his sides. A man next to the shaft pulled a bell cord, the new worker felt a small bump, and then the man-car slid beneath the shaft collar. The car briefly flashed by rock walls and timbers, before descending into darkness. The rider heard the clickety-clack of the wheels over the rails, and felt a strong

The Cost of Copper 117

wind in his face. As the man-car went deeper, the wind got warmer. After three to five minutes, the car slowed and then stopped. Suddenly, it was 80 degrees and humid. The immigrant watched small patches of light, cast by miners' lamps, flicker and bob across the mine walls as the men got off the man-car and walked down the drift.

In time, riding the man-car became just another part of the job. Although an unusual means of commuting to work, travel on the man-car almost always passed without incident. The equipment had a good safety record because the companies that maintained and operated it, and the men who used it, treated it with more than the usual amount of caution.

As the region's wealthiest mine, C&H could afford to build an extra measure of safety into its man-cars. They had their own hoisting ropes and engines, dedicated solely to raising and lowering men.[13] Calumet and Hecla deemed this practice safer, because the ropes were spared the strain of raising rock, and the engines were smaller and slower than normal hoists, and therefore less likely to run out of control.

At other mines, man-cars and rock skips shared the same hoist ropes and engines. The companies had strong incentives, above and beyond any concern for human life, to see to it that this equipment was in proper condition. A disabled hoist halted copper rock production at its shaft, and the breaking of a hoist rope was always a serious matter, even if rock was being transported and not men. A free-falling rock skip could tear up timbers and track for hundreds or even thousands of feet. Such smash-ups led to expensive shutdowns and repairs, so companies took measures to avoid them. The steps taken to assure that rock was hoisted without incident also made the transporting of men a safer proposition.

At Quincy, a typical hoist rope (about 1½ inches in diameter) raised some 100 loaded rock skips per day. The rope had a working life of up to three years. Workers inspected the wire rope several times per week; they treated it with special lubricants to cut down on corrosion and abrasion. Halfway through the rope's expected life, they turned it end for end on the drum to help even out wear and strain along its length.[14]

The rope was not the only component to receive special attention. The man-car was equipped with special safety catches, intended to brake the car's fall if it separated from the rope. It seems to have been common practice, before putting men on board, to run an empty man-car down and up the shaft, to make sure it was free of obstructions. And whenever men were being transported, an assistant joined the engineer at the hoist's controls, so he could help in an emergency and double-check the engineer's attentiveness.

The engineer raised men at about one-third the speed of rock, or no more than 1,000 feet per minute. He was to take special care to set the hoist in motion only after receiving clear and proper bell signals, and he was responsible for starting and stopping the hoist especially smoothly when men were aboard. The men riding the man-car were instructed to get on and off in an orderly fashion, to never stand up while the car was in motion, and to never carry any tools with them that might trip someone or get snagged by a shaft timber. While the man-car was in motion, they were to sit quietly, with only their lunch pails perched on their laps.[15]

As a result of such precautions, men were quite safe while riding conveyances intended for transporting workers. They were less safe when hitching rides on other vehicles. Between 1860 and 1929, at least 83 men died riding rock skips, water-bailing skips, or kibbles.

Companies did not operate man-engines or man-cars throughout the work day, but only at the start and end of a shift. Any man—common laborer, miner, captain, or even mine agent—who needed to go up or down in the middle of a shift had to improvise, and when he rode a skip or a bucket intended for rock, water, or tools, he placed his life at greater risk. It was particularly dangerous to ride on the outside of a rock or water skip, to lie on its top or sit on its bail. Clearances in many parts of a shaft were tight, and a low timber or piece of the hanging wall could crush a man instantly. Rock skips traveled fast and sometimes jumped their tracks. Any passenger who lost his grip could be pitched off and fall to his death.

It was safer to ride inside a rock skip rather than on it. If men crouched down on a half-full load of rock, or sat in the bottom of an empty skip, they were more likely to make it intact to the top or bottom of the mine. But there were no guarantees. One man died at Quincy by playing it safe. He got in the bottom of an empty rock skip, and then a filler, unaware of his presence, put a load of rock on top of him.

The worst accident involving a skip or kibble occured at Calumet and Hecla in 1893, when 10 men died at the bottom of the unfinished Red Jacket shaft, one of the few vertical shafts in the district. The men got in a large bucket to "go to grass." Up at the hoist, a small chain slipped off its sprocket; this chain controlled movement of the hand on the dial that indicated where the bucket was in the shaft. The hoist engineer failed to notice what had happened. He did not stop the bucket at the shaft collar. He over-hoisted the bucket and tried to pull it through the top of the shafthouse. The bucket separated from the hoist rope, and the men fell 3,300 feet.[16]

Men did not have to ride rock skips to be endangered by them. From 1860 through 1929, at least 127 men died by being run over by skips, which made this the most common type of machinery-related fatal accident.[17] The pedestrians who entered active hoisting shafts did so at their own peril, particularly in later years, when the companies installed faster hoists, and many switched to balanced hoisting. They ran two skips on separate tracks; one went down while the other came up. These changes made the technology less forgiving of any negligent or unwary worker caught in a shaft. He had less time and less room to get out of the way. After 1890, skips colliding with men accounted for 8 to 10 percent of all underground deaths.

Electric tramming and faster skips led to an increase in machinery-related deaths after 1890. At the same time, explosives became a less significant contributor to fatal accidents. The high explosives used in later years were safer than the less powerful blasting powders used in earlier decades. From 1860 to 1929, explosives killed at least 252 men and accounted for 16 percent of all underground deaths of known cause. In the 1860s, 1870s, and 1880s, however, explosives had been responsible for fully one-third of all fatalities. From 1890 through 1909, that fraction was halved to 16 percent, and from 1910 through 1929, it was halved again, to only about 8 percent of underground fatalities.

The handling of explosives entailed obvious dangers and called forth many safeguards. The mines received and stored their explosives in powder magazines located some distance from other mine buildings or dwellings. Miners did not take their own explosives underground. Instead, the companies took one or two weeks' worth of powder underground at a time, so the work could be closely supervised.[18] Mining teams kept their explosives in iron or steel boxes, located a safe but convenient distance from where they worked. They kept explosives in one box, and fuses and dentonators in another. When charging or tamping explosives, miners were never to use a metallic tool that might cause a spark and set off a premature blast.[19]

From the start, the mines used taped, waterproof safety fuse that promised a more dependable and even rate of burn. Miners varied their fuse lengths, so that explosions were staggered and not simultaneous. Once the fuses were in place, one miner stayed to light them, while his partner retreated a safe distance and watched to keep unwary persons from wandering into the blast area. When the fuses were lighted and all men were in retreat, they counted the explosions to make sure that all holes had fired.[20]

If miners detected a misfire, something had to be done about it, because an unexploded charge was an accident waiting to happen. If the men had time, they refired the missed hole themselves, after waiting 20 to 30 minutes. That interval allowed gases to dissipate and lessened the chance that a faulty fuse was still burning toward the charge. If they lacked time to deal with the misfire, the men reported it to a captain, so he could inform the next shift that they had a problem to take care of.[21]

Despite all precautions, men still died using explosives. Miners were at greatest risk when charging and firing shot holes. Premature blasts—when the powder exploded early, before miners had retreated to safety—accounted for half of the deaths due to explosives. Late blasts were another form of tragic surprise. Men returned to a blast area, thinking all charges had gone off, or that all fuses in unfired holes were surely dead. Just as they got back to the scene, a retarded explosion went up in their faces. Death by suffocation was somewhat similar; it befell men who had returned to the area of a recent blast. Instead of being blown up, they passed out and died, overcome by noxious gases and a lack of oxygen. Missed holes killed men who drilled into a rock face containing a hidden, unexploded charge. The percussion of the drill finally set the explosive off. Finally, flying debris occasionally killed miners who had not retreated a sufficient distance, or killed other workers who happened to enter the blast area at exactly the wrong time.

The percentage of underground deaths attributable to explosives dropped sharply after 1890, and then dropped sharply again after 1910. The first drop can be attributed to the inherent qualities of the recently introduced high explosives. Compared with black powder, they were harder to detonate (an act that usually required a blasting cap), and less likely to deteriorate or become unreliable because of environmental conditions, such as dampness. In short, high explosives were less likely to go off prematurely and more likely to fire when supposed to.[22] The second drop in the percentage of fatalities due to explosives was seemingly of different origin. For the first time, the companies promulgated and enforced stricter rules for dealing with various hazards, including explosives. Between 1910 and 1915 they accepted a more active role in accident prevention and no longer held the men accountable for learning and following safe practices on their own.[23]

The man who went out with a bang represented the quintessential mine accident victim. His death was particularly sudden and gruesome. The very handling of explosives symbolized the dangers of mine work. But a far more prosaic kind of accident killed as many men as explosives: falling down.

Mine ladders were often in poor repair, damp and slippery. Broken rock, scattered tools, pieces of timber, and empty powder boxes littered the mine floor. Tram rails could catch the foot, and loose rock underfoot could suddenly give way. Safety barriers and guardrails were unheard of. Men worked and walked in dim light. The work surface was often steeply pitched, and if a man fell, the landing zone was always hard and often a great many feet down the mine. It is small wonder, then, that at least 250 men fell to their deaths between 1860 and 1929. Decade after decade, falls claimed one-eighth to one-fifth of all fatalities.

An important underground technology that protected workers from one hazard put them at greater risk from another. The timbering that guarded against rock falls provided the fuel for deadly fires. Except for the wood found in stulls, square-sets, and roof-lagging, as well as the wood used for shaft stringers and crossties, there was virtually nothing combustible underground. The mining companies had strong incentives to keep these timbers from flaring up. First, there was the threat of a large-scale loss of life. Smoke, not flame, was the greater killer. Men did not have to be in a fire to die; they just had to be where the smoke could reach them. Second, major mine fires were very expensive in terms of lost revenues and large recovery rates. Fighting a stubborn fire could halt production for several months, and then a company faced removing the debris and rebuilding the mine.

Calumet and Hecla had the most to lose from fire, in terms of workers' lives and investment. Because C&H used infinitely more timber than other companies, it was also the mine most likely to go up in flames. On one occasion, Michigan's commissioner of mineral statistics wrote about C&H: "The mine is a network of fine timber; millions of feet of dry pine; the best fuel for a great conflagration, and to such a calamity there is, inevitably, constant danger." A decade later the commissioner wrote that "the Calumet and Hecla mine is the greatest single insurance risk in the world, and one on which no insurance company would place a dollar of indemnity. It devolves on the management of the mine to furnish its own [fire] protection."[24] Calumet and Hecla did protect itself from fire. It erected elaborate and expensive systems to prevent and fight fires, and to keep them from spreading across all its works on the Calumet conglomerate lode. But C&H did not devise its fire protection plans well in advance of actual need. Instead, it became more cautious after being burned by experience.

On 4 August, 1887, a fire started on the sixteenth level near the No. 2 Hecla shaft. This forced the closing of all shafts on the company's Calumet and Hecla branches for several weeks. Because men could not go into the smoky underground to extinguish the fire, the company fought the blaze by trying to smother it.

C&H sealed off all shafts to prevent the inflow of oxygen and pumped steam and carbonic acid gas into the mine. The company had only briefly reopened its Calumet and Hecla branches, when, on 20 November, fire erupted again. Once more the company sealed its shafts and injected carbonic acid gas, but the fire refused to die. C&H could not fully reopen the interconnected Calumet and Hecla branches until June 1888. Fortunately, neither fire took any lives, but they caused a production loss of ten million pounds of copper.[25] What saved C&H from total disaster was the fact that its third branch of operations, South Hecla, was not connected underground to the two older branches. As a consequence, South Hecla stayed in operation while the rest of the mine was sealed and shut.

Calumet and Hecla's fire troubles were not over. On 29 November, 1888, yet another fire burst out, this time in the No. 3 Calumet shaft.[26] C&H followed its standard routine. South Hecla remained open, but all Calumet and all Hecla shafts were closed. When the fire doors were shut in this instance, they sealed the fate of eight men trapped below ground who were asphyxiated before they could scramble to safety. Calumet and Hecla thought the blaze was out and tried to reopen its closed branches in February 1889, but as soon as air reentered the mine, it rekindled the blaze. Once again, C&H capped the shafts. Not until five months after the fire's start did the Calumet and Hecla branches reopen safely—and it was some time after that before the bodies of the eight victims were finally discovered.

The fires of 1887–88 led to a number of changes at Calumet and Hecla. Extensive fire damage prompted the company to eliminate three hoisting shafts, instead of rebuilding them. The fire destroyed the wooden man-engine and pump rods, so C&H turned to man-cars and new electric pumps. It also took several measures to guard against future fires.

C&H hired more watchmen to patrol the underground. On a regular basis, it sprinkled water on shaft-timbers to keep them damp. It treated underground timbers with zinc chloride and whitewash to make them more fire-resistant. Once a level was stoped out, laborers now closed it off from the shaft with a poor-rock wall, intended to keep any blaze from spreading across the underground. C&H installed fire alarms and telephones. While the fire alarms signaled most men to leave the mine, they summoned others to come fight the blaze. If men responded quickly enough, and if they had the right equipment, they might be able to douse a fire. So Calumet and Hecla installed water pipes, hydrants, hoses, and chemical extinguishers on its working levels.[27]

The last "fire insurance policy" that C&H underwrote for itself was by far the most costly: the sinking of the new Red Jacket shaft. This shaft demonstrated just how afraid the company had become of catastrophic fires. C&H wanted an entirely new shaft to gain entry, at great depth, to the northern end of the conglomerate lode. It wanted to open a new branch of the mine that, like the South Hecla branch, could stay in operation even if the older Calumet and Hecla branches went up in flames.

Red Jacket shaft stood considerably west of C&H's other works on the conglomerate. This vertical shaft intersected the lode 3,260 feet beneath the surface and then bottomed out at 4,900 feet. Miners began sinking Red Jacket shaft in 1889 and did not complete it until 1896. The shaft, reportedly the deepest in the world when finished, was also very large in cross-section. It measured 15½ feet by 25 feet, and carried six compartments (four for raising rock and two for transporting men).

Calumet and Hecla never publicly reported the cost of Red Jacket shaft, but it was immense. President Alexander Agassiz supported the endeavor because he would not risk the fortunes of C&H to more crippling fires. However, once the work was under way, stockholders became critical of it, and Agassiz expressed private misgivings to mine superintendent S. B. Whiting: "It now looks bad enough when you figure up its cost, the cost of the Engines, Boilers and Buildings, the cost of the cross-cuts, all of which are idle and likely to remain idle for a long time to come and in fact it is getting to look to me as if this whole point was an absolute waste."[28] Red Jacket shaft became an embarrassment to Agassiz, and yet C&H pushed it to completion. It became a symbol of the fact that Calumet and Hecla was anything but callous and reckless when it came to protecting its future. It is probable that no other metal mine in the world had ever expended so much money to protect itself from fire.

Other Lake Superior copper companies took no such precautions, because they felt they were unnecessary. Unlike C&H, they did not have continuous timbering running down and across their mines. They had no fears about losing drifts and stopes to fire, and focused their concern on shafts. That was where most of the timber was, and a shaft fire could cause considerable damage. Shafts were giant chimneys with strong drafts, and a fire could quickly spread up or down, depending on the direction of air currents. Still, in mines other than C&H, any fire would most likely be limited to one shaft, and that gave workers and managers a greater sense of security. Ironically, this feeling of security led to the worst fatal mine accident in Copper Country history. In 1895, 30 men and boys died of asphyxiation in the Osceola mine. Many perished because they did not think there was enough wood to burn in the mine to do them any harm.

At 11:30 in the morning of 7 September, 1895, men discovered a small fire of unknown origin at the twenty-seventh level of Osceola's No. 3 shaft. Since the fire was discovered early, Captain Richard Trembath and a small crew of men tried to extinguish it with buckets of water. The No. 3 shaft carried the mine's pump, and a second captain, Richard Edwards, went to the surface to get a hose to take underground to connect to the pump-works. When he tried to return down the No. 4 shaft, Edwards learned that both No. 3 and No. 4 were already too smoke-filled to be used. He then sent men down the Nos. 1, 2, and 5 shafts to warn others to get out of the mine at once. The company ran skips slowly up and down the shafts, so men could scramble into them.

Underground, the outbreak of the fire had not caused tremendous alarm. While some men fought the fire, others (many with long years of service) sat down to eat their dinner. They knew that the timbers in the No. 3 shaft were damp, so they did not expect the fire to spread vertically. They knew there was no timber in adjacent stopes to spread the fire horizontally. They also figured they could get out if they had to; the nearby No. 4 shaft was downcast and brought fresh air into the mine. It would remain clear. Two of these assessments proved deadly wrong.[29]

The fire got out of control in the No. 3 shaft, damp timbers or not, and the heat and rising smoke and gases reversed its normal air flow. No. 3 was usually a downcast shaft. Now, smoke from the fire rushed up No. 3. Most of it billowed out of the shafthouse, which quickly drew a crowd of anxious onlookers. Near the top of the mine, however, much smoke drifted through old, open stopes over to No. 4. This shaft remained downcast. Instead of taking fresh air underground, it now sent smoke and gas back into the mine. The men and boys who waited too long to start their escape, and who tried to get out through No. 4, were doomed by smoke and gas. Ultimately, their bodies were recovered from the seventeenth up to the fourth levels of No. 4.

Mine superintendent W. E. Parnall arrived at the No. 3 shaft an hour after the fire had been detected. In the midst of a nervous and frightened throng of mine workers and their families, Parnall faced a difficult decision. He had only one way of checking the fire's destruction: capping and sealing the shafts. But it was known that many men and boys were still underground. If they remained alive and were scrambling to get out, capping the shafts would end any chance of escape. They would die, even if workers initially capped only No. 3. This act would impede the fire by slowing the draft of air that fed it, but it would cause more smoke to pour into other parts of the mine. Parnall opted not to cap any of the shafts for several hours, even though he personally believed, after only a short time, that all workers still underground had to be dead. His delay let the fire do greater damage and apparently spared no additional lives. But the superintendent had another important consideration in mind:

> I fully realized, the moment I saw the smoke rolling out of the shaft in such volumes, that a first-class case for suits of damage could easily be made by an inconsiderate act. Mr. Watson thought of nothing but covering the shafts when I got there at half past twelve. At one o'clock he was a little touched because I would not consent. I am satisfied he was right at one o'clock so far as all hope of further escape was concerned.[30]

Even though Parnall believed all the men were dead by one o'clock, he held off on sealing the No. 3 shaft until five o'clock. He left the others open and did not even seal them off the next day. That would have helped choke out the fire, "without endangering a thread of existing life," but the families of the missing workers were still expressing their "ifs and buts." So Parnall, to avoid incurring the families' wrath, still deferred sealing the rest of the mine.[31]

After the shafts were all capped and the fire smothered, search teams went underground and recovered the bodies of one captain, 19 miners, five trammers, one laborer, and four boys. Others began surveying the damage done to the No. 3 shaft. It had burned from the twenty-seventh up to the sixth level. Ten days after the outbreak of the fire, Osceola resumed normal operations for the first time in shafts 4 and 5. On the same day, after three days of hearing testimony, the coroner's jury brought in its verdict: "That the deceased came to their deaths by suffocation caused by smoke and gas from a wood fire. . . . We believe that this fearful loss of life is due to the fact that deceased did not realize the seriousness of their danger, although from the evidence given this jury, we find that said deceased were duly notified. We exonerate the mine officials from all negligence in this sad affair."[32]

While the Osceola fire raged in No. 3, superintendent Parnall did not want to infuriate the families of the missing men. He did not want to commit any act that might encourage future lawsuits and cast the Osceola company as guilty of killing 30 workers. He acted cautiously, to assure "that no blame can in any way attach to the Company." Shortly after the fire died out, Parnall's caution paid off, when the coroner's jury determined that Osceola had not been negligent in any way. Nevertheless, many widows from the fire later sued Osceola to receive compensation for their husbands' deaths. The lawsuits served notice that near the turn of the century and beyond, workers and their families were less willing to accept mining deaths stoically and without protest.

As the death toll increased and reached 60 per year by 1905, worker's fears and complaints about underground hazards were certainly warranted. It seemed that too little was being done to cut the loss of human life. The Copper Country in this era had no other large-scale catastrophes like the Osceola fire, but men kept falling, one or two at a time, at all the operating mines. In truth, if one applied two standard measures of risk—the number of fatalities per 1,000 men employed underground, or the number killed per million pounds of copper produced—the mines at 1910 were no less safe than they had been in earlier eras. The death rate per thousand employed fluctuated up and down somewhat from decade to decade, but it was not inexorably going up as the mines went deeper and the companies turned to new technologies. As for the death rate as measured against a million pounds of new copper produced—that figure had steadily *fallen* for each successive decade, and was less than half as great in the 1900–1909 decade as it had been in the 1860s.[33] Workers' families, however, did not know fatality rates in terms of cold statistics. They just knew that more men were dying in the mines than ever before. And besides facing the threat of death, mine workers ran a very high risk of injury.

In 1911, the first year for which the newly formed U.S. Bureau of Mines published accident data, these data showed that the Lake Superior companies accounted for almost 30 percent of all copper mining deaths in the United States, and for nearly 11 percent of all metal-mine fatalities. In terms of fatalities per 1,000 men employed, the Michigan mines were actually somewhat better than the average. For all copper mines, 6.1 workers died for every 1,000 men employed underground; for all metal mines, the figure was 5.5. The Michigan copper mines lost almost 5 men per thousand working underground. The Lake Superior mines had a poorer record than other districts, however, when it came to less-than-fatal injuries. Copper Country workers received over half of all reported serious and minor injuries in the U.S. copper industry. If a man were to work a year underground, the odds were one in 200 that he would be killed—but they were better than one in three that he would be injured at least seriously enough to lose time from work. In 1911, when 60 Michigan copper workers died, another 656 suffered serious injuries, and 3,974 suffered minor injuries.[34] The pain and suffering inflicted by the industry had reached proportions that workers could no longer tolerate, that companies could no longer ignore.

8

Weeping Widows, Generous Juries

[Four months after 30 victims perished in the Osceola mine fire of 1895:]

There is no further doubt as to [our] company being sued by some of the widows from the fire. We are in for it. I have been notified by one of the worst blood sucking villains of a lawyer that ever lived that unless the company is disposed to compromise he is prepared to enter suit at once. Compromise is out of the question. In the light of justice I do not see how any jury on earth can award damages. But it will be a rich corporation (!) against a weeping widow, and under those circumstances no one can predict the end. Six of the widows have signed a release, more would have done so had they not been preyed upon . . . by the lawyers that are working them up to believe that ten thousand dollars is certain for them.

Osceola mine agent W. E. Parnall to
Thos. Nelson, 4 January 1896

"J.A.P." boarded the train in Chicago, bound for Hancock. He arrived in the evening of 12 September 1906, and started walking the streets, looking for a room. He carried a well-worn suitcase, and his clothes, too, had seen some hard wear. Stopping at a corner, asking directions for a good place to put himself up, J.A.P. conversed in Finnish or English. He looked and acted just like all the other men coming to the Copper Country to land a job at the mines. That, after all, was the whole idea.

J.A.P. had worked in mines before. Now he worked for the Chicago office of the Thiel Detective Service Company. His assignment was to get himself hired at the Quincy mine. Quincy had recently had a three-week strike, and the company had contracted with Thiel to have a spy planted in its labor force.[1] Quincy wanted to know what was going on in the minds of its workers, particularly the Finns. It wanted to know who the agitators were. So in finding a place to stay, J.A.P. needed a private room, a safe place to retreat to late at night, when he wrote up his reports. Almost every day he would send a report back to Thiel's office manager in Chicago, Mr. Seagrove. Seagrove would summarize J.A.P.'s activities and findings in a neatly typed letter and mail it to Quincy's superintendent, Charles Lawton.

J.A.P. worked under deep cover. Nobody at Quincy, not even Lawton, knew who he was. He had to get his job at Quincy without any friendly assistance and play his workingman's role well, for fear of being detected by the men he surveilled. J.A.P. was of modest build—a bit undersized, in fact, for the heavier kinds of mining work—and he did not want to take on any group of angry, betrayed trammers behind some Hancock bar. J.A.P. would have to get by on guile. As it turned out, the Thiel detective stayed at Quincy for two months. During that time, he took some lumps, delivered not by "fellow" workmen, but by the mine itself.

While not at work, J.A.P. prowled the local watering holes and attended socialists' meetings. He frequented seven or eight local bars where Finns hung out, as well as Rajala's pool hall. He went to hear speakers at the Finnish Temperance Society and at Jausi Seura, the Finlander Workmen's Society. The spy faithfully reported who he talked to and what he heard. He named names. He recorded if a man had anything good to say about Quincy, or if he was one of those who thought Charles Lawton an uncaring son of a bitch. J.A.P. presented Lawton with a window to the working class. He passed along all the gripes and complaints—about wages, abusive bosses, faulty equipment, underground working conditions, and hazards. The spy did not merely pass along hearsay evidence. He experienced the mine himself, first working as a laborer mucking out rock in the No. 7 shaft, and then as a waller, copper picker, and chute-man at No. 6.

The men had nothing at all good to say about No. 7. Even before going down to it, J.A.P. heard it was Quincy's worst. It was particularly hot and uncomfortable. Workers called it "the men's slaughter." Somebody got seriously injured or killed there every week. Because he had heard so many complaints, when he first went down No. 7 on 21 September, J.A.P. "was careful to examine conditions." Once there, he sided with the men. The hazards of No. 7 "were found to be as represented."

J.A.P. reported that the men considered the stopes at No. 7 to be traps: "No one knows when he enters one of them whether he will come back alive as there are stones weighing from 100 pounds to tons which are liable to come down on them at any time." The spy saw "spaces 50×75 feet and 75×100 feet without any timber," and where there were stulls they were "so far apart that they really are no protection." During his first day underground, on the fifty-eighth level, J.A.P. scurried a number to times to escape falling rock. He reported that captains and foremen explained the hazards this way: "Timbering costs money while men do not cost us anything." On top of the physical risks incurred, the men had to endure Captain Conn and foreman Nichols, who "were very abusive to the men in the way of cursing them." The foreman, "instead of giving the men instructions as to how to do the work . . . , only swore at them." The work, too, was no treat. After two days of it, J.A.P. laid off for a day, because he was "so sore" from mucking rock.

On 26 September J.A.P. had another narrow escape when a piece of the hanging wall came down without warning. He reported that in town he had met a lot of men who had quit Quincy, particularly No. 7, because it was too dangerous. Quincy always needed new men at No. 7, because older, wiser ones wouldn't stay there. He also reported the men's fear of the "air blasts" that rocked the Quincy mine. They didn't understand these "explosions" but knew they were of terrific force. They jarred the ground all the way to the surface and aroused a fear that kept the men continually disturbed. On 28 September J.A.P., "not being able to work longer under Nichols," and having had enough of the hazards at No. 7, quit there and started the process of getting hired at No. 6, which the men said was Quincy's best shaft. This took some time. He applied to Captain Kendall several times for work, but was told there was nothing to be had.

On Saturday, 13 October, J.A.P. again trekked to the mine to hunt up Captain Kendall. As he approached No. 6, he saw a crowd milling about a large hole. Between the Captain's office and the No. 6 dryhouse, the surface had collapsed. Somewhere at the bottom of a 400-foot-deep hole lay the body of a dryhouse man, John Shea, who had worked forty years for Quincy before the mine literally swallowed him up. J.A.P. wasn't able to meet that day with Captain Kendall, who had his hands full with the cave-in. But he spoke to a number of men at the scene. They "expressed the opinion that the mines were getting worse every day and as they became deeper and more hollow the work was correspondingly more dangerous."

J.A.P. spent another nine days down in Hancock before getting hired at No. 6. During his lay-off, several times he heard mysterious air blasts go off at Quincy. On the night of 22 October he finally started work at the fifty-first level of No. 6, laying up walls of broken poor rock. The bosses were better in No. 6, and the men had fewer complaints. The major complaint seemed to be the man-cars. They did not stop at all the levels, and sometimes men had to climb 200 to 400 feet at the end of their shift to get a ride to grass. In some places, the ladders were old, worn, and dangerous. During his first shift in No. 6, J.A.P. heard nearly half a dozen air blasts. When he got to the dryhouse, the hot water was gone so he "had to wash with cold water in a dirty bucket, without soap." Back in Hancock, at nine in the morning he felt another Quincy air blast that slightly shook the houses in the town.

J.A.P. worked at No. 6 for two weeks. One day the man-car let him off above his level, so he had to climb down the shaft ladder. Some of the steps were missing, and he fell off the ladder onto a pile of iron, injuring an ankle and knee. While J.A.P. was still recuperating, a man at one of the Finn saloons revealed his suspicions about him—it seemed Quincy employed him in a special capacity. The spy took that as a good sign to leave Hancock, while he was still in one piece.

In the privacy of his office on Quincy Hill, Charles Lawton read J.A.P.'s reports. After scanning the contents, he neatly refolded each report, put it back in its envelope, and hid it away a bottom drawer of his oak rolltop desk. So the men thought that Quincy was becoming more dangerous? They were scared of rock falls and air blasts? They were more prone to agitate about working conditions than before? For Lawton, none of this was news. He knew—and admitted to Quincy's president, if not to the workers—that accidents were coming "thick and fast these days."[2] He knew that mine accidents were becoming a more contentious—and expensive—social issue. He knew that those air blasts were a particularly threatening and even spooky phenomenon, a veritable plague brought down on his head by the sins of his forefathers—the agents and mine captains who, for half a century, had failed to leave proper supports under Quincy's hanging wall. Largely because of the air blasts that he would ultimately have to fight for over two decades, Charles Lawton, more than any other mine manager in the district, keenly felt the hazards involved in taking these old mines deeper and deeper.

In a sense, the Quincy mine played a mean trick on Charles Lawton. He had wanted to get out of Utah and back to Michigan, where his father had once been the state's commissioner of mineral statistics. He had attended the mining college in Houghton, and his brother, who served as Quincy's attorney, had an office in Hancock. Being named Quincy's superintendent late in 1905 had brought Charles Lawton back to his roots, and it should have been a real boost to his career. Quincy was a highly regarded firm. It had been tagged with the moniker of "Old Reliable" for paying dividends every year since 1868. But right after Charles Lawton moved to the mine, "Old Reliable" became very unreliable, in terms of its structural soundness. The air blasts confronted Lawton with a complex conundrum, in which mine economics, technological limits, worker safety, and the psychology of fear all played an important part.

In 1906, J.A.P. and ordinary workers little understood the air blasts, but Lawton was already certain of their cause. They were "man-made earthquakes." Since the late 1850s, Quincy had taken out all ground showing good copper value.[3] This practice maximized Quincy's production in the nineteenth century, and seemingly did the mine no structural harm. But in 1906, the massive weight of the hanging wall began shattering the too-few pillars left in the mine. Large sections of the hanging collapsed, and the foot wall even shifted. The collapsing roof compressed air in stoped-out ground and shot it onto other parts of the mine. Hence the local term for these large rock falls: air blasts. On 8 February 1906, only months after taking his new position, Lawton wrote to Quincy's president, W. R. Todd: "To my mind there is no mystery about the so-called 'air blasts.' It is simply the caving in of the mine."

As he began combatting air blasts in 1906, it was perhaps just as well that Charles Lawton did not know what they held in store for him. He did not know that between 1906 and 1920, these unpredictable cave-ins would cost the company $4.5 million in damages, due to crushed and closed shafts, twisted tram tracks, severed trolley wires, and smashed pipelines. He did not know that 400 separate air blasts would be counted just between 1914 and 1920, or that an air blast in 1927 would kill seven men in the No. 2 shaft, making it Quincy's worst fatal accident.[4] He did not know that as an engineer, groping for solutions, he would often be rebuffed by company officers in New York who were unwilling to pay for his suggested remedies. He did not know that it would take the 1927 catastrophe to force Quincy's directors to pay for a major fix to the problem—a switchover to the retreating system—that Lawton had first suggested just as soon as the air blasts had begun, 21 years earlier.[5]

Charles Lawton did know from the start that air blasts raised hell with workers. Thankfully, the vast majority of these collapses occurred in parts of the mine stoped out many years before. When the hanging wall collapsed in long-abandoned stopes, it didn't threaten injury or death, as long as the collapse did not extend to a shaft and crush it. Air blasts at Quincy claimed surprisingly few lives. Still, they frightened the men, drove many good ones away, and ruined Quincy's reputation as a good place to work.

Lawton grew frustrated with the men who were rattled by Quincy's "shake-ups," "puffs," "old-rousers," and "radiator rockers," as the air blasts came to be called. He was not wholly unsympathetic, however. Air blasts woke him up at night, too, and he often got out of bed to telephone an underground captain to find out if the last tremor had done any harm. Lawton was a hard-nosed manager, working in a dangerous industry. He expected occasional accidents, killing one or two men at a time. But these air blasts presented a threat of unusual magnitude. Quincy's hanging wall might finally reach equilibrium only by collapsing all at once while an entire shift was underground. After five years of air blasts, Lawton wrote to W. R. Todd that "they are matters of very grave concern and shake one's nerves when he realizes the number of men that are underground and the great possibilities for disaster."[6] And Lawton himself was not always tucked behind his desk, safe from harm. Sometimes he had to go underground into an air blast area to inspect the damage and reassure the men that the danger had passed. Lawton, just like the frightened miner or trammer, had to screw up his courage:

> I assure you it is not a very pleasant situation where I have to take the lead and go to make an examination when all the timbers are cracking and snapping and the hanging wall [is] going off in reports like cannons, and especially when you really do not know how quick you may or may not "go to glory."[7]

For Lawton, the onset of the air blasts could not have been more poorly timed. They spoiled his arrival at the Quincy mine and agitated more trouble in the worker safety controversy that was brewing on the local scene, and nationally as well. Muckrakers made much ado over the fact that wealthy, industrial America had grown fat at the expense of its own poor workers, too many of whom had been consumed by on-the-job accidents.

In the first decade of the twentieth century, newspapers printed gruesome accounts of accidents that snuffed out hundred of lives. Many of the worst tragedies occured at American coal mines: 201 dead in Utah; 239 killed in Pennsylvania; and 361 victims in West Virginia.[8] Such large-scale disasters caused a spotlight of concern to be shone into the dark corners of American shops, factories, and mines. Workers, lawyers, juries, and legislators seemed bent on illuminating sinister hazards and exposing huge companies as uncaring villains.

In the Copper Country, long-held, fundamental beliefs about risk and responsibility were being challenged. Mine managers heard the dissent from bosses and captains who passed up labor complaints about working conditions. They heard it from lawyers, who were being called upon more often to defend the companies in cases involving personal injury, death, and negligence. Superintendents had been accustomed to handling accident victims in their own way. Now, workers, courts, and lawmakers intruded into the process. They wanted to amend the "old-time rules" of mine safety that in the nineteenth century had served employers well and employees poorly.

In the nineteenth century, it was unfortunate when a man died at work, but his death usually did not trigger public protest. Death was never distant or rare on the Keweenaw. Infant mortality ran high. Year after year in Houghton County, children under five died in far greater numbers than the members of any other age group.[9] Disease swept children away, and the perils of childbirth claimed many young women. There was nothing exceptionally tragic about a man's death under a pile of mine rock, and communities did not single out mine accident victims as worthy of extraordinary sympathy.

Mine accidents usually killed one or two men at a time, and random, individual deaths did not cast a collective pall over large segments of the population. A death at one mine had little or no meaning to residents of another. Also, each mining community was subdivided into various ethnic elements, and ethnic identification largely defined for whom a person should mourn. In the nineteenth century, no unified working-class consciousness existed when it came to mine accident victims. People mourned for their "own kind" more than for others.

Often, residents of the Keweenaw did not even learn of accidental deaths, unless they hit particularly close to home and were communicated by word of mouth. The formal reporting of such deaths was haphazard. Of fatal accidents occuring before 1900, no more than half made it into official county death records or into local newspapers. Newspapers ran no obituary columns, and when they reported a death at all, they often did so in just one or two sentences buried on a back page. They announced that some anonymous Irishman or Finn had been killed.

In early decades of mining on the Lake, many workers, and particularly the dominant Cornishmen, came out of long mining traditions. These men differed greatly from later immigrants entering the mines, who largely came from agrarian traditions, and who tended to complain about underground conditions. Early in the twentieth century, mine managers harkened back to the days when "real" miners took pride in their work, and all the more so because of its challenges and dangers. Managers had preferred those proud men, who laid claim to their dignity by following a hazardous profession and accepting the risks that came with it.

Under the old rules, the companies were not responsible for providing employees with a safe environment. A mine was a dangerous place, and there was nothing the companies could do to change this cruel reality. Under the old rules, companies looked after the safety of their tools and equipment, but the men looked after themselves. Underground captains and bosses occasionally scolded workers for following unsafe practices, but the person most responsible for a man's safety was the man himself. Men primarily learned their jobs, and their attendant safety practices, from other more experienced workers. If a man did not follow safe practices and got killed, or if he died after a fellow worker put him in jeopardy, that was not the company's fault.

Under the old rules, local society was more prone to believe that "accidents do happen." People acknowledged that in cases of accidental death, you often had no villain, and managers and workers alike were less driven to affix blame. All the copper mines suffered "unavoidable" accidents. At Quincy, a shift captain went into a stope and warned workers of some bad ground overhead. A man sounded one piece of the hanging wall with an iron bar and deemed it safe. The captain then started to pinch down a second piece of the hanging. While he did so, the first piece fell and killed the man who had just tested it. In another instance at Quincy, one day superintendent Charles Lawton and Captain Kendall rode up from the bottom of the No. 6 shaft in a rock skip. They both held on to the hoist rope, just above where it connected to the skip's bail. Neither man noted anything wrong. On the very next day, the hoist rope broke just where they had held it, and the skip fell down the shaft and killed two men.[10] In neither accident was anyone at fault. Nobody could be held accountable for the sometimes freakish behavior of hanging walls or hoist ropes.

Under the old rules, if anyone could be blamed for a fatality, it almost always was the victim himself. Men were negligent, careless, and paid the price. The miner who started to drill a new hole only six inches away from a missed hole (still containing dynamite)—even though he had been "repeatedly and everlastingly cautioned" not to do so—would pay for the error with his life.[11] The men near the Osceola fire, who nonchalantly sat down to eat instead of racing out of the mine, died because of their lack of respect for a life-threatening hazard. Mine officers believed that carelessness was the chief killer of men. And, in insidious fashion, men became more careless the longer they worked underground. After encountering numerous hazards and escaping harm, workers stopped looking for them. Then, one day, the overly confident mine worker died.

In his annual report of 1899, Michigan's commissioner of mineral statistics wrote, "Familiarity with underground dangers often breeds contempt, which is apt to be followed by serious consequences." Quincy's Charles Lawton subscribed to the same idea: "It is well understood that perpetual contact with conditions in which lurks the element of danger, tends to breed carelessness and contempt of danger on the part of those at work in such surroundings." At Calumet and Hecla, James MacNaughton also believed that his company could not be held accountable for the way the men themselves responded to danger:

> If a man gets rattled and runs into danger or if a man fails to use good judgment and deliberately stands in the way of danger it would seem to me that these are

contingencies we cannot provide against as long as we continue to use men in the mine. . . . Indeed I think I am safe in saying that so far as I have observed two thirds of our underground accidents are the result of carelessness and the longer some of our men work in the mine the more careless they seem to become.[12]

Companies put a heavy burden on individual workers. They had no truck with timid men who complained. At the same time, they faulted others for being too confident or brave, because they became foolhardy and careless. The companies left workers to find and walk their own safe line between paranoia and personal courage.

One early (and quite ineffective) challenge to the old-time rules covering mine safety was launched in 1887, when the Knights of Labor were active in Michigan and not without influence. The state legislature considered a liberal bill calling for elected county mine inspectors, an eight-hour workday in all mines, and a halt to the underground employment of boys younger than 16 years of age. By the time the house and senate passed this legislation, its provisions against child labor and for an eight-hour day had been dropped, and the law called for mine inspectors to be selected by county boards of supervisors (which were controlled by the mining companies). The key man who stripped the bill of its more liberal provisions was state senator Jay A. Hubbell: chairman of the Committee on Mines, Minerals and Mining Interests; resident of the Copper Country; and friend and supporter of local mine managers.[13]

The act creating county mine inspectors required them to visit each mine at least once per year. The inspector was to "condemn all such places where he shall find that the employees are in danger from any cause, whether resulting from careless mining or defective machinery or appliances of any nature." When the inspector discovered a dangerous place, he was "to order the men engaged in work at the said place to quit work immediately" and to notify the mining company of "the work to be done or change to be made to render the same secure, ordinary mine risks accepted." Each inspector was also required to produce "an annual report of his acts and proceedings . . . , specifying among other things, the number of mine accidents occuring during the preceding year causing either death or injury to persons . . . , and so classifying said accidents as to show what occured through the fault or negligence of employers and those occurring through the fault or negligence of employees."[14]

Mining interests dominated the Houghton County board of supervisors, so that board did not err by appointing a bleeding-heart mine inspector. The first appointed to the post was 61-year-old Josiah Hall.[15] Hall, a Cornishman, had arrived in the Copper Country in 1851. At the Cliff mine he had served as a miner, timberman, and then mining captain. He also served as captain at the Northwestern and Pewabic mines. His last position, before accepting the $1,000-a-year inspector's job, was as superintendent of the Centennial mine. Hall was a man the mining companies could trust, a man who understood the "old-time rules."

Captain Hall and his successors did not bother the mining companies with too-frequent visits. They seldom, if ever, condemned any part of a mine or demanded that companies alter their practices. The Houghton County mine inspector's annual reports never lived up to the full letter of the law. They did not record the mine visits made or safety problems discovered. The reports were supposed to specify "the number of mine accidents . . . causing either death or injury," but they almost wholly ignored the thousands of less-than-fatal accidents. The mine inspectors did, however, do a thorough job of recording fatal accidents. They reported what had happened and summarized the verdict of any coroner's jury impaneled to rule on the accident. Also, the mine inspectors affixed fault when possible. They often concluded that an accident had been unavoidable. When they could point to negligence, they invariably pointed to the victim. Meanwhile, in their annual summaries, the mine inspectors often commended the local companies for their fine safety record.[16]

Mine inspectors were not alone in consistently blaming workers for their own accidents. Courts, applying select common-law doctrines, did the same thing in the second half of the nineteenth century. During this era—when industrialization boomed and industrial accidents rapidly multiplied—torts (covering negligence, liability, and personal injuries) became a significant branch of the law.[17] As more personal injury cases went into the courts, the law of torts rapidly developed in a manner heavily favoring employers over employees.

Common-law doctrines provided the mining companies with three excellent defenses to use in personal injury suits. The first was "assumption of risk." Companies argued, and the courts often ruled, that when a man assumed a job he simultaneously assumed the risk of injuries common to that occupation. The second doctrine, related to the first, was the "fellow servant rule." Large employers argued that they could not control all men on their payroll, so they could not be held responsible if the negligence of one worker injured or killed another. An employee assumed the risks of his occupation and the risk of being endangered by co-workers or fellow servants. The third doctrine, "contributory negligence," allowed companies to be exonerated completely if they could prove that the victim had, in even the slightest way, contributed to his own accident.[18]

During the second half of the nineteenth century, the Michigan copper mines grew to maturity just when the courts were stacking the legal deck of torts in favor of business and industrial interests. Since employers held all the cards, many workers did not try to get in the game. With little hope of winning a lawsuit, most elected not to sue at all. It is small wonder that the copper companies held the "old-time rules" of mine safety to be near-sacred. Under those rules, they either won in the courtroom or they won by default. By 1900, however, new challenges confronted the established rules and doctrines.

At the turn of the century, industrial accidents in America killed 35,000 workers per year, and injured another two million. While many flinty-hearted judges still stuck to narrow principles of the law, jurors, who could overlook doctrine and facts if they got in the way of fairness, more and more sided with blinded or crippled workers, or with impoverished widows and children. The public mood shifted on the issue of worker safety. Many believed that workers had borne the burden of industrial acccidents alone for too long, and that companies now had to take up part of that burden.[19]

Traditionally, the mining companies treated injured miners and the families of deceased miners as charity cases. They investigated the needs and living conditions of injured parties, and then decided unilaterally whether to provide financial assistance. This practice survived into the first decade of the twentieth century. At Quincy, a rockhouse worker got killed by a skip, leaving a widow who had borne 18 children, 10 or 12 of whom were still living. Superintendent Lawton wanted his company to pay off the balance of $200 due on her "little home." In another instance, Lawton discovered that the widow of a man killed underground at No. 8 had three young children and "little or nothing to live upon." Lawton wrote Quincy's president, William R. Todd, that he thought it "well that we do something for her," along the lines of a contribution of $400 or $500.[20] But the companies began backing away from such charity, once local courts began giving accident victims more sympatheic hearings. While charitable overtures had once been interpreted by workers as acts of corporate goodwill, they were now seen as acts of guile and guilt: the companies were buying off widows with a pittance, because they knew they might pay out far more money in damages if taken to court. The companies became more chary, and less charitable, when their offerings seemed to encourage, rather than discourage, the filing of lawsuits.

Relations between companies and injured employees or widows grew far more formal and antagonistic after the late 1890s, when Copper Country attorneys began filing and winning more lawsuits on behalf of workers. Two attorneys in particular, Patrick H. O'Brien and Edward F. LeGendre, solicited personal injury cases and commonly confronted the mining companies in court. O'Brien set up his law practice in Laurium in 1899, and LeGendre joined him as a partner by 1910. O'Brien's participation in these legal battles was tinged with poetic justice. O'Brien, the son of an Irish miner who had emigrated to Lake Superior in 1856, was born in 1868 near the Phoenix mine. As the father moved from company to company, the son was educated in local schools, first in Allouez, then in Osceola, and finally in Calumet's high school, while his father worked for C&H.[21]

After high school, O'Brien left Lake Superior to attend law school in Indiana from 1889 to 1891. In 1890, while the son was in the midst of his law school education, the father fell to his death in Calumet and Hecla's No. 11 shaft. Upon completion of his schooling, Patrick O'Brien first set up practice in 1891 at West Superior, Wisconsin, where he established his specialty in personal injury cases. Eight years later, he moved to Laurium, right on the doorstep of his father's former employer, to set up a similar private practice.

With attorneys such as O'Brien in the area, mine managers were no longer surprised when an accident resulted in a suit for damages. Upon hearing of a potential suit, a mine agent typically investigated the facts of the case and huddled with his company's attorney. They evaluated how the facts squared with the doctrine of assumption of risk, the fellow servant rule, and the doctrine of contributory negligence. They reviewed the recent behavior of judges and juries and the amounts of damage awards. The agent and lawyer then decided if it seemed wise to settle the case out of court, or let it run its full course.

Out-of-court settlements usually cost a mining company from $250 to $2,000. In 1904, for example, Olinto Rocchi settled with Quincy for $1,000, after losing both legs below the knee in a skip-riding accident.[22] In another instance, Lawton settled with an underground worker at Quincy who had suffered a broken arm and a severely bruised back and hip. Lawton noted that "the accident was one for which, according to old time rules, we were not the least responsible." Still, the man had been unable to resume his job for eight months, and now he threatened suit. He also wanted to give up mine work to become a peddlar. The man agreed to Quincy's settlement offer: $103 to cover his house rent and fuel costs since his accident, and $225 to cover the cost of a horse, wagon, and peddlar's stock.[23]

When mining companies could settle claims for modest sums, they often did so, because such settlements proved cheaper than litigation. At Calumet and Hecla, "the general plan of the Company . . . has been to settle cases having any degree of merit, if reasonable settlements can be had."[24] Settlements, however, were not always easily achieved, and mine managers sometimes found themselves entangled in nasty negotiations. At Quincy, Charles Lawton sought to settle with the son and daughter of John Shea, the old dryhouse man at No. 6 who was swallowed up by a cave-in at the surface. Regarding John Shea's accident, Lawton wrote Quincy's president:

> I have looked the matter up very carefully in detail and am very strongly of the opinion that the Mining Company is in no way whatsoever responsible. . . . Our Attorneys also are of this opinion, still law-suits are expensive things, and I was in hopes that I could bring matters about so that it would not be necessary; that is, I was rather inclined to give them something, rather than go into a law-suit, but in doing so I do not like to admit that the Mining Company is in anyway responsible.[25]

Lawton wanted to settle this case but could never catch John Shea's daughter in a "conciliatory mood." She upbraided him and stated "very strongly that the Quincy Mining Company was extremely to blame for her father's death." Under the circumstances, Lawton decided not to offer any settlement terms. She would have taken the offer as an admission of guilt, and it would have intensified her resolve to make the company pay dearly for her father's death. Lawton in this instance determined to "let the matter take its course." If it went to a jury, so be it. By 1910 to 1912, it became company policy not to initiate settlement negotiations. Lawton had discovered it unwise "to seek out and make overtures of settlement to the injured," because "any advances on our part are generally considered a weakness."[26]

By 1910, Calumet and Hecla also believed that any show of corporate weakness stimulated more personal injury cases. James MacNaughton thought these lawsuits were "becoming altogether too numerous," and his company adopted a more combative, less conciliatory stance.[27] C&H's chief attorney, Allen F. Rees of Hougton, became more aggressive toward those attorneys he commonly faced in court:

> My inclination is strong to fight any case that has any possible defense, which might get into the hands of O'Brien and LeGendre. They are extremely active in looking up and obtaining these claims for prosecution, and I understand have agents who are on the look out for business for them, and whose duty it is to urge anyone having an accident of any kind to put the claim in their hands for suit.[28]

Rees noted that other Houghton County attorneys, "envious of the business which O'Brien and LeGendre are doing in this line," were following in their footsteps and seeking personal injury cases for themselves. C&H and the other mines were now in a widescale and costly battle, because "the situation will soon be that every accident of any nature whatsoever will be looked up by attorneys in an effort to obtain a chance to prosecute for damages." What had started as a trickle of cases in the 1890s had become a flood.

By 1910, when company attorneys went into courtrooms they did not just argue the merits of individual cases. They defended and tried to sustain the law of torts that once had been such a dependable ally. Now the law seemed unpredictable and uncertain, and companies hoped to prop up old, friendly doctrines before they fell altogether. Swaby Lawton defended Quincy in a $25,000 damage suit over the death of a timberman. The plaintiffs claimed the company had not provided a safe workplace, and that a timber boss had been responsible for the accident. For Quincy, there was more at stake here than $25,000. The company wanted to make certain that the court still recognized a "boss" as an accident victim's "fellow servant," even though he held a supervisory position. Under older interpretations of the fellow servant rule, a judge likely would have taken this case away from the jury, saying it had no merit. Quincy wanted the judge to throw this case out, too. Also, Quincy could not back down from the plaintiffs' assertion that it was the company's duty to provide "a safe place for the men." That claim, according to Charles Lawton, struck "right at the very foundation of mining." A company could never provide a safe environment for timbermen, because a timberman, by definition, entered unsafe places for the purpose of shoring them up. When the local judge failed to dismiss this case and let the jury take it, Quincy immediately jumped to the conclusion that it would have to appeal it all the way to the state supreme court. As soon as it went to a jury, Charles Lawton became "fearful of the results, simply because jurymen are apt to give a judgment regardless of the facts."[29]

When local judges threw personal injury cases out of court, the companies won on several levels. They paid no damages in the individual case; the judge's ruling bolstered old doctrines; and the ruling discouraged other plaintiffs and made them more willing to dismiss expensive lawsuits and accept modest settlements. But when judges let cases go to juries, the companies stood to be big losers, because public sympathy typically rested with the plaintiffs, particularly if socialists or union men served on those juries.[30]

Just when fatal and nonfatal accidents peaked in the Copper Country, local lawyers actively sought out accident victims as clients, and workers proved willing to litigate. The common law was wavering, and perceptibly moving away from the companies' interests. Judges were proving unpredictable, but juries were all too predictable: they sided with employees. Personal injury cases proved more and more expensive, divisive, and problematic to the companies. Within just a few days in 1912, the Quincy Mining Company won one important case, when the judge took it away from the jury. That encouraged two other plaintiffs to settle inexpensively, for $1,000 and $2,000. Then Swaby Lawton went ahead with another case, when he "thought that he had a pretty fair-minded jury." The jury "gave a verdict against the Quincy Mining Company of $17,050."[31]

For companies, the legal situation showed clear signs of getting worse, not better. In Houghton, Circuit Judge Patrick O'Brien took his seat on the bench in 1912.[32] The liberal lawyer parlayed his victories over the companies in damage suits into a popularly elected judgeship. After several years of legal setbacks, the companies were not displeased when Michigan passed workmen's compensation legislation in 1912. Now, they could see predictability returning to the whole issue of mine accidents, their victims, and their costs.[33]

When the act "relating to the liability of employers for injuries or death sustained by their employees" went into effect on 1 September 1912, the copper companies all elected to go under the workmen's compensation plan. It rendered them responsible for providing monetary awards to the dependents of all fatal accident victims, and to all injured employees who lost more than two weeks' work because of on-the-job accidents. As late as 1905 to 1910, the mining companies would have vociferously opposed any such plan because it set aside the time-honored notions of fault and negligence, and said that all injured workers were entitled to compensation, even if personal carelessness caused the accident. Now, however, they endorsed the workmen's compensation act because it ended recent confusion in the court and struck a compromise between the interests of employers and employees.

Workers gained a sense of security, knowing the law provided their families with a buffer against financial disaster should the breadwinner fall to an incapacitating or fatal accident. The act stipulated just how much a laid-up worker would receive (depending on the severity of the injury, the duration of the recovery period, and his ultimate ability to resume his former work), and how much the dependents of a deceased worker would receive. For example, the widow of a miner in the Copper Country would receive a death benefit amounting to about $2,700. If an accident stopped short of killing a mine worker, but rendered him incapable of future work, he and his family received four to ten dollars per week for 500 weeks, up to a maximum of $4,000. The law specified the compensation to be received by workers suffering partial incapacity due to the loss of thumbs, fingers, toes, hands, feet, arms, legs, and eyes. (Under workmen's compensation, "costing an arm and a leg" had real meaning. An arm was worth 50 percent of a worker's wages paid over 200 weeks; a leg, 50 percent over 175 weeks. An arm and a leg together equaled total incapacity: 50 percent for 500 weeks.)[34]

For the worker, these payments were entitlements to be received quickly, without resort to lawyers, trials, or complicated hearings. The compensation schedule was certainly not as generous as many jury awards had been, but it was more generous than many out-of-court settlements—and it was far more generous than receiving nothing at all. What workers or their families gave up in return for receiving workmen's compensation benefits was the right to sue the employer for any greater award. When beneficiaries accepted payments, they released employers of future claims.

The mining companies gained a cap on the amount an employee could receive and no longer had to concern themselves with the effects that pitiful plaintiffs had on sympathetic juries. They now had some basis for estimating just how much money accidents were likely to cost them in any given year, and they could budget accordingly. What they lost were all the old-time rules that once made it so hard for employees to collect damages from employers. They lost the right to say that when a man took a job, he had to suffer any tragic consequences that came with it.

Two years before Michigan enacted workmen's compensation, federal legislation created the U.S. Bureau of Mines.[35] The new bureau was charged with using scientific and technical expertise to make American mines safer. It collected data on accidents from operating mines, researched hazards and the means of preventing accidents, disseminated its findings in numerous publications, conducted on-site training programs for mine employees, and, when asked, sent experts in to help companies in the midst of an emergency, such as a mine fire. The Bureau of Mines had no regulatory power over the practices and operations of the Lake Superior copper mines, but the mining companies strongly sensed that the federal government was now looking over their shoulders.

Only one year before workmen's compensation, the Michigan legislature had passed a bill significantly changing the state's system of county mine inspectors. Since 1887, inspectors had been appointed by county boards of supervisors. Now, each county selected its mine inspector by popular election.[36] Furthermore, the inspector was required to visit each operating mine every 60 days, instead of a minimum of once per year. The creation of the U.S. Bureau of Mines, the new mine inspector law, and the initiation of workmen's compensation effected considerable change in company attitudes about mine accidents. The mining firms now espoused that many accidents were preventable, and they actively participated in the "Safety First" movement sweeping across their industry.

Earlier, the mining companies had never formalized rules or regulations for employees. Now, in 1912–13, they rushed to print little booklets such as "Rules and Regulations for the Protection of Employees." Besides providing lists of "dos" and "don'ts," these booklets tried to engender a cooperative spirit among employees. After a decade and a half of growing antagonism between managers and employees over worker safety, the companies now preached mutual interest: it was in everybody's interest to reduce the frequency of accidents. The companies espoused a deep and abiding concern for every human life involved in their industry. While the booklets usually reached all workers, companies directed them particularly at men in supervisory positions, from captains down to trammer bosses. These people not only had to know the rules; they had to enforce them.[37]

For the very first time, the mines began requiring physical examinations of all new employees. They started putting belt guards and other safety appliances on surface machines. They fenced off dangerous areas underground. They posted new signs above and below ground, warning workers of hazards.[38] Mine agents secured the assistance of mine doctors and Bureau of Mines experts in teaching first aid to numerous employees. (C&H aimed to have one in five of its underground workers versed in treating and handling the injured.) They placed comprehensive first-aid kits in their shops and at numerous underground stations. They requested periodic visits from Bureau of Mines personnel, who taught workers how to avoid and fight mine fires.[39] Finally the companies hired their own experts, college-trained mining engineers, to serve as safety engineers. These men helped codify, communicate, and enforce safety regulations. Since their job entailed close observation, problem analysis, and the application of new techniques to solving problems, the safety engineers were not unlike the other mining engineers recently hired to deliver the benefits of scientific management. At Quincy, in fact, one engineer filled both bills—he was the "safety and efficiency man."[40]

Between the mid-1890s and 1912, mine safety became a contentious issue at the mines. The death toll reached more than one man per week. Even though most underground workers still died only one or two at a time, these small-scale fatalities had a cumulative human cost simply too big to be ignored. And for every man who died underground, another 75 suffered injuries that cost them time and wages, or left them incapacitated for life.

The mines expanded tremendously in size during this era, hiring many new men who had never worked underground before. Traditional workers, such as the Cornish, had been more willing to accept mining hazards. The risks of mining had for them given their hard occupations more dignity and respect. The new mine workers, to the contrary, found the mines a hostile and alien environment. Employees agitated more about working conditions and safety. They became convinced that the mines were becoming more dangerous all the time.

In fact, in terms of the level of risk encountered by each underground employee, the mines were not becoming more dangerous—they simply were not getting any safer. As a consequence, the larger the companies grew, the more men who invariably died. Just as the mines suffered ever more accidents, so did the great increase in industrialization across the American economy lead to an epidemic of work-related accidents of all sorts. The local mines became caught up in a nationwide reform movement, an important element of the Progressive Era, that sought to protect employees not only from the physical pain of accidents but also from the financial hardships that wreaked havoc on working-class families.

The battle over reform was largely fought in courts and in legislatures. The law of torts, as it developed between 1850 and 1890, made it difficult to prove an employer's negligence and to win damages in personal injury suits. But the gross inhumanity and unfairness of the common law came to be challenged. After 1890, the public put pressure on judges to be more liberal, and when put on juries, average citizens often made large damage awards to plaintiffs. The law of torts entered a 15- to 20-year period of difficult and uncertain transition. The law was muddled and the scales of justice, from case to case, tipped back and forth from employer to employee.

Finally, new legislation returned predictability to the system and simultaneously offered workers greater protection. In Michigan, a historical watershed that redefined relations between employers and injured employees was the Workmen's Compensation Act of 1912. In the Copper Country, the mining companies and virtually all their employees elected to place themselves under the provisions of this law. By no means, however, did workmen's compensation end all disputes over safety at the mines.

Workmen's compensation guaranteed a measure of financial security in the event of an accident. But the law in and of itself did not stop accidents, did not spare men the physical anguish of injuries. Workmen's compensation did not change the minds of the men, who still believed they toiled for too little money in an environment that was too dangerous. The mining companies published their rules and regulations and preached cooperation and compliance, but this did not end a long period of dissention and antipathy. Mine managers, after all, still believed careless men caused the vast majority of all accidents, while workers pointed a finger at the mines themselves.

When the mining companies geared up their Safety First programs, workers did not perceive them as altruistic. The new safety regulations emanated from the main office, where they were drawn up by "experts" without worker input. Down in the mines, every captain, boss, or foreman was expected to know the rules, to keep much closer scrutiny over his charges, and to come down hard on violators. Because the mining companies held that careless men caused the injuries that now had to be paid for, they believed the best way to control accidents and their costs was to enact stricter controls over worker behavior.

Safety First quickly came to wear a punative rather than protective demeanor. It arrived when men already felt they were watched and pressed too hard in the name of efficiency and productivity. Instead of alleviating conflict between labor and management, the sudden inauguration of Safety First campaigns became another source of irritation. At Quincy, Charles Lawton reported that the men were making "great complaint" because "our discipline about the mine had become so rigid."[41] Safety First gave bosses and foremen more justification and opportunity for riding their men, and the men did not like it.

9

Homes on the Range

> How pleasant it is to see taste and comfort consulted in the arrangement of our mining locations. . . . We would like to see the agents, in laying out village or location lots, leave a reasonable garden plot to each house. Every family might have from 25 to 125 feet for garden and yard to make their houses attractive to themselves and others. We believe that stockholders, by consulting the comfort of their workmen, are consulting their own interest in the long run. Men who have spent long hours several hundred feet below the reach of sunshine must have recreation. And many who now become disorderly would not frequent the bar-room if they had a garden to cultivate or a comfortable house to bring themselves about.
>
> *Mining Magazine*, 1856

The first decades of mining on the Lake conjure up images of isolation and hardship. Of lone cabins in the woods; of desperate groups of unemployed men trying to walk from the Keweenaw to Green Bay in the depths of winter; of infrequent arrivals of the U.S. mail by dogsled. Early travelers' accounts emphasized the rawness of the region and romanticized the hardy individuals who endured it. But over half a century, everything changed. The mining companies came to take almost equal pride in two new images of their region that later travelers reported.

The first was what visitors saw when looking down a company's row of shafts: boiler stacks billowing coal smoke into the air; huge, polished hoists rapidly winding up copper rock from the depths; tall shaft-rockhouses receiving thousands of tons of the red metal; vast stockpiles of supplies; and busy rail yards filled with the coming and going of locomotives. This was the image of corporate wealth and power.

The second impressive view could be found only a few blocks from the first and still on company property: rows of single-family houses, neatly arrayed and substantially built, each with space for a vegetable garden; streets that had "the apperance of having been swept every morning"; schools filled with workers' children and qualified teachers; a library with books in over a dozen languages; ballyards and other recreational facilities; a hospital filled with modern equipment; and a multitude of churches. This was the image of a company's wisdom and "soul." These large organizations were not "grasping, grinding" ones. They exploited copper, but not workers.[1] Here, capital presented immigrants with opportunities. The opportunity to work at an honest job, live in decent housing, and to marry and raise a family in a wholesome environment. The opportunity to educate one's children in the American ways that would allow for enhanced prosperity from one generation to the next.

Early in this nation's history, many Americans had shunned or feared industrialization. In a sparsely populated republic dominated by villages and farms, the rise of industrialism seemed to portend much evil.[2] Selfish, uncaring companies would erect "satanic" workplaces within communities populated by large numbers of poor families. When drawn off the land, these families would surrender their independence and become trapped in permanent squalor. Republican harmony and unity would dissolve as the gap between the wealthy owners and poor workers inevitably widened. Within crowded industrial slums, law and order would disintegrate, morality decline, and ignorance and pestilence flourish.

Some of these evils did come to pass in select American coal fields, where generational poverty and substandard living conditions became infamous, and in metal mines in the West, where scoff-laws wore as a badge of courage their rough-and-tumble ways. They came to pass in many urban slums, where adults and children crowded in at night after twelve-hour days in the local sweatshop.

For a brief time, the Copper Country, too, appeared headed down the wrong path. In their formative years, the mine locations claimed none of the usual symbols of civil authority: no police, courts, or jails. Mine managers were loathe to keep much cash on hand for fear of theft. Agents laid down and enforced whatever rules they believed necessary to maintain law and order. They fired offenders, or levied "fines" or "damages" that they deducted from a man's pay.[3] Still, drunkenness flourished and violence often erupted between rival ethnic groups. The Civil War era represented a nadir in terms of the social development of the region.[4] Order and authority broke down, and violent disturbances escalated. God-fearing individuals formed armed militias and drilled in secret, preparing themselves to defend Houghton County not from Southern rebels but from their own neighbors.

But shortly after the Civil War the Copper Country took on a new cast. Several successful companies settled in for a long run at profitable mining. Social order and control became more important as the mines consciously shucked off the trappings of fledgling ventures to wear the dress of permanent fixtures on the landscape.

At this juncture, Calumet and Hecla rapidly became the region's dominant industrial force. Modern mines were now to be built, requiring control over more capital, technology, and people. C&H functioned as the premier symbol of advanced mining techniques, and of an advanced society dedicated to mining. Previously, "good" mining companies had been extolled for erecting picturesque villages amidst the wilderness. But such rough-hewn settlements no longer sufficed. Major mining firms, with Calumet and Hecla at the fore, sought to obliterate the frontier by building harmonious, well-populated communities dedicated to rigorous and systematic copper mining.

Top mine managers believed they succeeded as society builders. Frank W. Denton, manager of the Copper Range mines, reported, "We who are in charge of mining operations in this district believe that the . . . living conditions are the best of any mining district in the United States."[5] James North Wright, agent first at Quincy and then at C&H, concurred: "From an experience of thirty years . . . , I can truthfully say that I know of no mining region where the relations between the companies and their employees have been more friendly and pleasant."[6] Wright admitted that the companies had not perfected an egalitarian society: "Social life at the mines partakes somewhat of the characteristics . . . of an army post, in that officers and men mingle but little together." Although some social distinctions remained, Wright believed that great strides had been made by the turn-of-the-century worker and his family:

> The younger miner of today never signs the payroll with an X, but with a clear, legible, and often handsomely written name. Should you chance to meet him at his home, you will find him a most hospitable host. His wife will lay a snowy napkin at your plate, and his daughter will entertain you at the piano, often with not a little credit to herself.

Wright's bragging over his industrial Camelot seems overdrawn, but numerous observers echoed his message over several decades. Michigan's commissioner of mineral statistics opined, "No mining company in the world treats its employees better than Calumet and Hecla." In 1882, a writer for *Harper's* visited Calumet and found it to embody high values and standards of behavior: "As a straightforward manly development of American civilization, the village of Calumet is without peer." The author noted that "two 'Molly Maguires' from the coal regions would make more noise than the two thousand employees of Calumet. . . . Here are Swedes, Norwegians, Danes, Finns, Scotch, Cornishmen, Canadians, Bohemians, Spaniards, Italians and Germans quietly and harmoniously developing into self-respecting American citizens."[7]

Some 30 years later, Claude Rice, writing in the *Engineering and Mining Journal,* concluded that labor conditions in the Copper Country "are probably the best in the United States. The companies take a wholesome interest in the welfare of the community and of their employees especially." The companies still turned out good Americans: "If the system through which labor conditions in this region have been worked out so satisfactorily verges on paternalism, it is the kind of paternalism that kills unionism and, in one generation, builds out of foreigners, ignorant of Anglo-Saxon institutions, citizens that any community can well be proud of."[8]

Rice understated the case. The system employed in the Copper Country mines did not "verge" on paternalism. More properly, it defined paternalism. Vernon Jensen, in *Heritage of Conflict: Labor Relations in the Nonferrous Metals Industry,* was more correct in observing that "the Copper Country in Michigan always was a tight little world isolated snugly in and around the Keweenaw Peninsula," and that "perhaps no more completely controlled paternalism ever existed in this country than that which developed there."[9]

In an industrializing America, many companies used paternalism as a bridge between management and labor. One hallmark of paternalism was the belief that life and work were not separate domains, but were interrelated and mutually reinforcing. Paternalistic companies provided workers with more than a job and a wage. They involved themselves in workers' private lives by engaging in health, education, and social welfare activities. In return, paternal employers generally expected more diligent and loyal service from their workers.

Paternalism took many forms and had different, highly visible champions at different times. In the 1820s and 1830s, Boston merchant-manufacturers made Lowell, Massachusetts, the premier industrial city in America by pairing paternalism with the new technology of factory production. They built mills to manufacture cotton textiles on impressive new machines. To operate the machines, the mills turned to young New England women out of school and not yet married. These mill girls had to be taught the discipline required by industrial work; they also had to be protected in their private lives. To achieve both ends, the manufacturers promulgated regulations that proscribed the mill girls' behavior both inside and outside the factory gates. They built boarding houses, run by proper matrons who watched over their charges. They required the mill girls to attend church and prohibited drinking. They instituted nightly curfews. They told the mill girls to deport themselves in a moral manner and to improve themselves by participating in educational and cultural activities.[10]

Half a century later, in the 1880s, the most renowned example of paternalism in America could be found on the outskirts of Chicago in Pullman, Illinois. Its founder, George Pullman, had made his fortune via the Pullman Palace Car Company, whose elegant sleeper cars had made long-distance train travel fashionable and popular. When he wanted to expand his car-manufacturing enterprise, Pullman bought 4,000 acres of land near Lake Calumet and turned a swamp into an $8 million planned industrial community that integrated life and work.[11] He hired architect Solon S. Beman to plan his town's buildings and designer Nathan F. Barrett to landscape its residential, industrial, and commercial areas. Pullman's guiding belief was rather simple: if you liberated workers from urban slums (like those in neighboring Chicago) and put them into a harmonious, refined, and attractive community, then you would be rewarded. The town would produce a superior class of workmen. The men would emulate their environment. They would be content, diligent, and loyal.

In 1914, Henry Ford sought essentially the same objectives as George Pullman, but through different means. He would obtain a superior workingman, but without building his own town. At the Highland Park plant that mass-produced the Model T, Ford's assembly-line technology required that thousands of men conform to the manufacturing system, that they always be at their work stations, never missing a beat. To enhance discipline and reduce labor turnover and absenteeism, Ford inaugurated the "Five Dollar Day." This wage and bonus plan effectively doubled the income of industrial workers, but they first had to qualify for the program by meeting standards of behavior set by the company and monitored and approved by a new sociological department. Ford used the promise of a 100 percent wage increase to get investigators into workers' homes, to examine and modify, if necessary, their mores and manners. At the same time, Ford schools instructed immigrant laborers in English and taught them how to get along in their new country: how to eat at the table, shine shoes, and open a bank account; how to greet people and purchase a home; how to get along in the workplace and become good citizens.[12]

The paternal bridges built by American companies were anchored at one end by a cooperative bond between managers and workers and at the other by a sound economy and solid profits. These bridges sometimes suffered spectacular falls, when one anchor or the other gave way. It is ironic that paternalism's most notable successes became some of its most notorious failures.

Lowell served as a model of a new, humane industrialism for perhaps two decades. Then, in the face of economic problems, mill managers initiated wage cuts and work speed-ups. These encouraged the mill girls first to strike and then to abandon Lowell, leaving it to Irish immigrants or whoever wanted it. Lowell became just another grimy mill town.

Paternalism brought harmony to Pullman for about a decade and a half. Then a severe slump in the railroad industry caused the company to enact wholesale layoffs and pay cuts. But the company did not lower the rents for worker housing in its town. Suddenly Pullman became home to one of the most serious labor strikes in American history up to that time, and President Cleveland ordered troops into the area to maintain peace. In the minds of labor, the strike transformed George Pullman from a steward of social progress into a "son of a bitch."

Henry Ford later suffered the same transformation. Ford's paternalism briefly achieved its intended purpose, but then collapsed in less than a decade. Economic and social changes wrought by the First World War caused Ford to abandon uplifiting talk of loyalty and good will between employer and employee. He abandoned paternalism and adopted a much more aggressive and domineering management style. Believing now that men just worked for money, Ford determined to pay them a good wage—and in return for it they would simply have to do what they were told.

As a means of handling labor-management relations, paternalism had a spotty track record that attracted criticism from both social conservatives and liberals. Conservative businessmen often espoused the belief that the business of business is business—and not philanthropy, social welfare, or social reform.[13] They faulted paternalism for diverting money and attention away from a firm's primary work. Capital should not go outside the factory gates to fulfill the social needs of workers, because workers' wages were supposed to pay for those needs. Managers should make important wage and employment decisions in strict accordance with business criteria, and not be influenced by broader community concerns. Finally, employers were misguided if they believed paternalism could achieve permanent harmony with labor. The more benefits workers received, the more they expected. But the paternal company could not always afford to be generous. It competed in an uncertain economy, and business fluctuations could force unpopular cutbacks. These would generate shock waves of dissent, when laborers were suddenly denied their customary perquisites.

Liberal objections to paternalism centered on the issue of whether it was proper for an employer in a democratic society to assume a "fatherly" role in relations with adult employees. That role presupposed that the employer was the superior actor, the source of wisdom, the keeper of values, and the judge of proper behavior. Workers necessarily became the inferior actors. They were always the led, never the leaders. Paternalistic employers, regardless of any benefits they provided, took something essential away: workers' dignity and independence.

Richard Ely, who wrote "Pullman: a Social Study" for *Harper's* in 1885, found much that was seductive about this well-planned, attractive town. Yet its social relationships concerned him. Here the worker "had everything done for him, nothing by him." Ely concluded that Pullman was "un-American" because of the inferior status assumed by workers and their lack of self-determination. Ely believed that "if free American institutions are to survive, we want no race reared as underlings." Some 30 years later, a member of the Auto Workers Union criticized Ford's "Five Dollar Day" in the same vein: "The workers are not little children who need some fatherly minds to guide them, to tell them how much they may earn and what they shall do with it."[14]

Critics saw paternalism as limiting freedom and independence. Conservatives believed it entangled managers in social problems and issues that impeded their practice of business. Liberals saw managers not as victims but as likely villains. Paternalism too often coaxed or coerced workers into doing the bosses' biding. It relegated labor to a servile status that violated egalitarian ideals. It will be seen that the paternalism practiced in the Lake Superior copper mines demonstrated that both sides could be correct. Nevertheless, the system worked quite well for half a century, because mine officials and many workers bought into it. Both paid a price for paternalism, but got something of value in return. This was particularly true in the case of company housing, the cornerstone of Lake Superior paternalism.

Mining companies entered the housing business out of necessity. On a frontier that experienced heavy winters and lacked transportation conveniences, it was essential to locate housing as close to a mine as possible. The companies invariably owned the adjacent land, and they had to erect the first houses on it, because nobody else was going to do it. Even if a company platted a village and tried selling lots, outsiders balked at buying them. Opening a mine was a high-risk venture. Only after a mine achieved some success could it expect to see much housing or commercial development on its margins. In 1858 the Minesota mine platted 80 acres into lots, creating the village of Rosendale (later Rockland), and it sold many of them.[15] Similarly, Quincy parted with land to support the growth of the village of Hancock. But before these villages saw any real boom, their "parent" companies first had to become productive mines. By the time that occurred, the mining companies had already built many houses of their own.

Early-arriving mine workers had no choice but to take company dwellings, because they were the only ones to be found. But for both companies and workers, what began as a necessity evolved into a mutual convenience. After half a century of mining, companies kept building houses, and laborers queued up to get them.

Workers paid a price for occupying a company house. They had less choice and selection at the mine than in neighboring independent villages. They had to take what their employers gave them, and what the ordinary worker received was a solid dwelling that expressed no individuality and had few frills. The worker knew that his job and house were inextricably linked. Any problems with the boss, if serious, could result in problems keeping the house. His house lease, then, encouraged reticence and compliance. On the other hand, the very fact that a man had been given a house signified that he was a valued employee. Not everyone got one. Those that did enjoyed a greater sense of security in their employment. There were other benefits as well. Men in company houses lived closest to work, and for comparable dwellings they paid the lowest rents.

Companies, too, paid a price for serving as landlords. They invested capital in structures that did not return a profit directly. They employed men to build and maintain the houses. As the standard of living went up, workers came to want more, so managers endured an onslaught of requests for modernization. They had to bother with leases and rent collection, with public works matters like streets, garbage collection, and water hook-ups. When a breadwinner died or became incapacitated, some company official had to deal with a sensitive social issue: Does the family keep or lose its company house?

Running a town complicated the running of a mine, but companies thought the benefits outweighed the costs. By renting company houses to some, and by leasing inexpensive building lots to others, companies hoped to lure and keep the men it particularly wanted, because of their marital status, occupation, or ethnic group. Inexpensive, company-provided housing offered competition to private housing and thus acted as a check on workers' cost of living. The payback to the companies came in the form of reduced agitation for higher wages. They could pay lower wages than many other mining districts, because copper workers paid less for shelter. The company house, or the employee-owned house on leased company land, also produced a more trustworthy and loyal employee. The house was a tangible expression of the bond between employer and employee.

Housing was indeed the most fundamental element of paternalism on the Lake. All companies provided it. By 1913, the mines rented out 3,520 houses, and another 1,751 employee-owned houses stood on leased company land.[16] Some 5,000 houses, of course, could not accommodate over 14,000 mine workers. This shortfall, this gap between supply and demand, meant that many occupants of the 5,000 houses considered themselves truly fortunate to be there. But for many on the outside looking in, company housing practices only proved that paternalism was highly discriminatory. Yes, it served some well, but it served others poorly. The housing situation proved that paternal companies treated some employees like favorite sons and others like bastards. While the one group remained loyal, the other grew more discontent.

The early history of the Pewabic mine demonstrates how a company used houses and land to shape its labor force and community. In the early 1850s, only about 800 men worked in the entire Lake Superior copper industry. The largest mine, the Cliff, employed fewer than 200; the next three largest—the Northwest, North American, and Minesota—employed about 100. Fifteen to 20 small companies trying to establish themselves employed only 5 to 30 workers, and Pewabic was on the low end of that scale.[17] But in 1855, the company discovered the Pewabic lode just north of Portage Lake. The lode showed great promise, so Pewabic began hiring more workers.

Pewabic opened its lode with a labor force composed entirely of single men (or, if married, the men arrived without their families). From November 1855 through February 1858, Pewabic built six boarding houses at a cost of $500 each and spent another $1,200 on bedding, stoves, and boarding-house furniture.[18] Each boarding house accommodated up to 18 men. Pewabic's 108 employees had no choice of living in individual houses, because the company had not built any.

By the spring of 1859, Pewabic's employment stood at 211 men. The company put additions on two boarding houses and erected five more. Apparently, married workers with families arrived for the first time, because Pewabic also built eight log houses at a total cost of $800.[19] On a per-capita basis, it cost Pewabic $100 to shelter a married worker and his family, while it cost only one-third of that to shelter a single man. Despite the cost differential in favor of single men put up in boarding houses, Pewabic built no more shelters of that type for miners or general laborers.

Instead, it added to its complement of log cabins, constructed frame houses, and provided special accommodation for men of select rank. In 1862, Pewabic put five "lodging rooms" for managers in the mine's new office building, and erected a frame dwelling "of the better class" opposite the hoist-engine house "for the accommodation of engineers." In six years, company employment had risen from perhaps 10 to about 430 men, and company housing at the location already drew distinctions amongst these employees: officers, skilled mechanics, married, and single men. All had different accommodations.

By 1862 Pewabic began phasing out boarding houses for single laborers. Some boarding houses were "subdivided so as to receive from two to three families, as it is deemed better, for many good reasons, to have more families and fewer large boarding houses." In support of making a fuller transition to family men, while minimizing its own costs in the process, in 1861–62 Pewabic introduced another housing alternative: it allowed workers to erect their own houses on company land:

> Last summer a portion of the Company's land . . . was laid off in half-acre lots in a regular manner. These lots have been leased, under proper restrictions and reservations, to such of the better class of our men who are desireous of making permanent homes. Some 20 cottages have already been built; more than this number will doubtless be built next season, to the great advantage of the Company, as well as of their employees.[20]

By 1866, a decade after starting its housing "boom," Pewabic had put up individual dwellings for its agent, chief engineer, and a handful of other officers. Another 80 double houses of frame construction and 26 log cabins, all built by the company, dotted the mine location, and many employee-built houses also stood on company land.[21] Noticeably absent were the boarding houses that once comprised the company's entire housing stock.

The demise of boarding houses did not signify that the location was devoid of single workers. By closing company boarding houses, Pewabic encouraged families at the mine to take remaining single workers into their homes. Many families welcomed boarders because they provided the family with additional income. Since there were no jobs with the company for women, and since wives shouldered the burden of feeding and sheltering boarders, this practice allowed women to contribute directly to the economic well-being of their families. Employees' families benefited, and so did the employer. Pewabic saw crowded boarding houses as potential breeding grounds for labor unrest. It was better to have single laborers diffused across the location in family settings.

Pewabic's experience was typical. For a company just starting out, the right worker was a young, adventuresome male, unfettered by family obligations and willing to endure a hard and tenuous existence. The company lavished no money on surface improvements until it thought it had a good chance of becoming a paying mine. In the meantime, living conditions for the pioneers remained extremely crude.

By the time a company started into real production, the right worker was somebody else. Along with the stamp mill and steam engines came doctors, churches, and rudimentary schools. Single-family dwellings sprang up; boarding houses went down. The right worker now was a male, approximately 20 to 40 years of age. He was married and accompanied by his wife and children. He was a man willing to put down roots and ready to link his family's future with the future of Lake Superior copper mining.

As the wilderness receded in the Copper Country, it became easier for companies to lure more women and children to their locations. By 1857, women comprised only 13 percent of the adult population of 618 persons at the Minesota mine; 16 percent of 181 adults at the Rockland mine; and 26 percent of 285 adults at the North America mine.[22] But overall gender imbalance in the region leveled out rather quickly. Between 1850 and 1860, women moved from 25 up to 33 percent of the total population. By 1870 they comprised 45 percent, and they stayed at that figure for several decades.[23]

The companies actively encouraged family men to settle at the mines, because they believed that marriage made men better behaved and less likely to run off in the event of a wage cut or the opening of some new mine. Also, women bore children. In the future the mining companies would not have to rely just on immigrants. They could employ homegrown boys and young adults, reared in a mining environment and educated in American ways.

When it built company houses for its agent, chief engineer, and other officials, Pewabic signaled that its paternalism was not aimed just at the working class. It also served managers, clerks, mining captains, boss mechanics, and mine physicians. Life in the remote Keweenaw, when not enhanced by the companies, attracted precious few Americans to fill top posts. To lure Americans, as well as the most sought-after immigrants (such as Cornish mine captains), companies universally provided superior housing and other perquisites to key managers and bosses. On the Keweenaw, the men who administered paternalism also lived under it and garnered its most generous benefits.

At a given mine, workers of equal occupation or status tended to live in dwellings of comparable quality, but these houses were not always identical. Companies built housing in periodic waves, as made necessary by employment growth, and they often changed styles from wave to wave. At a long-lived company, this led to considerable variety in housing.

From the 1840s until the 1870s, log construction dominated. Companies built one- or two-room cabins and log houses of four, six, and even eight rooms. During the 1870s, machine sawn and planed lumber became more available, and companies switched to frame construction. Two-story, rectangular houses with double-pitched roofs were very common, with their ridge lines running either parallel or perpendicular to the street. Often the kitchen extended off the back, covered with a shed roof. The companies also built T-shaped houses, and telescope houses, where, as the house receded from the street, each successive room became narrower. By the 1890s, several companies built their version of the New England saltbox, and by the First World War era a few purchased worker housing from the catalog of Sears, Roebuck and Company. Sears shipped precut building materials by rail, and mine carpenters put the pieces together in accordance with standard plans.[24] Most houses were clapboarded, but some were covered with cedar shakes. Most were detached, single-family dwellings, but to save on costs the companies built a considerable number of duplexes, joined side-to-side or back-to-back. Communal, kitchen and utility spaces occupied the first floor, while bedrooms filled the second.

Worker housing improved over time. Newer dwellings offered more square footage and more rooms. Papered lath and plaster interior walls became standard. Houses founded on cedar posts gave way to those on masonry foundations. Bare exteriors were painted, often white or grey. More windows pierced the walls, making interiors brighter. The barest of entrances—a mere door and some wooden steps—evolved into an entrance landing covered with a storm roof. At more prosperous mines, piped-in water replaced the well, and, wonder of wonders, an inside toilet occasionally replaced the privy.

Although diverse housing stock could be inventoried at a given mine location, within one block all houses were usually alike. The companies eschewed the ploys that twentieth-century surburban developers later used to conceal uniformity. They did not create a false appearance of individuality by mixing together a small number of standard model houses. They did not trim them out just a bit differently, make one house the mirror image of another, or paint them different colors. The mining companies put their highly regular houses on lots of a uniform size—50 feet by 100 feet was common—and used the same setback and house positioning on each lot.[25] Each house had its privy in the back, and if a company put a small 12' by 12' barn behind one house, it put the same barn behind the others. Each housing tract assumed the visage of orderliness run rampant.

This orderliness was barren of adornment. Most mine housing went up during the Victorian Age. Elsewhere in America, architects, builders, and carpenters celebrated their violation of the notion that form must follow function. They exuberantly nailed up capitals, dentils, brackets, pendants, finials, and spindles. They took the products of the planer, scroll saw, and lathe and hid their houses behind gingerbread. They threw up turrets and towers, wrapped fancy porches all around, busted out walls with bay windows, and made the light dance through stained glass. Yet looking along a row of ordinary workers' houses in Calumet, Quincy, or Osceola, not a stick of any such embellishment was to be had. A house was truly blessed with fine detail if it had turned posts supporting the small roof over the front door. No ordinary worker's house, for instance, ever merited a front porch to be used for relaxation or socializing with neighbors.

Uniform housing within neighborhoods suggested that an overzealous American egalitarianism was at work in the mining towns, but this was not the case. Contrasts between neighborhoods made it clear that a hierarchy of housing mirrored the occupational hierarchy at the mine. While the Quincy Mining Company built workers houses for a few hundred dollars each, in 1880–81 it built and furnished a house for its mine agent at a total cost of $25,000.[26]

The agent's fine Italianate house embodied a concern for style not seen at all in ordinary Quincy dwellings. The only return the company could expect from this investment was attracting and keeping good agents (and not losing another one, like James North Wright, to Calumet and Hecla). The company got no monetary return at all. While common workers paid rent for small log or frame houses, Quincy's agent, and as many as 15 other managers, bosses, and doctors, lived rent-free in more spacious and refined dwellings. The company adopted this policy early on, "when good men were hard to obtain and salaries comparatively small." Besides their housing, until 1900 these officials also received, free of charge, medical service, fuel, light, the personal use of company horses and vehicles, and the use of company men or boys to run errands or do odd jobs.[27]

Company housing did not separate mine employees into just two neat categories, "managers" and "workers." Finer gradations existed on the occupational and social scales. One can see how status translated into housing by examining three plans executed by C&H architects between 1898 and 1903. One was a generic "Miner's House," one a generic "Captain's House," and one a custom-built house for an upper-level manager.[28]

The two-story miner's house, having a simple rectangular plan, measured 18 by 26 feet. It had a masonry foundation and a full basement with a concrete floor. The façade carried no porch, but the front and rear entries had storm roofs over the doors. The first floor divided into a living room (with fir wainscoting) and a kitchen (with pantry). The second floor contained three bedrooms. Floor to ceiling heights were 6′6″ in the basement, and 8 feet on the first and second floors. The house lacked central heating, electrical service, and an inside toilet. Architecturally, this five-room house had but one novel feature on the exterior. In a mining town where simple double-pitched roofs were ubiquitous, this house displayed a gambrel roof.

The two-story "Captain's House," also rectangular in plan, measured 25 by 48 feet. It had a concreted basement and masonry foundation. The house carried a roofed veranda across the front, and a single-story storage area with a shed roof off the kitchen in the back. The first floor divided into a vestibule, parlor, sitting room, dining room, and kitchen with pantry. The second floor contained five bedrooms and a bath with toilet and tub. Although equipped with plumbing, no central heating or electrical services were indicated for this ten-room house. The ceilings were 8′6″ in the clear on the first floor, and 8′ on the second.

Calumet and Hecla residence no. 1138, for an upper-level manager, was a two-and-a-half-story structure having a somewhat irregular "L" shape. It measured 42 feet along one leg of the "L" and 65 feet along the other. Its concreted basement divided into cold cellar, storage, and laundry areas and contained a half-bath. The first floor carried a broad porch and clapboard siding; the top floor carried cedar shingles. The first floor contained a vestibule, entry hall, parlor, sitting room, dining room, den, pantry, kitchen (with range and McCray ice-box refrigerator), storage area, and a half-bath. Wainscoting and woodwork were oak. The second floor divided into four large bedrooms, a dressing room (with fireplace), and a full bath. The attic or third floor provided finished space for a fifth bedroom, and unfinished storage space. The ceiling heights were 7′6″ clear in the basement, 9′ on the first floor, and 8′4″ on the second floor. Electrical service was available throughout, and steam radiators provided heat. The house contained no furnace and boiler—the steam was piped in from one of C&H's nearby industrial lines.

The turn-of-the-century "Miner's House" paled in comparison with some of its contemporaries, but it was a substantial dwelling, intended for a class of good workmen that C&H wanted to keep. Many men hoped to get their families into such a house. It is just as certain that many of them came away disappointed, because C&H and other large companies never guaranteed housing to all workers who wanted it.

As landlords, the company had great discretionary power in reaching out the paternal hand. Evidence from the Copper Range Company's Champion mine clearly indicates that housing went to the best-paid workers the company wanted most to keep, and not to the lowest-paid workers most in need of inexpensive housing. It went to ethnic groups holding the favored occupations, while others were locked out altogether. The "best" housing went to the "best" workers, generally meaning those from the United States, the British Isles, and western Europe.

In 1916, the superintendent of Copper Range authored the most explicit statement of housing discrimination on the Keweenaw. A year earlier, his company recognized a need to build houses for an additional 25 to 30 families. A number of young men who had done good work for the company for four or five years wanted to get married. Copper Range deemed it "good business" to accommodate them with some new five-room houses, most of them to go up at the Champion mine in Painesdale, where the housing shortage was most acute.[29]

In the spring of 1916, as company carpenters erected the new dwellings, the housing problem took a turn. Champion, along with other mines in the district, now faced a difficult time, not so much in keeping miners, but in keeping unskilled laborers. In Calumet, James MacNaughton scrambled to stabilize his force of trammers: "We have lately been trying to give preference to trammers because we feel that that class of workmen have not had equal sharing with the miners in getting Company houses."[30] At Copper Range, Frank Denton coped with the same problem: how to house unskilled workers who might otherwise leave. The houses under way in Painesdale would not do. As Denton wrote in a letter to president William A. Paine, they were "adapted to our best families . . . , and they will not accommodate the Croatians, Italians, Finns and Armenians that we want to accommodate." For the Croatians, Italians, and Finns, Denton suggested that Champion rush the construction of "four-room houses built like those in Seeberville," one of the poorest neighborhoods near Painesdale. As for the Armenians, "we need something in the nature of a camp located out of sight, and yet near enough the mine to keep them from being afraid of being molested."[31]

The effort to accommodate unskilled workers with housing in the years surrounding the First World War was new. Earlier, companies had treated unskilled laborers as expendable transients. The traditional pattern of discriminating against unskilled labor, and against the ethnic groups filling that rank, is clearly documented by the Champion mine's allocations of company houses in 1905 and 1912.[32] In 1905, Champion employed 1,027 adult males, but had only 214 houses to rent out. On a purely random basis, a worker had about a one in five chance of obtaining a company house. However, there was nothing random about the way Champion distributed this benefit.

To qualify for housing a worker had to be married and accompanied by his wife and any children. Applying that criterion eliminated slightly more than half of Champion's workers, and left 450 men vying for 214 houses. Occupation and skill determined the final distribution. In 1905, Champion rented 130 houses to underground workers. From a pool of 150 married miners, 118 lived in a company house. Six of 16 married timbermen occupied a Champion dwelling. The 170 married trammers, pickers, and wallers did not fare nearly as well. Only six of them lived in a company house. The remaining 84 houses went to those filling managerial and skilled surface positions: clerks, doctors, mining engineers, captains, hoist engineers, machinists, electricians, and the like. In 1912, the situation was little changed. Champion then rented 162 houses to underground workers. Some 140 went to miners; 16 to timbermen; three to pickers; and three to wallers. Of a total of 207 trammers working at Champion, not one received the keys to a company house.

This occupational bias in the allocation of housing served some ethnic groups well, and others poorly. In 1905, although few in number, Irish, Scottish, French Canadian, German and American employees got far more than their proportional share of housing, because they clustered in supervisory or skilled positions on the surface. Of the larger ethnic groups, the Cornish fared best in obtaining housing, because on the surface they comprised over half the skilled tradesmen, while underground they held all the mining captain positions and a plurality of miners' positions.

The Finns represented Champion's largest ethnic group in 1905. When the Finns arrived on the Keweenaw in large numbers in the late nineteenth century, they were relegated to tramming and other unskilled jobs. Then select Finns worked themselves up into miners' positions. At Champion, those Finns who achieved upward mobility in their occupation also worked their way into company housing. Just over half the married Finns at Champion lived in a company house, and those who did were miners. Finns still engaged in tramming or other unskilled work were left on their own to find housing.

The Finns at Champion formed an unusual group in 1905, a group half in and half out of the social order. By arriving in substantial numbers (which seemed to give them an advantage over some other groups, like the Italians), and by staying on at the mines in the face of much discrimination, 65 Finns had earned their keys to a Champion house. The Finns were by no means as favored or sought after as the Cornish, and they undoubtedly lived in something other than Champion's best houses. But at least they were no longer at the bottom of the socioeconomic ladder. In 1905, that dubious distinction was held by the "Austrians," diverse immigrants coming from the Austro-Hungarian Empire of east-central Europe.

The Austrians, in effect, were the "new Finns," stacked into the poorest jobs and locked out of company housing. Of 181 Austrians working underground in 1905, not one was a miner, and not one received a company house. But like the Finns, they kept coming, persevered, and became more valued employees. By 1912, they outnumbered the Finns and all other ethnic groups on Champion's employment rolls. A good number worked as miners and 35 lived in company houses. Meanwhile, yet another new group, 74 Lithuanians, found themselves without housing or esteemed jobs.

Copper Range, heavily capitalized and rapidly developed at the turn of the century, did not have 40-year-old log cabins in one corner of its operations and

Sears, Roebuck houses in another. Nevertheless, by 1913 its 607 houses gave mine managers considerable leeway in matching the "right" house to the "right" worker. Eighteen had but two or three rooms, while 101 had only four. Some 182 working families lived in five-room houses, and 112 resided in six-room dwellings. Five- and six-room company houses were most common on the Keweenaw, and occupying any house larger than that set the man and his family apart from the ordinary worker. Copper Range rented 93 houses of seven rooms and 89 having eight. At this size house, one hit another important break in the social and occupational hierarchy. Going above an eight-room house, one left the domain of skilled tradesmen, bosses, shaft captains, and mid-level managers, and entered the domain of upper-level managers: the general superintendent, agents of the Champion, Trimountain, and Baltic mines, head mining captains, chief engineers and clerks, and physicians. Copper Range had but 12 houses of nine to twelve rooms.[33]

One can get a sense of how the hierarchy of work, combined with discriminatory housing practices, dealt ethnic groups into different standards of living by examining a 1909–10 survey of 493 foreign-born workers and their families.[34] All resided north of Portage Lake, at Calumet, Quincy, Mohawk, and Wolverine. About a third owned their houses; the others rented. Overall, the average family's house had 4.6 rooms, but there was considerable deviation from this mean. Beneath the average resided Finns, Croatians, Italians, Hungarians, and Slovenians—the ethnic groups that held the most unskilled positions. Above the average resided the Cornish, French Canadians, Irish, Norwegians, Swedes, and Poles—groups that eschewed underground work for skilled surface positions or, if underground, commonly worked as miners, not trammers. The larger houses meant more convenience and allowed residents, quite literally, more breathing space—a health factor in terms of avoiding contagious diseases—and more privacy, particularly in the additional bedrooms.

Mine managers had definite ideas about which workers belonged in which houses, but they could not keep absolutely everyone in his proper place. A mine had a social dynamic. Because of the ebb and flow of workers, companies could not strictly segregate occupational or ethnic groups by setting them off in permanent enclaves. True, the companies batch-built identical houses, intending them for a specific class of worker. Sometimes this classification was evident in the new neighborhood's name: both Calumet and Hecla and Quincy had a "Swedetown," and Quincy had a "Frenchtown" and a "Limerick" as well. But the neighborhoods stayed, long after the Swedes, French Canadians, and Irish had become an extremely small percentage of the mine's labor force, or had left altogether. By 1910, Calumet's Swedetown indeed had six Swedish households—but it also had three houses headed by Cornishmen, three by Germans, and 10 by Slavs. By this date, Swedetown should have been "Finntown," because 25 Finnish families resided there.[35]

Competition for company housing was keen. Workers watched for other men to leave and put in bids for their houses. James MacNaughton reported "a dozen applications for every vacant house."[36] This put mine managers in a delicate situation. While satisfying one worker, they risked alienating 11 others. Still, this was exactly the situation the managers wanted. They wanted demand to outstrip supply, so that those in company houses would feel fortunate (and grateful) to be there. They wanted housing to be seen as an earned privilege, and not as an automatic benefit bestowed upon employment.

Company housing was indeed a benefit, in terms of its cost. Housing costs at the Lake Superior copper mines were perhaps the lowest to be found in any major metal-mining district in the country.[37] Typically, employees paid $1 per room per month.[38] Sometimes the rate went even below that, for a company's smallest, oldest houses. By 1913, Quincy charged only $1 to $2 per month for its log houses, regardless of size, and C&H rented its log dwellings for as little as 50 cents.[39] These exceptionally low rents, however, came with a string attached. The companies had written these houses off as unworthy of improvement, and their occupants could not expect painters, carpenters, or roofers to pay them regular visits.

Alexander Agassiz had a company house at Calumet, where he stayed during his one or two short visits to the mine each year. In 1895 he mentioned to "John the Cook" that he wished his mattress hardened, but he left town without being more specific. John, not wanting to err in getting this done, mentioned the mattress to C&H's general manager, S. B. Whiting, and Whiting turned the problem over to engineer Preston West. West wrote Agassiz for more definitive intructions, and Agassiz responded that he "should like to have my mattress restuffed so that it should not be quite so flabby as it is now, adding some hair."[40]

James MacNaughton, C&H's general superintendent after Whiting, lived year-round in his company house. MacNaughton received numerous letters from workers (or their wives) requesting house maintenance. In 1915, MacNaughton wrote such a letter himself to the company's boss carpenter: "Will you please have the shingles on my house examined; Mrs. MacNaughton says the roof is leaking."[41] It is fair to say that letters written by MacNaughton were acted upon somewhat more quickly than those he received.

The thousands of houses standing at the mines represented innumerable maintenance problems, particularly as the housing stock grew older. Struggling firms often let their houses go to ruin, as managers conserved capital by halting all maintenance work. Profitable, ongoing companies proved more conscientious in tending to workers' houses. They wanted the dwellings to last the life of the mine, so they tended to leaky roofs, crumbling chimneys, and broken windows. Still, even the wealthiest companies acted as frugal landlords when it came to general maintenance. While Copper Range painted its better houses regularly, hundreds of dwellings "occupied by miners and employees of that grade" went without paint for up to 15 years. Finally, Copper Range decided to tackle the work, because it would "make a tremendous difference in the general appearance of our location and please our employees very much." To cut costs, the company hired a small army of Tom Sawyers—local school-boys on summer vacation—to lay on the paint.[42]

Early in the twentieth century, workers wanted greater access to the modern marvels of science and technology. They wanted home improvements. Suddenly, the chamber pot and privy would not do; workers wanted inside toilets. Families wanted electric lights and telephones. Because wood-burning stoves were dirty and inconvenient, more workers wanted radiators. They wanted additions that provided more space and privacy. Elite employees requested that their old barns be renovated, because they had no use for a horse or milk cow. They wanted stalls and feed bins removed and new doors and sills installed, so that a newfangled automobile could be driven into its garage.[43]

After 1900, workers in the Copper Country wanted more improvements in their housing than their employers were prepared to make. At C&H, James MacNaugton was not about to modernize over 700 houses all at once at company expense, but in the name of preserving harmony with labor, he had to make some accommodations to their rising expectations. The telephone company could run lines over C&H property, but the workers had to pay all installation and service charges. C&H started wiring old houses for electricity, but then backed off. The houses were too numerous, the demand too great. The company told employees to engage approved local contractors at their expense to do the work. When employees moved out of their electrified company houses, the wiring they paid for had to stay, but they could remove any chandeliers or other fixtures they had put up.[44]

As C&H ran new sewer lines along select streets, residents there received hook-ups and inside toilets, which were usually set into a basement corner. The company did not put bathtubs into every house, but workers could avail themselves of this "luxury" at a C&H communal bathhouse. MacNaughton or the land agent turned aside most requests for major renovations or additions, but sometimes allowed tenants a new floor, ceiling, or wall covering.

Employers blanched at retrofitting hundreds of houses with modern conveniences, because these dwellings did not generate a profit. By 1913 the mining companies reported about a 5 or 6 percent per year return on the capital invested in their houses, before any taxes or maintenance costs were deducted.[45] Calumet and Hecla had built its 764 frame dwellings at a cost of $1.2 million; between 1902 and 1912, its rent receipts averaged $61,863 per year, or a 5.15 percent return.[46] But over that same span, C&H spent $61,228 per year maintaining and upgrading its housing. These figures suggest that Calumet and Hecla absorbed the initial cost of house building as part of its overall costs of doing business. Collected rents did not pay off the houses. Instead, they set the company's annual budget for house repair and modernization.

Since the companies profited little or not at all from their rents, it is not surprising that they encouraged workers to build their own houses on company property. By 1913, a total of 1,751 employee-owned houses stood on company lands, with more than half of them at Calumet and Hecla.[47] These houses constituted the ultimate expression of a strong bond between employee and employer, and of a worker's dependency and loyalty. In economic terms, the security of the employee's housing investment was wholly dependent on his company's continuing success. As fate had it, it did not prove foolhardy to build a house on property owned by the five or ten largest companies, and this was particularly true for employees at Calumet and

Hecla. C&H posed no threat to the market value of a worker's house, because for decades the company was in no jeopardy whatsoever of a possible shutdown. Here what posed a threat was the ground lease a man signed to obtain his lot. C&H, like other companies, used its leases—both for company houses and for building lots—to exercise broad social control over its domain.[48]

Leases for company-owned houses at Calumet and Hecla stipulated more than a rental rate and payment schedule. With management approval, portions of the house could be sublet to boarders or lodgers, but they had to be company employees. The occupant could not sell "spirituous or intoxicating liquors." More important, the tenant had to vacate the house within 15 days of quitting his job, or of being dismissed. But the lease did not necessarily remain in effect as long as employment continued: C&H could terminate the tenancy on 15 days' notice "for any cause or reason whatsoever."

The lease for C&H company houses stacked the deck in the employer's favor, but was a model of restraint when compared with the company's ground lease for building lots. This lease was liberal in just one sense—the rent, which was only $5 per year. But that tempting carrot dangled from a very thorny stick. A C&H ground lease ran for only five years. It was renewable, of course, but renewal was not guaranteed. The lease-holder's house occupied the surface, but the company maintained the mineral rights underground and could drift or stope to within 15 feet of it. The lease-holder also had to share his lot with any open excavations or pipes the company deemed necessary for mining operations.

The homeowner could not operate his dwelling as a bar or brothel. He could not transfer his lease, sublet his premises, or even sell his house without the written consent of the company. He had to pay all taxes on the land and structures within 60 days of their being due. As long as the owner complied with all lease provisions, he could "peaceably and quietly have, hold and enjoy the said premises." However, if the employee violated any provisions, quit the company, or was fired, then his lease became void. After 90 days, the company had the right to reenter and repossess the property, including the house. During a second 90-day period, a man could not occupy the house, but he could pick it up and move it, or sell it to a company-approved buyer. If he failed to do either, any structures on the lot became the property of Calumet and Hecla "without any liability . . . to pay for the same or any part thereof."

Ground leases written by other companies were nearly as stringent. In 1914, the Bureau of Labor Statistics commented on the "drastic nature" of these ground leases, and found it "strange" that any person would sign one, and "most astonishing" that 1,000 employees had done so at Calumet and Hecla.[49] From the perspective of a strike-torn era, it seemed inconceivable that an employee would erect a house on land rented under such unfavorable terms. But examined from the perspective of earlier decades, erecting employee-owned houses on company ground made more sense. The prevalence of this practice testified to how well company paternalism had then worked, particularly at Calumet and Hecla.

A C&H employee in the early 1890s had three housing options. He could hope to rent a company house, and if fortunate enough to get one, his housing costs amounted to about 5 percent of his total earnings. If he had adequate savings and income, he could build his own house on a lot leased from his employer. If he did so, 8 to 9 percent of his earnings went toward housing. An employee could also turn to the housing available in the adjacent villages of Red Jacket and Laurium. There, shelter typically claimed 10 to 14 percent of his income.[50]

Employees who built their own houses on leased land often boarded fellow countrymen or extended family members. This common practice supplemented their income and afforded them the opportunity to build and occupy more house. As Calumet and Hecla admitted, employees who owned their own houses were generally "better housed" than those who rented from the company.[51] Still, one must look beyond housing costs, and the attainment of modestly superior dwellings, to understand why so many employees signed such a one-sided ground lease.

Calumet and Hecla was a careful and conservative company that honored orderly progress and abhorred breakdowns. It wrote extremely harsh provisions into its ground leases, so if a social breakdown occurred it could expel any troublesome members of the community. Still, C&H would exercise its right to evict only as a last resort, and it had absolutely no desire to effect its lease powers ruthlessly. Doing so would have destroyed its reputation as a model employer—and would have discouraged workers from building their own houses. The company's policy was to do exactly the opposite. It wanted workers to build more houses, so that it could build fewer. That saved the company construction costs, lessened its maintenance obligations, and increased the number of employees who had a very large stake—the family home—invested in the company's present and future operations.

By 1893, when C&H rented out 570 houses, employees had already erected 782 houses of their own on leased land.[52] These men trusted their company and wanted to strengthen their bonds and pledge their loyalty to it. C&H was one of the richest mines in the world. For "good workers" it offered steady employment at higher wages than those offered at other local mines. The ground-lease arrangement seemed of mutual benefit. The company got a stable force of dependable employees in important positions. As for the workers, as one observer noted, "Like Ohio office-holders, 'few die and none resign' among C&H employees, and jobs are handed down from father to son, so that the transactions in houses or leased lands are not numerous."[53]

The mining companies entered the housing business because they had to. To shelter the work force they desperately needed, firms erected rental housing and made it possible for men to build their own houses on company property. These activities hastened the establishment of stable mining communities, and allowed thousands of workers to live in decent, albeit rather plain, accommodations.

The companies did not receive a profitable return on their housing investments, at least not a direct return. They did, however, intend for their housing and ground-lease practices to help lower their labor costs. By providing houses and building lots, they offered competition to private developers in adjacent villages who might have gouged workers in need of shelter. By holding down their workers' cost of living, employers hoped to hold down wage scales. The mine managers believed they succeeded in this effort. By the early twentieth century, it cost a mine worker considerably less to occupy a dwelling on the Keweenaw than in Butte, Montana. Local managers figured the difference for comparable dwellings at $12 per month, or nearly 50 cents per shift worked.[54] That was 50 cents per day, per worker, that they did not have to pay out in wages, because of the housing services rendered.

Besides using housing as a check on labor costs, the companies manipulated housing to shape and control their communities. They eschewed boarding houses, seeing them as breeding grounds for trouble, and invested in single family dwellings. They wanted to attract as many married men as possible, because they deemed the husband to be more reliable, sober, and industrious than the single man. In this, too, they were quite successful. As frontier conditions on the Keweenaw eroded, more women and children came into the area. All mature companies boasted a sizable number of men who had come with their families and who had stuck, the men completing decades of service, and their sons following after them.

Yet even before the turn of the century, serious limitations or flaws could be detected in the companies' paternalistic housing practices. In the long run, the promise of inexpensive housing could not compensate for the fact that the mines offered thousands of jobs that were dirty, dangerous, and otherwise undesirable. When the mines expanded tremendously after 1890, they signed up relatively few men from the British Isles or western Europe. They lost their ability to pick and chose their laborers. Instead, they employed hosts of men from Finland and southern and eastern Europe who had never worked in mines and who were unversed in industrial discipline or skills. The companies never reconciled themselves to this situation. They became dependent upon Finns, Italians, Austrians, and other immigrants whom they considered inferior—and they only begrudgingly moved any of these newcomers into better jobs and into company houses. Meanwhile, the companies reserved their best housing for managerial and working-class elites, for men in supervisory and skilled positions.

In serving as paternal landlords, the companies did not get into labor trouble by treating their tenants in a high-handed, dictatorial manner. They did not err by forcing their will too overtly on their workers' life-styles, mores, and manners. True, the companies dragged their feet in repairing and modernizing their nineteenth-century houses. Many tenant families undoubtedly chafed at the delays, and at their dependence upon their employer to fulfill housing needs and wants. Many others, however, happily accepted less-than-perfect shelter, as long as the rent was low. Altogether, those under the care of the paternal hand seemed relatively content.

The problem was that the paternal hand reached out only to some, and not to others. The men performing the worst jobs underground got no relief when they got to grass. While preferred workers left the mine for their own houses on company ground, or for their company-owned house, the trammers and other unskilled laborers went off to whatever housing they had arranged for themselves, where they usually paid more for less. When payday arrived, the companies explained that the wages might be a bit low, but the benefits of working in the Lake Superior copper district were high—particularly the benefit of good housing. That argument still carried weight with the miners, mechanics, and bosses who lived on company property, but it rang hollow for the unskilled "beasts of burden" who did not.

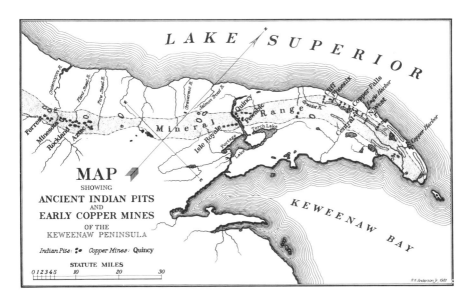

The early mines congregated around ancient Indian pits. Two later companies of great import, Calumet and Hecla and Copper Range, were located about ten miles northeast and southwest, respectively, of the Quincy mine. (*Quincy Mine Hoist Association*)

The Minesota mine, quick to become profitable, erected a substantial church, company office, and store/warehouse on a clear-cut hillside before 1860. (*Michigan Technological University Archives and Copper Country Historical Collections*)

A Calumet and Hecla smelter worker completes the last step in the production of marketable copper—ladling the molten metal into ingot molds. (*Smithsonian Institution*)

In the last quarter of the nineteenth century, Calumet and Hecla, atop its Conglomerate lode, built one of the most extensive and expensive mine plants in the world. (*Michigan Technological University Archives and Copper Country Historical Collections*)

Not much movement on 5th Street in the village of Red Jacket (also known as Calumet)—at least not until work crews cleared it of snow. (*Michigan Technological University Archives and Copper Country Historical Collections*)

Calumet and Hecla dedicated a beautifully landscaped park to the memory of Alexander Agassiz, the firm's longtime president, in 1923. (*Michigan Technological University Archives and Copper Country Historical Collections*)

Before they had rock drills driven by compressed air, miners used sledgehammers and hand-held drill steels to drive shot-holes for blasting rock. (*Quincy Mining Company*)

A mine captain and a timber crew lean against square-set timbers in the Calumet and Hecla mine. (*Michigan Technological University Archives and Copper Country Historical Collections*)

For about thirty years after being installed in the 1860s, this man-engine carried workers into and out of the Quincy mine. (*Michigan Technological University Archives and Copper Country Historical Collections*)

The main drive shafts of the Superior Engine at Calumet and Hecla came from the Krupp works in Germany. The massive engine powered several different mine operations. (*Michigan Technological University Archives and Copper Country Historical Collections*)

Quincy erected this Nordberg cross-compound, condensing, four-cylinder hoisting engine at its 9,000-foot-deep No. 2 shaft in 1920. It was the largest direct-acting steam hoist in the world. (*Michigan Technological University Archives and Copper Country Historical Collections*)

Thirty men, their dinner pails hooked over their arms or sitting on their laps, are set to ride a man-car from the shaft-rockhouse down into the mine. (*Michigan Technological University Archives and Copper Country Historical Collections*)

At Tamarack, the taller portion of this structure (ca. early 1890s) is the shafthouse; grafted to it is a rockhouse with crushing machinery inside. The giant tree sections in the foreground will serve as stulls, or mine supports, when taken underground. (*Michigan Technological University Archives and Copper Country Historical Collections*)

Inside a rockhouse at the Isle Royale mine, laborers still have to stoop to pick up copper rock that has tailed off the grizzly. They will put it into nearby crushers. (*Michigan Technological University Archives and Copper Country Historical Collections*)

The "Little Giant" drills manufactured by the Rand Drill Company brought tremendous productivity increases and cost savings to the mines in the early 1880s. (*Smithsonian Institution*)

Tramming was the most arduous underground occupation—and trammers were wont to complain about being "beasts of burden." (*Michigan Technological University Archives and Copper Country Historical Collections*)

Rock falling from the hanging wall was the most frequent killer of men. As a precaution, a Calumet and Hecla miner bars down a loose piece of the hanging. (*Michigan Technological University Archives and Copper Country Historical Collections*)

In September 1895, a crowd gathered on the surface near an Osceola shafthouse billowing smoke. Underground, thirty men and boys were dead in the Copper Country's worst mine disaster. (*Michigan Technological University Archives and Copper Country Historical Collections*)

The log cabins that Calumet and Hecla erected in the late 1860s were still inhabited half a century later. By that time, though, they were distinctly inferior to all other company houses at the mine. (*Michigan Technological University Archives and Copper Country Historical Collections*)

Near the Calumet and Hecla mine stands a row of company houses intended for a "better class" of workers. (*Michigan Technological University Archives and Copper Country Historical Collections*)

Most of the mine and mill towns had company-sponsored bands. Here, the Calumet and Hecla band poses on the sidewalk in Calumet. (*Michigan Technological University Archives and Copper Country Historical Collections*)

After being reared at Calumet and Hecla, James "Big Jim" MacNaughton came back to rule over the community in 1901. (*Michigan Technological University Archives and Copper Country Historical Collections*)

In 1913–14, thousands of miners and trammers joined the Western Federation of Miners and struck to protest poor wages and working conditions and to oppose technological change. (*Michigan Technological University Archives and Copper Country Historical Collections*)

Young girls head up a union parade during the Copper Country strike in 1913. In the background, bivoacked National Guardsmen stand watch. (*Michigan Technological University Archives and Copper Country Historical Collections*)

In the midst of the bitter 1913–14 strike, tragedy struck at Italian Hall (Calumet) during Christmas Eve celebrations when a cry of "Fire!" started a stampede to the exit of the hall. Seventy-four, mostly children, died in the mad rush. (*Michigan Technological University Archives and Copper Country Historical Collections*)

10

Cradle to Grave

On paternalism:

It is a ruse that has been employed on many different occasions by those who are vainly attempting to substitute charity for justice in the relations of employers and employees. . . . It is an attempt to preserve the slave's feeling for his master in the modern machine age.

Miners' Bulletin, 28 August 1913

In the second half of the nineteenth century, the mining companies nurtured a man's body, mind, and soul. They protected his income by encouraging competition among merchants. They gave him land for food production and kept his cow in a company pasture. They hired the doctors, built the hospitals, and administered medical and benefit programs. They erected the schools and libraries and set aside the parks and athletic fields. They supported the churches and maintained the cemeteries. They comforted widows, gave work to orphans and fuel to paupers. They pumped up the water, pumped out the privies, and picked up the garbage.

Paternalism succeeded for decades at the Lake Superior copper mines because it served the interests of managers and workers, as well as the interests of the broader industrial society being formed around the mines. Paternalism speeded the development of stable companies and stable communities. Yet once the mining companies and the local society reached maturity, the system began to unravel. It did not cope well with social change and the influx of new peoples. By 1900, despite the companies' continuing presence in a wide range of social and community affairs, the paternal bonds between employer and employee grew noticeably weaker.

When the mines first opened, they operated farms to produce food for employees and animals.[1] The environment put some restraints on this activity. Keweenaw summers were short, and although summer days in this northern region were long, average temperatures remained low. Also, much of the soil was not prime for agriculture, so some crops, such as corn and wheat, did not flourish. Nevertheless, company lands yielded hay, oats, timothy, onions, cabbages, squash, and potatoes. The "much cherished" potato did particularly well and became a staple in the workers' diet. A visitor to the North American mine in 1853 reported that he saw "their green and vigorous tops in every little patch attached to the miners' dwellings, as well as spreading clear along the face and upon the very summit of the bluff, wherever a potato could be stuck along the roads, disputing possession with the blueberry and the bramble."[2]

The grain and grasses raised on company farms went mostly to animals, while the vegetables were sold to employees. Companies engaged a "head farmer" to oversee operations, or they leased the land out in sizable plots to contractors who supplied "the employees of the Company with the produce, at reasonable rates."[3] Companies did not get wealthy off their farms, but they did expect to generate profit from them. In 1854 the Ohio Trap Rock Company received $5,100 for its 30 acres of potatoes, and in 1860 the Cliff mine reported that its farm paid "a handsome profit on amount expended."[4]

Company farming was a temporary activity that nurtured new mining villages in their early years. Later, many companies continued to produce livestock feeds and to maintain pastures, but they quit the business of producing food for human consumption. Despite profits made from their farms, companies retreated from their role of "provider." Large, company-managed farms gave way to smaller plots, tended by the workers themselves. In 1862, the Pewabic mine raised 24 tons of hay and oats, and seeded 11 acres of land in grasses. At the same time, "a large portion" of its cleared land was "farmed out" to employees, and 70 newly cleared acres were to be "leased in the spring to our people for planting." In the same manner, in 1913 the Tamarack mine still had 19 lots in the midst of its settlement set off as "potato fields."[5]

The companies did not shift any "burden" of food production onto employees. More correctly, they allowed workers the opportunity to till the soil, if they wanted. On the smallest scale, a worker's family could maintain a small vegetable garden behind the company house. If a family wanted to do more, it could lease an acre-sized garden plot and, at little or no cost, keep milk cows on company pasture land.[6] A man who wanted to produce cash crops might lease 40 or more acres from his employer. By opening up land for agricultural use, the companies provided a benefit that many immigrant workers wanted—particularly those who arrived after 1890.

Many of the Finns, Italians, Croations, and Slovenians who came to Lake Superior had worked as farmers or farm laborers in Europe. In Michigan, they were able to recapture some of their agrarian past by farming company lands. In winter, they had little choice except to labor in the mines. But in summer, disenchanted mine workers looked for surface work. For many, and particularly the Finns, this meant heading off to the fields. Three-fourths of the foreign-born Finns in the Copper Country had been agricultural workers in their homeland, and they became the dominant farming group on the Keweenaw. By the 1910 era, at least 500 Finns leased 40 acres or more from the copper companies.

The history of company stores paralleled that of company farms. In their first decade or two, many firms involved themselves in running a store. Then they largely withdrew from the provisioning business, turning it over to shopkeepers who leased stores on company property, or over to independent merchants located in adjacent villages. Finally, just as many mine workers provided food for themselves through agriculture, some also secured and distributed foodstuffs and other goods by forming workers' cooperatives.

When no commercial businesses operated near the remote mines, the companies filled the breech. They built a store and took the proceeds. Workers bought on credit, and companies withheld the amount due from a man's pay. This system offered the potential for great abuse. If the company store set high prices and caused employees to run up large bills, then on settlement day, they might be left with little money to take home. The agent of the Osceola mine once faulted the Allouez mine's company store for credit manipulation. He noted that "scarcely a man of family [is] not more or less indebted to the store at that mine."[7] Nevertheless, such abuses seem to have been uncommon. The observer for *Mining Magazine* was correct in noting in 1853 that "each working mine has a store, at which the men are supplied at reasonable prices by the companies."[8]

Several factors limited company store abuses in the Copper Country. First, mine officials realized that their "bread and butter" profits came from selling copper, not provisions. They preferred to concentrate on their real business, while minimizing the amount of attention paid to a store. Second, these companies offered similar jobs and competed with each other for laborers. In this competition, it did a company no good to earn a reputation for being greedy. Finally, the companies were consistent in wanting to hold down their workers' cost of living. This is why they stayed in the housing business—and why they did not abuse company stores. They did not want monopolies and high prices, but competition and low prices, so that workers would agitate less for wage increases.

Companies sometimes opened stores to break monopolies and compete with established merchants. The Quincy Mining Company, through the 1850s and early 1860s, opted not to run a store. Ransom Sheldon and other early merchants in Hancock and Houghton catered to workers from all the Portage Lake mines. With the onset of the Civil War, Quincy faced keen competition for laborers, and wartime inflation drove up the cost of living. Quincy wanted to stabilize its work force without resorting to ever higher wages. Because it felt local merchants were profiteering, Quincy opened its first store in 1863 and ran it for about three years, selling goods at or near wholesale.[9]

The Osceola mine, like Quincy, initially relied upon independent merchants to provision employees. In its case, these merchants were in nearby Red Jacket village, where they had come, primarily, to sell to Calumet and Hecla employees. Osceola's agent, John Daniell, was particularly opposed to credit operations. He did not want his mine involved in holding "a man for a merciless creditor."[10] Yet at the same time he believed it "would be an advantage to our people" to open a store at Osceola in 1878. Wages were tending downward in the region, including at the Osceola mine. But Red Jacket merchants had not taken notice of that fact, largely because of the more liberal wages paid by the richer Calumet and Hecla. A store at Osceola, then, would be good for competition and lead to lower prices.[11]

By the early twentieth century, only the Copper Range Company operated its own stores. It had four: two at its mines and two at its mill towns. The stores carried groceries and meats, dry goods, boots and shoes, men's clothing, furniture, hardware, kitchen utensils, and animal feeds and hay. In 1914, its stores at Trimountain and Beacon Hill together netted a profit of $13,800 on gross sales of $205,000.[12] Besides netting modest profits, the stores made shopping more convenient for employees and offered competition to the independent merchants of South Range. The Copper Range Company and the South Range merchants, after all, had dissimilar interests. The former wanted low prices; the latter wanted high profits. This conflict became apparent in 1914, when war in Europe began affecting prices in the United States. The South Range merchants quickly marked up their goods, while Frank Denton, superintendent of Copper Range, wanted the company stores to serve as a buffer against the first wave of inflation:

> Fortunately our stores had a fairly good stock of food supplies, especially of flour, and all this has been sold at the same price as existed before the war was expected. It is estimated that our customers will be supplied with flour for nearly sixty days at the old price. My instructions to the stores are to raise prices only when an increased price has to be paid for purchases. Some of the South Range stores have resented our actions but that does not matter.[13]

Other mining companies helped provision employees, established competition, and controlled prices without running their own store. Instead, they leased out stores on company property. They received some income from the stores, but escaped the bother of operation. Meanwhile, the shopkeeper had to run his business under the terms of a lease that protected the company's interests. For example, the lease could keep out any "demagogue" who had "extreme views on the rights of the poor and likes to air them." A lease could stipulate the goods to be carried, the prices to be charged, and could forbid the sale of beer and liquor.[14] The North American mine wrote leases to a butcher and a shoemaker in 1857; the shops brought in some money and were "of great advantage to both our people and Company."[15] In 1886, Calumet and Hecla's board of directors authorized James North Wright, the agent, to sign ground leases "for stores . . . as he may deem it for the best interest of this Company to retain upon the location."[16]

Companies did not develop full-scale commercial districts on their properties. A mine location had no "downtown." Often, it had but one or a few company-operated or company-leased stores. To find specialty shops and to broaden their selections, residents of the locations journeyed to adjacent villages or towns.

The companies believed their store policies did a good job of controlling prices, both at and away from the mine proper. The former Quincy and Calumet and Hecla agent, James N. Wright, opined in 1899 that "competition is open and abundant, and the miners of the Upper Peninsula can shop as cheaply as the laborer of Detroit, excepting only the added cost of transportation on his goods."[17] In 1913, mine managers trotted out figures demonstrating that a Michigan copper miner paid one-quarter to one-third less for groceries than his Montana counterpart.[18] But the mining companies were not solely responsible for this; they shared credit with workers who had started cooperative stores.

The first cooperative opened in 1890 as a joint venture involving labor and capital. The idea originated with a mine worker, Robert Bennett, who had once organized a cooperative in his native England. Bennett believed a co-op needed influential friends to help it stand against unfriendly merchants, and considerable capital, so it could be well stocked and truly competitive. He turned to Captain John Daniell, then the Tamarack mine agent, who had long espoused that workers should get a fair deal in the marketplace. Daniell talked other mine officers into investing in the cooperative; workers invested in it themselves; and the Tamarack Co-operative Association opened a large store in Red Jacket that carried clothing, dry goods, shoes, hardware, furniture, kitchen utensils, groceries, and meats.

The Tamarack Co-op catered to workers from the adjacent Tamarack, Osceola, and Calumet and Hecla mines, and it extended its market by delivering goods some 15 miles north and south. Co-op wagons made weekly deliveries to consumers as far away as Houghton, Hancock, and Mohawk, and daily deliveries to families within three miles of the store. After 1890, the Tamarack Co-operative, more than any other store, set the level of retail prices in the Copper Country. By establishing low prices and selling to a broad market, this store generated more business than any other cooperative in the United States.[19] From 1890 through the end of 1912, its total sales registered $9.5 million, and annual sales reached nearly $850,000. Following the success of the Tamarack Co-op, at least four other co-ops were opened by 1910—one by Croatians and three by Finns.[20]

An important family expense on the Keweenaw was fuel—wood or coal to heat the home over long winters, and fuel throughout the year for cooking. Again, to help hold down their workers' living costs, the companies used their bulk purchasing power to obtain lower fuel prices from local merchants. Often, mining families did not have to shop fuel suppliers on their own. Their employers negotiated with the firms, securing the best prices for home-delivered wood or coal. In 1896, Calumet and Hecla reported, "We have made a contract for the delivery of wood and coal to our employees, retaining only the supervision of the prices—a condition of things which cannot fail to be satisfactory to both employees and the company."[21] Similarly, when a local utility firm wanted to run service to C&H houses around 1910, it had to negotiate with the company. In return for getting to build up its load by reaching out to C&H workers, the utility had to lower its rate to these employees from 12 to 8 cents per kilowatt hour.[22]

Shortly after the companies began building single-family houses, they started erecting schools, libraries, churches, and hospitals. These social institutions were by no means equally well endowed from location to location, as their fortunes were tied to those of the parent company. In some instances, these social institutions in the Copper Country were as fine as those found in any industrial setting in America. Others, however, limped along pathetically, in step with an impoverished company.

At the mine locations, no hard and fast line separated civil and corporate authority, or public and corporate interests. Residents of an unincorporated mine location had no mayor, no councilmen. Instead, they had an agent who set the course of their community. Adult male workers, if citizens (and many were not), voted in township, county, state, and federal elections. But within the boundaries of a mining company's property, where hundreds or thousands of people lived, elections hardly mattered. Nobody voted on a bond issue to build a water supply system, to pave the streets, or to build new sewers. That was the company's prerogative. So, too, the company largely determined the quality of the location's schools. Nominally, schools were the provenance of local government. But the public schools at a mine were, in effecct, just another extension of the company's paternalism. They were as good or bad as the mining companies made them.

Schools appeared at the Keweenaw and Ontonagon county mines in the 1840s and 1850s, just as soon as they attracted married men with families. Houghton County was the last to see the development of schools, but because its mines flourished while others declined, after the Civil War era, the vast bulk of Copper Country schools stood in Houghton County mine and mill towns, and in the independent villages near the mines. In 1861, Houghton County had four school districts, seven teachers, and 455 enrolled students. By 1910, it had 34 districts, 127 schools, 639 teachers, and 19,557 enrolled students.[23] The tremendous increase in public education was wholly due to the growth and success of local mines.

The Cliff mine offered instruction as early as 1848, when ministers or priests did the teaching without the aid of textbooks.[24] By 1860, the mine had a permanent school, apparently a two-room affair, supported by a two-mill tax collected on property in the district. Some 90 to 100 "scholars" received their instruction from "two very competent fellows."[25] On the southern end of the peninsula, the Minesota mine claimed one of the earliest schools. In 1852, when the mine had 33 dwellings and a population of about 312 persons, including 100 women and children, it erected "a new and comfortable building designed and regularly occupied as a church and school house."[26] One of the earliest Houghton County schools went up in 1861 in Franklin Township, where it served children from the Pewabic and Franklin mines.[27]

Early schools were ungraded, poorly equipped, and offered primary education only. Teachers often met with 50 or more students of mixed ages in their classrooms. In 1873, the superintendent of the 12 schools in Keweenaw County reported, abashedly, that only four had wall maps, only three had a Webster's Unabridged Dictionary, and only one "exalts in the possession of a respectable globe."[28] Education further suffered from the fact that in Keweenaw County, workers constantly moved from mine to mine. They were "almost as migratory as birds of passage," and children often attended three different schools in a single year.[29] Finally, there was the problem of foundering companies failing to support the schools. One, probably the Conglomerate mine, claimed "the right to govern the school board in every particular," because it paid the greater portion of the school millage in its township. When in financial straits, the company appointed its surface boss as the school's teacher. The school paid him $70 per month, the mining company only $25. The man's first obligation, however, remained with the company. The teacher abandoned his students whenever the surface boss was needed. School seldom began before 10 a.m.; it usually recessed from noon to two or three o'clock; and sometimes, when the surface boss was extremely busy, the teacher did not show up at all.[30]

Conditions were not much better at the early Houghton County schools. In 1870, a slump in the mining industry caused several mines—and their schools—to shut down. In 1872, the county's school superintendent bemoaned the lack of globes, maps, and other teaching materials. He noted that "our mining officers and businessmen are not in the habit of employing good workmen at high wages without providing all the necessary appliances for good efficient work, but as school officers they are continually doing this." That year, one school held forth in an unplastered attic over a mine office, while another convened in a room over a saloon.[31]

There was one area, however, where the mining companies ultimately could not skimp in supporting schools: teachers' salaries. A male teacher who taught at the Cliff mine in 1863 quickly escaped from the Copper Country, deeming it the "most God-forsaken place" in the United States.[32] American-born, he did not like the region's environment, its deprivations, or the foreigners he lived with and taught. Those company-controlled school boards that wanted educated American men and women to serve as teachers had to pay well to get them. In 1880, out of all 77 Michigan counties, Houghton ranked second in the average salary paid to 23 male teachers, and first in wages paid to 57 women.[33] The county also ranked first in terms of having the longest school year, which was probably a tribute to its long winters and short summers. On the debit side, it ranked only 71st in its percentage of school-aged children (56 percent) who actually attended school.

Although poor companies were chary about school funding, successful companies did a more respectable job of nurturing "public" education. Quincy Township created its first school district in 1867. The Quincy mine, for all practical purposes, was Quincy Township, so it ran the school. The school board that drafted the first "Rules and Regulations of Quincy School" consisted of James North Wright, mine agent; Thomas Flanner, mine physician; and J. M. Foster, chief mine clerk. The mining company designed and built Quincy School and then rented it to the school district. The first school accommodated up to 150 youngsters; by 1877 it held 300. Quincy's school was one of the early graded schools in the Copper Country; it sought out qualified teachers; by at least as early as 1876 it offered instruction in Latin, trigonometry, and algebra; and in 1884 it added a high school department. The school itself was a rather nondescript frame building, and the student-teacher ratio was high, reaching almost 100 to one in 1877.[34] Nevertheless, it was one of the better schools found at a mine location.

One could find more elaborate schools within the villages of Houghton and Hancock, but as a mine school, Quincy was outdone only by the Copper Range's school at Painesdale, and by Calumet and Hecla's schools at its mine, mill, and smelter communities. In Painesdale and Calumet, the schools lost the appearance of small, rural affairs and adopted the demeanor of substantial, urban-like institutions. Painesdale High School, erected adjacent to the Champion mine in 1910, was a three-story structure handsomely crafted from red sandstone. Nothing about its façade indicated that it was a mine school. The truth became apparent, however, when the students arrived from Copper Range's more far-flung operations, such as the stamp mill at Redridge. Few American youths rode passenger trains to and from school, but those at Painesdale did. The Copper Range Railroad ran school trains into Painesdale, on tracks usually dedicated to carrying copper rock.[35]

Calumet and Hecla was the company most heavily engaged in school affairs. By 1910, ten company-built schools, erected at a cost of over $350,000, stood at the mine location or near C&H's mill and smelter at Torch Lake. The rents received from the local school districts amounted to a 3 percent per year return on the money invested in the buildings, which C&H also maintained. Calumet and Hecla paid 85 or 90 percent of the taxes in its township. Since it put up the money, it got the schools it wanted.[36] A pipeline from the company's top officials ran straight to school administrators, just as surely as pipelines from company boiler shops ran to school radiators.

On his periodic visits to Calumet, president Alexander Agassiz checked school conditions, just as he inspected all other company business. When Calumet rebuilt its manual training school and high school in 1905 because of fire, no time was lost in endless school board debates over what to do, or in soliciting citizen comment. Superintendent MacNaughton and president Agassiz simply got their own architect and went to work on the new, fireproof schools, to be paid for largely from the company's insurance fund.[37] While Calumet and Hecla wielded a strong hand in school affairs, it also received credit for building strong schools.

C&H earned its reputation for supporting good schools from the start. By 1870, Houghton County's school superintendent already credited Calumet for securing the services "of a thoroughly competent and experienced principal" and for erecting a new primary school. He faulted the general level of support given education, but lauded Calumet for manifesting "a highly encouraging degree of interest" in students' welfare.[38] As the company expanded rapidly, so did the local schools. While many mine towns made do with a one-room school, or at best with two- to four-room schools, C&H went its own way in designing, erecting and furnishing its Washington School in 1875:

It is said to be the largest and best schoolhouse in the State. Its length is 192 feet; its width one hundred feet. Its basement is of cut sandstone, and the superstructure, two stories in height of wood, is heated by steam throughout and seated with the Sherwood single seat. Its seating capacity is ample for 1,200 pupils. To Mr. Alexander Agassiz . . . and to Mr. J. N. Wright, superintendent, are we indebted for this costly and commodious structure, complete in all its appointments, and as perfect as the best skilled modern school architect, with an unsparing use of money, could make it.[39]

C&H was consistent; it ran its schools like it ran its mine. With Agassiz at the helm, C&H had a penchant for centralization and impressive edifices. Seeing that corporate success would spawn a burgeoning population, C&H quickly erected the largest grammar school in the state, rather than opting for myriad small schools nestled in different neighborhoods. Only later, when the population outstripped even Washington School, did C&H build satellite schools around it.

Calumet and Hecla also had a penchant for naming things, whether they were stationary steam engines, pumps, railroad locomotives, mine shafts, or neighborhoods. Often, the names drew allusions to Native Americans, but when it came to naming its grammer schools, C&H turned to American presidents and men of letters: Washington, Jefferson, Lincoln, Irving, Longfellow, Whittier, and others.[40] The names carried a message for the children of immigrants; they attended schools in Calumet to learn the English language and American ways. They also attended school to prepare themselves for work. In 1875, the year the Washington School opened, the supervisor of the Calumet Township schools noted, "Owing to the fact that there are only a few in this community who expect to give their children anything more than a very limited education, our classes in the 'high school' department are as yet, and henceforth will be, necessarily small."[41] Calumet strengthened its high school offerings by 1880, and yet for many years it emphasized a very practical curriculum geared toward students who would not be staying through to graduation.

By the turn of the century, Calumet's high school still attracted only a few hundred students, but it was becoming more popular. C&H anticipated a high school enrollment of about 600 students by 1910. In 1905, James MacNaughton believed the recent enrollment increase was due to one factor: manual training.[42] Under the company's direction, the Calumet schools became more vocationally oriented, more geared toward providing the skilled labor that C&H needed in surface shops. More boys stayed in school after 1899 because the manual training schools taught them skills C&H would pay for.

In shop classes at Calumet, high school boys did not fabricate decorative or useless trinkets: "The shop-work here differs quite materially from that of other schools . . . , in that the work is all directed to utility. Pupils make things to be used. This is especially true of the work in iron and steel. Every piece made is for use in the mining work. . . ." The students did not feel that this orientation diminished their education. On the contrary, the manual training school provided just what they needed:

> Here . . . was found the most intense interest in the shop-work. There seemed to be a feeling of ownership in every piece of work being made. . . . They have been reared in the midst of a world by itself, for Calumet is not like any other city in Michigan. All the work they had ever seen was in the line of mining. Their fathers and brothers are miners, and they too expect to be connected with the industry in some of its departments. All these conditions give to their work a reality not found in the ordinary manual training schools.[43]

Calumet and Hecla and the other major mining firms manipulated local schools to serve their own ends. The schools served as incubators. They took in children from a dozen or more ethnic groups, and each year produced a new brood of English-speaking and -writing mine workers. The companies, however, did not have to force their particular view of education on local parents and students, because these parties largely shared the companies' view. They believed in a practical education and a quick dose of reading and writing. Many working-class parents wanted their sons to leave school early, so they could work for the mining companies and contribute to their family's income for eight to ten years before getting married in their mid-twenties. In the context of their own time and place, the schools fitted in. They had limited range in terms of what they did, but at the more prosperous mines, at least they did it well.

At smaller mines all social institutions suffered in comparison with those sheltered by a Quincy, Copper Range, or C&H. In terms of paternalism on the Keweenaw, the larger mines exerted more social control over their communities, yet they were the preferred employers. These sterner "fathers" had the deeper pockets and provided more benefits, such as libraries to accompany the schools.

The adjacent Pewabic and Franklin mines, acting in concert, created two early libraries in 1861. The president of both companies, E. L. Baker, exhibited "thoughtful kindness" in donating some personal books to form the nuclei of a School Free Library and a Miners' Free Library. Pewabic's *Annual Report for 1861* solicited stockholders "to contribute to this most useful object any works of interest which they can conveniently spare."[44]

Despite this pioneering effort, well-stocked libraries remained rare until the turn of the century and thereafter, when Calumet and Hecla (in 1898), Copper Range (in 1903), and Quincy (in 1916–18) established libraries for employees' families. It is interesting that in each case the library shared its attractive stone or brick building with small meeting or lecture rooms and with bathing facilities. Within the paternalistic structure of the mining companies, this odd coupling of facilities made social sense. The library served as a company-sanctioned focal point of wholesomeness, of the ideal of having a clean body and a clean mind. Here workers and their families could use their free moments in ways that brought them personal benefit.

At Quincy, president William R. Todd vetoed the idea of coupling a library with billiard or card rooms, or with a bowling alley. He did not see Quincy's new facility as a place for "amusement." He did not want such a place on the location because it would "affect good discipline and interfere with work." "This matter of amusement," he wrote Charles Lawton, Quincy's agent, "is entirely personal and a thing of business I do not believe the company would be justified in encouraging, it promises only trouble without any good results." He wanted a library, bathing facilities, and quiet, civilized reading rooms—all under the watchful eye of a church-going teetotaler who would banish card playing and other frivolous pursuits, including the intermingling of young men and women.[45]

At Painesdale, the Sarah Sargeant Paine Memorial Library served workers at the Champion, Trimountain, and Baltic mines and functioned as a social and cultural center. William Paine paid for the library (named in honor of his mother) and then deeded it over to his company, Copper Range.[46] William Paine was less paranoid about "amusements" than Quincy's president, and the library hosted many events. Of course, they were of a proper kind. The Painesdale Social Club held its May party at the library in 1904. Members listened to vocal soloists and to musicians on piano, mandolin, and cornet. Then the evening's speaker, O. J. Larson, perhaps the most prominent Finn in Houghton County politics, presented his oration on "American Citizenship." Finally, the members got up from their chairs for the concluding dance program, which included such local favorites as the "Trimountain Two-Step" and the "Stamp Mill Waltz."[47] Surely such an evening bred no socialistic or anti-company sentiment.

At Calumet, the library, with its highly decorative exterior masonry and handsome interior, became one of the gems of the community. The woman who headed the library oversaw its bathing facilities in the basement, its sizable collection of books, magazines, and newspapers, and its lecture rooms. She consulted with James MacNaughton and Alexander Agassiz on which books to buy, which towels were best, and how they should be monogrammed with the company's initials.

The entire facility was well stocked and well used. The downstairs bath strictly segregated the sexes: men bathed in one area with 18 tubs; women and children in another with six tubs. In all likelihood, the men had a common locker room for changing clothes, while the women and children disrobed in private changing rooms at each tub. Initially, C&H made no charge for the facility; later, men alone had to pay 2½ cents per bath. In 1906, men took 24,675 baths courtesy of their company, while women took 7,951. It is not surprising that Friday and Saturday nights were prime times for bathers, and attendants had to turn many men away when the hot water ran out.[48]

When the Calumet and Hecla Library opened in 1898, it contained 6,800 books, including some 3,000 sent out to the mine from Boston. By 1900 it had 25,000 volumes "in more than a score of languages," and the shelves held 35,000 books by 1913. Claude Rice, writing on labor conditions at C&H in 1910 for the *Engineering and Mining Journal,* noted that "the books in the foreign languages are made up of the best of their native literature, books on the history and government of the United States, and as many biographies of the great men of this country as can be obtained." The selection proved popular. In 1912, there were 8,700 individuals who held C&H library cards. Adults checked out 6,937 books in foreign languages and 74,696 in English. Children took out some 63,000 titles, and 24,506 used the library reading rooms.[49]

Nominally, the Calumet and Hecla Library was controlled by the library board of the Calumet Township school board.[50] In reality, if any controversy regarding the library arose, the president and general superintendent decided it. Unquestionably, the local population benefited from its access to the C&H Library—yet the institution was a quiet means of social control. The employee at the card catalog perhaps did not sense this control, but it was there nonetheless, in terms of what he could and could not get at the company's library. In 1912, James MacNaughton and Quincy Shaw, Agassiz's successor as president, had all the normal problems of overseeing a large company, plus the very special problems of the declining copper values found in the Calumet conglomerate lode and the troublesome introduction of the one-man drill. As paternal employers, they were also plagued by their own insistence upon social control. Paternalism meant that at the mine location, *everything* mattered. They could not be careless and sow any seeds of unrest in their own backyard. Only in a highly paternalistic company would the president and general superintendent worry about the contents of a book on woodworking.

Marie Grierson, the librarian, had asked Quincy Shaw to find a book entitled *Woodwork Safeguards,* put out by the Aetna Life Insurance Company. Shaw found the book, but in an era characterized by debates over workmen's compensation and worker safety, he did not send the book directly to the librarian. Instead, he wrote to James MacNaughton:

> I have no doubt that there are many valuable points to be gained and perhaps some of them to be adopted but I question the advisability of putting such a book in the Library at the present time unless someone goes through it, more or less competent to judge the wisdom of the conclusions and the attitude in which the book is written.[51]

MacNaughton concurred. The book would not go to the library unless and until it had been "looked over carefully" and passed inspection.

The mining companies sanctioned other types of entertainment and recreation, besides reading. Most mine and mill towns boasted a musical band that was often company-supported. At Quincy, the company provided a cushy job for the director of the "Quincy Excelsior Band" and paid him an additional wage for his musical services.[52] The bands often played free outdoor concerts in the summer, and many local organizations booked them for dances and other special events. The Copper Range Band was made up almost entirely of Finns, but they performed for numerous ethnic societies whose members were Austrian, Lithuanian, Polish, or Italian. The Copper Range Band played for pay, without discrimination. They entertained both socialist clubs and Republicans alike.[53]

The mines also supported athletics and often set aside a portion of their lands as sports fields. C&H had baseball diamonds in a large common near the mine, and Tamarack laid out a lot for cricket. In the winter, the companies flooded open ground to create skating rinks. At least two of the mines, Calumet and Hecla and Baltic, operated bowling alleys for employees. C&H also erected a 26' by 40' indoor swimming pool and contributed generously to the erection of a new YMCA in Red Jacket in 1908.[54]

In the case of athletics, C&H believed that what was good for the men was also good for officers and influential friends, albeit in separate facilities. In 1902–3, the company took one of its larger houses near the main office building and converted it into a club house "for officers and other persons connected with the company." Equipped with bowling alleys, a billiard room, handball courts, reading and social rooms, this structure became headquarters for the Miscowaubik Club—an all-male, elitist organization whose elected membership was set at 250 Keweenaw residents and 50 nonresidents. This is where Agassiz, MacNaughton, managers, doctors, lawyers, and officials from other local mines came to play.[55]

Neither the well-to-do nor the common workers were limited to the amusements provided by the mining companies. Red Jacket, Laurium, Lake Linden, Hancock, Houghton, and South Range offered entertainment and recreational opportunities of their own. In these villages, numerous ethnic societies erected halls for meetings and social events. Calumet and Hecla's Miscowaubik Club had its counterpart on Portage Lake, in the equally exclusive "Houghton Club." In support of music and theatrical events, at least 21 theatres operated in Houghton County before 1910, including the Lake Linden Opera House, the Lyric Opera House in Laurium, the Kerredge Theatre in Hancock, and the Calumet Theatre in Red Jacket.[56] The last, a municipally owned structure erected in 1900, was a grand and glorious facility seating 1,200. Here, national touring companies and bands staged professional performances in a fitting environment. Between 1900 and 1910, the Calumet Theatre booked over 800 events, including John Philip Sousa's band, popular American melodramas such as *Uncle Tom's Cabin*, 16 different Shakespearean plays (including *Hamlet, Macbeth, The Merchant of Venice,* and *The Taming of the Shrew*), the operas *Faust* and *Carmen,* and Ibsen's dramas *Ghosts* and *A Doll's House.*[57]

In terms of music, entertainment, and culture, the Copper Country had come a long way in 50 years. In the winter of 1853–54, one of the big social events had been a "dancing party" held at the mining captain's double log house at the Ripley mine on Portage Lake. Guests trekked in on snowshoes, coming from as far as Ontonagon to the south and Eagle River to the north. The crowd spilled over; those not dancing stood outside in the snow. Everyone thoroughly enjoyed the feasting and fun, even though the band consisted of only one violin played by the Webster mine's blacksmith.[58] By 1902, residents could ride the Houghton County Traction Company's trolleys out to

Electric Park, located on the route between the Quincy and Osceola mines, where bands and vaudeville acts performed in the pavilion, children rode the merry-go-round, and families picnicked.[59]

Summer and winter sports flourished on the Keweenaw. Hunting, fishing, cross-country skiing, boxing, and wrestling were very popular, as were the team sports of baseball, football, and especially hockey. Large arenas with inside rinks and spectator seating testified to the fact that the Keweenaw was one of the early centers of ice hockey in the United States. The schools fielded teams, and numerous adult leagues were formed. Some of the hockey teams achieved semi-pro or professional status. Local teams played each other, and also toured other northern regions.

The Copper Country was hard-working, hard-playing, and hard-drinking territory. Bars and brothels did not want for business. Compared with other mining districts, perhaps, there was a "marked absence of saloons" on the Keweenaw, but it was far from dry.[60] The mining companies banned liquor establishments and bawdy houses on their own property, but did relatively little to shut them down in neighboring villages. In 1907, in the midst of a local population of about 40,000, nearly 100 saloons operated in Red Jacket and neighboring Laurium. In 1910, Houghton County boasted 88,000 residents and 179 bars.[61] Houses of prostitution are less easily identified and counted, for obvious reasons. Operators did not register their businesses in city directories, and few were as forthright and honest as John Cabus of Hancock, who listed his occupation in the 1880 census as "Keeper of a House of Ill-Fame," or Laura Lytton (age 22), Anna Whitcomb (21), or Anna MacDonald (19), Cabus's employees, who listed their profession as "Prostitute."[62]

When made aware of indiscreet behavior on company property, mine officials quickly put a stop to it. In 1908, Frank Denton, agent at Copper Range, threatened to terminate the lease of Frank Santoni, unless he quit selling beer out of his company house and stopped encouraging visits by disorderly men.[63] Companies rarely reached beyond their own property lines, but in 1888, Quincy's agent, Samuel B. Harris, took it upon himself to write to a Hancock resident:

> Whereas a complaint had been made to me that you are harboring on your premises certain women of bad character,—particularly one known by the name of "Black Mag":—and that you are keeping a disorderly house, much to the disgrace of yourself and the annoyance and danger of your neighbors,—you are therefore hereby notified to put a stop, at once, to all such disgraceful proceedings:—and if any further complaint of a like nature is made of you, or your house, immediate steps will be taken to punish you to the full extent of the law.[64]

More typical, mine managers tended to look away from the "seamier" sides of life in adjacent villages, leaving oversight to local officials and police. In 1898, Red Jacket village had ordinances against disorderly persons, defined not only as "all gamblers, drunkards and tipplers" but also as fortune tellers, vagrants, prostitutes, and "all jugglers, common showmen and mountebanks who exhibit or perform for profits any puppet shows, wire rope dancing or any other idle show, acts or feats." Ordinances forbade gaming houses, dancing, or "boisterous entertainment" between midnight Saturday and noon on Sunday, and houses of ill-fame. Saloons were permitted to open at 6 a.m. and to remain open until 11 p.m., except Sundays, election days, and legal holidays.[65]

In 1898, John Ryan, the president of Red Jacket village, wrote Alexander Agassiz to complain about a C&H employee. James Renwick had been appointed a deputy sheriff at the company's request, for the purpose of protecting company property. But Ryan believed Renwick was overstepping his bounds. He had "been causing a good deal of trouble throughout the town all summer by making complaints against the better class of our saloon keepers." The "better class," it seems, served alcohol discreetly on Sundays, behind drawn blinds. But Renwick kept citing them for violations. Ryan asked Agassiz for help and got it. Agassiz instructed C&H's superintendent to inform Ryan that he was no longer to interfere with the village of Red Jacket in any way.[66]

The mining firms were not anxious to serve directly as moral police. They left that role to civil authorities, and in large measure to local churches, which they encouraged and aided from the start. Company directors and agents were not evangelical missionaries. They did not seek converts to any preferred branch of Christianity, or mandate that workers financially support or attend a given church. Rather, they accommodated the religious beliefs and customs their workers brought with them to the Keweenaw.

Managers and workers both sought benefit from churches. For workers, the church served as an important focal point of immigrant culture. Here, worshipping alongside fellow countrymen, they recaptured traditions and established some semblance of continuity between their lives in the old and new worlds. For managers, church steeples served as beacons that first attracted hard-working family men, and then helped get them rooted into the region. Ministers and priests also served the mining companies' interests by encouraging proper behavior. The attitude of Quincy's president, W. R. Todd, toward one particular clergyman was typical:

> We believe that Rev. Father Krone may prove a desirable resident and may be of some value to us in cases of misunderstanding with our men, and if he can do as he writes, "train the minds and hearts in the fear of God and in respect for authority, law and order" he should prove a good man to keep among us.[67]

In the 1850s, and perhaps the 1860s, companies were directly involved in erecting churches, bearing much if not all of the cost and supplying labor and materials. In that era the mines employed a relatively small and homogeneous labor force. Because a mine location included men from only a few ethnic groups, it needed only a few Protestant or Catholic churches, depending on its mixture of Cornish, German, and Irish laborers. The Minesota mine built one church in 1852. Since it doubled as a school and must have been plain and nondenominational in design, the building may have served more than one congregation. The company also erected, in 1857, a "substantial and comfortable" house for a resident minister. By 1860, three churches stood at the Cliff mine, Protestant Episcopal, Methodist, and Roman Catholic.[68]

The mining companies grew larger, their labor forces more diverse. One or two Protestant or Catholic churches no longer sufficed. Protestant denominations multiplied as Episcopal, Congregational, Presbyterian, Baptist, and Lutheran churches joined Methodist ones. Each denomination further subdivided on the basis of ethnicity. In Calumet alone, the Germans, Swedes, Norwegians, and Finns established separate Lutheran churches. The same pattern held for Catholics. In the 1850s, one priest who spoke English, French, and German accommodated all Catholics at the mines near Eagle Harbor. By the turn of the century, these ethnic groups had their own parishes and priests, as did the Italians, Poles, Croatians, Slovenians, and Austrians.[69] When an ethnic group was newly arrived and small in number, it often joined an already established church. But as soon as their numbers and resources made it possible, they established their own.

While early Christian settlers made do with meeting in a carpenter shop or school, later residents often raised large and impressive churches of stone, complete with carvings and stained glass. By this time, the mining companies had withdrawn from actually building churches. They no longer provided construction materials or bore the cost. Still, they supported the march of Christianity in more modest ways.

The companies made land available, through sale, lease, or donation, to virtually every church group that requested a building site. Relatively few churches stood on Quincy Hill above Portage Lake, but numerous churches located just down the hill in Hancock, on land once owned by the Quincy Mining Company. In 1908, Quincy's president, W. R. Todd, noted that "we have always thought it desirable to encourage the erection of churches, [and] in fact, I think every church located in Hancock was given free the ground on which it stands."[70] Quincy, by deeding land over, supported at least two Catholic churches; Congregational, Methodist, and Lutheran churches; and one Jewish synagogue. It also provided the land for the first Catholic and Protestant cemeteries in Hancock.

In the summer of 1898, Calumet and Hecla hurried the construction of a new company dock at Torch Lake by having the construction crew work on Sundays. Word of this breaking of the sabbath reached president Agassiz, who received "a good many complaints" from "people who dislike to see Sunday work going on." Agassiz wired superintendent S. B. Whiting to stop the Sunday work unless it was absolutely necessary, because "we do not want to antagonize anybody on that account."[71] In giving this order, Agassiz was being consistent with his longtime policy of having C&H support Christianity by accommodating the religious preferences of its large work force. By the turn of the century, 30 churches stood near the mine, "in which the Gospel is preached in English, French, German, Italian, Swedish, Norwegian, Finnish, Slavonian and other languages."[72] Calumet and Hecla, besides finding lots for these churches, contributed some $36,000 toward their building funds.[73]

Calumet and Hecla wanted to see its men and their families in church on Sunday morning. In support of this, Agassiz allowed for a great multiplicity of churches, even though they violated his personal preference for strong, centralized facilities and institutions. In 1905, he and James MacNaughton agreed that C&H workers had built about twice as many churches as necessary. If half as many existed, attendance would be the same, and yet each church would be better supported financially. Both men believed, for example, that it was "foolish" for workers to support two Catholic churches in Lake Linden. The Holy Rosary Catholic Church had just burned down, and its German and Irish members wanted to rebuild it. Agassiz and MacNaughton believed it wiser for Holy Rosary, instead, to combine with the French-speaking Saint Joseph's Catholic Church. They recognized, however, that the two groups wanted no merger, so they did not push the issue. C&H had contributed $2,000 to Holy Rosary upon its founding and gave an additional $1,000 to help rebuild it.[74] C&H quietly and privately bowed to the workers' desires to hold on to their own traditions.

Although Calumet and Hecla bent to the workers in allowing them numerous, diverse churches, in other matters it held firm. The company's support of religion had practical limits. C&H was absolutely committed to maintaining hegemony over its property, so new churches could not receive, through purchase or donation, deeds to their property.[75] Instead, they settled for ground leases for their building sites. Also, churches could not expect regular offerings from the company. C&H believed that those who founded a church had to sustain it. Only in exceptional cases of need, such as the loss of the Holy Rosary Catholic Church to fire, could a minister or priest pry loose additional monies from Agassiz or MacNaughton. When the Swedish Methodist Episcopal Church petitioned for help in replacing its heating plant, MacNaughton replied that "after a church is once built it must of necessity be kept in repair and operation by the congregation. If we once let down the bars in this respect we will be importuned on every occasion by each of the churches in this locality."[76]

Because of Calumet and Hecla's reputation as a generous and wealthy employer, it received a barrage of charitable requests, which its top officials often had to deflect. MacNaughton, besides having to say "no" for his company, often had to say "no" for himself as well. Paternalism put additional pressures on the mine's top officials to contribute to worthy causes, including churches. MacNaughton, for one, chafed at the many solicitations.

When the Reverend P. W. Pederson asked Calumet and Hecla's general superintendent for a personal contribution of $150 to help defray the costs of painting and repairing the Norwegian-Danish Methodist Episcopal Church, James MacNaughton replied that such a donation was "out of the question." He explained that "the demands of this character that are made on me are probably more numerous than those made on any other individual in this community." He already contributed to "three English-speaking denominations," and that was enough. The Reverend Mr. Pederson begged to disagree, and tried to shame MacNaughton into a contribution: "Of course you are expected to donate to churches and charitable institutions according to your means. . . . In my opinion a man who is so selfish that he can not give a part of his earning for the welfare of his social surroundings is not good enough to dwell among civilized men." If MacNaughton could not contribute $150, the church would kindly accept whatever contribution he chose to make. James MacNaughton still chose not to make one. He wrote the good Reverend back:

> As to your remarks on the obligations of a citizen to his fellow man and to society in general, they fully coincide with my own views in the matter, otherwise I should feel indebted to you for your observations. It is a pleasure to me to note that you approve of my donations to English-speaking churches. However, the added pleasure of your approval should I make a donation to your church is one I must forego.[77]

The Reverend Mr. Pederson had not learned the system at C&H. Officials meant "no" when they said it, whether speaking as individuals or for their company. These men set their own limits to paternalism and would not cave in to external pressure. No churchman, like no dissident trammer, could challenge, scold, or lecture them and expect to win.

11

The Social Safety Net: Patients, Paupers, and Pensioners

> I think the time is coming, if it is not already here, when the employees will expect from the hands of their Company special medical and surgical service. Your working men of today would not be content to live in the small log houses that were furnished them forty or fifty years ago, nor would they today be satisfied with the mine physicians that practiced forty or fifty years ago. The standard of living has been raised.
>
> A. B. Simonson, Calumet and Hecla's chief physician, 1913

Compared with workers at smaller, more isolated companies, by the 1870s and thereafter the men employed by the larger Houghton County mines were distinctly advantaged. They found steadier work. They traded at a wider variety of stores and partook of more diverse amusements. They had a much higher likelihood of finding a church that met their preference exactly, offering the right denomination and language. Also important, they availed themselves, in times of need, of a more extensive social safety net, erected in part by their companies, and in part by fellow workers. The more successful companies had the wherewithal to provide employees' families with greater protection against the harsher shocks of life. Company-sponsored medical services made up a key part of the mines' social safety net. All operating mines provided at least a minimal level of medical care. Larger, more prosperous companies could afford to hire several physicians and build more comprehensive medical facilities.

The companies included medical service as part of their paternalism from the 1850s on. They did so, in part, because their early employees expected it; the provision of a mine surgeon was a Cornish tradition transplanted to Lake Superior.[1] The mining companies, however, willingly accepted the obligation to provide medical treatment to settlers in new mining locations. They saw the employment of doctors as a humanitarian move that also made sound business sense, in terms of attracting laborers. One "laborer" who definitely needed inducements before coming to the mines was the American-born doctor himself. Physicians did not flock to the Keweenaw as medical missionaries bent on healing the sick in this newly developing corner of America. They came because the mining companies offered salaried positions, free housing, and high status. Companies treated their doctors as important officers.

The companies hired the doctors, built and owned the hospitals and dispensaries, ordered the medicines, and purchased the medical instruments. They operated compulsory health insurance programs, administered by them but largely funded by mandatory, monthly payments from the men. By the early 1850s, the standard fee was set that remained in effect for upwards of 60 years: single men contributed 50 cents per month; married men, $1.[2] In return for this payment, workers and their families received outpatient care and medicines without additional charge.

The Minesota mine had a single "qualified physician and surgeon" in residence by as early as 1852, and the Cliff mine employed at least two doctors by 1860. Small or very new companies, which alone could not support a doctor, joined together to share one. At the small Keweenaw and Ontonagon county mines in the early 1850s, "a physician is supported by way of two or three mines, within a convenient ride." When the Pewabic, Franklin, and Quincy mines developed on Portage Lake in the late 1850s and early 1860s, they initially shared a single doctor.[3] Usually, though, a company obtained its own physician once it employed more than 200 or 300 men.

Some doctors treated company employees only; others established private practices on the side. Quincy generally restricted its physicians' outside work, believing that it opened the door to workers' complaints about getting less than full-time attention from their doctors. The company wanted its physicians to concentrate on the needs of the population that supported their salaries.[4] Calumet and Hecla followed the opposite policy, but for equally valid reasons. It encouraged doctors to have an outside practice, because "the mine doctor who is working for tips on the outside has to keep . . . more in touch with modern practice . . . , and this additional knowledge and work on his part must help in his mining practice."[5]

Within a few years of hiring its first physician, a mining company typically built a wood-framed hospital for surgical operations and treatment of the injured. The Minesota mine erected one of the region's first hospitals over the winter of 1857–58. Quincy obtained its first physician who worked solely for its employees in 1859, and between 1862 and 1865 completed a hospital with room for 35 patients. Pewabic and Franklin jointly erected a hospital in 1862, when "Humanity, justice, and the true interests of the companies all demanded that this good work should no longer be delayed." In 1870–71, C&H turned what was originally intended as an agent's house into its first hospital. Early in the twentieth century, after Copper Range had been in operation for about five years, president William Paine wrote his general superintendent: "I feel that our operation is now of such a magnitude that we should have a hospital—and hospital service, which shall be first class in every particular."[6]

The Copper Range hospital for employees at the Champion, Trimountain, and Baltic mines opened in 1906.

The mining companies seemingly funded their hospitals on their own, without tapping the monthly fees paid in by the men. Perhaps for this reason alone, mine hospitals did not provide absolutely free care. Hospital patients did not pay for medicines or treatment, but they did pay for their ward or private room. However, in no instances did the mine hospitals charge exorbitant fees.[7] Their daily charges were less than those set by several public and Catholic hospitals in the area.

Calumet and Hecla, for one, claimed it ran its hospital at a loss and never recouped its construction costs. From 1871 to 1914, fees paid to the C&H hospital reportedly fell $463,000 short of its operating expenses. This shortfall had medical implications. Because C&H covered any deficits, the mine superintendent served as the hospital's gatekeeper. To check costs and to control the demand for doctors, nurses, rooms, and facilities, he limited hospital admissions. He let in patients in need of major surgery and the victims of serious mine accidents, but kept many others out. Persons having a serious illness or disease, and maternity cases, were generally excluded from hospital care and treated at home. As MacNaughton noted, "At times, if we have room, we take care to extreme sickness but at no time do we bind ourselves to do so."[8]

C&H physicians did not have full authority to decide who or what to treat. Agassiz and MacNaughton sometimes made important decisions for them that were informed not by medical science but by economic considerations, by notions of loyalty (owed or not owed) to certain classes of people, or even by morality. On direct orders from Alexander Agassiz, C&H physicians did not treat venereal diseases.[9] Agassiz apparently viewed venereal disease as a self-inflicted wound incurred off-the-job by persons leading a profligate life. His company did not aid those suffering the wages of sin. Similarly, Agassiz was none too keen on treating liquored-up men injured in street fights. In 1903, the mine's chief physician felt obliged to argue that since the company did not make temperance a condition for employment, it should not refuse to treat the wounds of intemperate combatants.[10] While Dr. Simonson won this appeal, he made no attempt to challenge Agassiz's edict on venereal disease.

Early company medical departments were understaffed and hard-pressed to treat the many illnesses and injuries that presented themselves. The physician served as general practitioner, obstetrician, and surgeon. When Dr. Fischer arrived at Quincy in 1890 and opened a drawer containing dental forceps, he realized "that I was to be dentist as well as doctor." An even greater surprise must have befallen one of Fischer's predecessors at Quincy, Dr. A. A. Shepard, who found himself thrown into the breech as company veterinarian. In 1874, Shepard ordered books on veterinary practice, veterinary surgery, and horse diseases.[11] Perhaps this same sort of pressure, to have to reach beyond one's competence, is what finally did in yet another Quincy physician in 1897. The physician, with the sadly appropriate name of Dr. Downer, was a "prince of a man" and an "ideal mine doctor," who always knew "when to rave and when to smile." Big-hearted and sympathetic, Dr. Downer was revered by the residents on Quincy Hill. Dr. Downer, however, was also a drug addict who took too much of his own medicine. Quincy's agent finally fired him when "he got on a 'toot' again and drugged himself into imbecility."[12]

Near the turn of the century, medical service improved as the number of doctors multiplied, specialists arrived, and hospitals and equipment became more modern. At the end of 1897, C&H opened a new hospital "fitted with the best laboratory and surgical apparatus procurable in Europe and America." The company's medical department included eight physicians, two interns, three pharmacists, and four nurses. In the 11 months following the hospital's opening, it recorded 2,795 patient days, while company doctors received 52,000 office calls, made 35,000 house calls, and wrote 78,000 new prescriptions. The doctors also met about 4,000 patients off the C&H payroll, including paupers "who are, by special arrangement between Calumet and Hecla Mining Co., and the Superintendent of the Poor, allowed to obtain their medicine at the Company's dispensary."[13] C&H received only a few hundred dollars per year for treating the poor, and its doctors used the money to purchase books for their medical library.

Despite its neglect of venereal diseases, Calumet and Hecla evidenced considerable concern over communicable diseases. Its medical staff functioned as a public health department, and from at least as early as 1897, the mine's chief physician provided the general superintendent with monthly "Sanitary Bulletins." These bulletins pinpointed on a village map all cases of scarlet fever, typhoid fever, tuberculosis, meningitis, smallpox, and diphtheria that company doctors encountered, as well as whooping cough, measles, chicken pox, and mumps.[14] The reports made it clear when any epidemic was under way, such as an outbreak of scarlet fever in 1900, or of diphtheria in 1910.

The sanitary bulletins sometimes prompted special action to protect the public health. By early June 1907, C&H doctors had reported nine cases of typhoid. The company, believing its water supply might be contaminated, sent samples to the University of Michigan for analysis. James MacNaughton also had the township health officer publish notices in local papers warning citizens to boil all drinking water as a precaution. By mid-June, 30 cases had resulted in six deaths. Since the tests on company water had shown it to be uncontaminated, C&H concluded that unsanitary surface conditions were at fault. The company organized crews that cleaned up accumulated filth and disinfected certain locales by spreading lime on the ground.[15]

The oddest public health "menace" to befall C&H was its single case of leprosy encountered in 1910–11. Leprosy was a greatly feared and misunderstood disease, and victims were viewed as dangerous social outcasts. A Norwegian, Marelius Jensen, lived with his wife and four daughters in the Centennial Heights neighborhood. Calumet and Hecla doctors first diagnosed Jensen's leprosy, and the company deemed this case so unique and important that it called in Dr. Shumway, of Michigan's health department, and Dr. McClintock, of the federal health service, to confirm the diagnosis. When these doctors left Calumet, James MacNaughton had a certified case of leprosy on his hands, the only case of its kind for years in all of Michigan.

Jensen's leprosy highlighted the complex and pivotal role that James MacNaughton played in local society, because of his position as general manager of a highly paternalistic company. As an individual, McNaughton showed some real compassion for Jensen and his family. Since Jensen could not work, MacNaughton informed the County Superintendent of the Poor that "it is going to be up to you to see that he and his family have the necessities of life. . . . I would thank you therefore to get in communication with him to the end that he is liberally supplied with what they may need." He urged that the case be handled with discretion: "It is bad enough for this man to be afflicted with leprosy and for his wife and children to realize that they may possibly have the disease, without having them pointed out as lepers by everyone, and I am asking the Doctors as well as yourself that the question of isolating them and indeed the question of the case in general be kept quiet."[16]

Personally, MacNaughton did not want the family to suffer. As a mine official, he did not want a public health scare on his hands that would agitate the labor force and blight his company's reputation. Finally, McNaughton wore yet another hat. He was a supervisor of Calumet Township. Once this case inevitably became public knowledge, he had to respond to it as a public official, and interact with other concerned officials, such as the health officer of Calumet Township and the local school board. Since these officers also tended to be C&H managers or bosses, it was never clear if Jensen, his wife, and his school-age children represented a public problem or a company problem. In a sense, the issue was academic, for either way the buck stopped at James MacNaughton's desk.

To MacNaughton's credit, he handled the Jensen case in a way that protected the public health and yet did not pander unduly to public fears. He got the local Superintendents of the Poor to provide food and clothing, but C&H continued to provide the Jensens with housing. To effectively quarantine Jensen, the company moved him to an isolated house about a mile out of town. His wife and daughters lived in a second house right by his, so they could care for him and yet minimize their own risks of contracting the disease.

MacNaughton viewed his housing of the leprosy patient as only a temporary solution to the problem, but it continued for at least a year and a half. He sought to have Jensen placed in a proper institutional setting for such cases. This need became particularly acute when Jensen became "ugly and morose in disposition," and threatened his wife and children with violence.[17] The problem was that nobody wanted to take Jenson off MacNaughton's hands. Houghton County was not prepared to do it. Michigan had no "leper colony," and a C&H attorney tried but failed to find him a home in other midwestern states. The federal government did not come forward with any help. While Jensen continued to live in his C&H house, his case spawned touchy legal and medical issues: Could they enforce an absolute quarantine and restrain him within the house, if necessary? When the schools opened in September 1911, the focus shifted to his daughters. Jensen's case had by then become notorious, despite MacNaughton's desire to hush it up, and public sentiment ran "very strongly against" allowing the daughters to enroll in school.[18] The school board, in fact, voted to ban their attendance. What was the real risk of their having or spreading the disease? Did the school board have the authority to institute such a banishment?

To resolve such issues, MacNaughton and other C&H employees, acting as company officers or as public officials, corresponded with Michigan's attorney general and with the secretary of the state board of health. Unfortunately, the downstate legal and medical experts failed to offer any definitive advice that MacNaughton could act upon. They were of little more practical help than the Christian Scientist in Iowa who read of the case in a newspaper and recommended that "you turn this most worthy case over to God's protection."[19]

A year and a half after his leprosy was diagnosed, Jensen apparently still resided locally in isolation, but his family had at least been allowed to return to Calumet, even though the children were not allowed back in school. MacNaughton seemingly arranged for their return to town and tried to make it possible for Jensen's wife to earn a living. Such gestures, it appears, were not universally endorsed, as a letter from the "People of Houghton Co." made clear:

> . . . you gave her the permission to move to town to spread the disease, rented her a house with dancing quarters so all of our youngsters goes there learning to dance and get that disease. Within few years all of us working class will have it if yous won't send her back, for the children goes to their father and what is the good of it. Children don't go to school, run from house to house spreading the leaphrasy. If the people round here get it will make it hot for the co.[20]

The leper was a unique social problem; others were more ubiquitous. For chief officials at paternalistic mining companies, paupers, like those who queued up at Calumet and Hecla's dispensary, posed such a problem. The companies sought stable, harmonious, hard-working communities. In the work they offered, and in the communities they ran, there was a suitable spot for vigorous men imbued with stamina and the work ethic, and for their families. But the system offered up casualties, men who no longer fitted into a productive niche.

No employees, not even the hardest working ones, remained young forever. They grew old and at some point could no longer perform arduous labor. Others lost their function when a rockfall or machine snatched away an arm or a leg, or when a premature blast stole their eyesight. Some were maimed in an instant; others weakened slowly due to illness or disease. Still others died, leaving behind widows and children. The mines relied on active men, but their communities produced and included old, maimed, and ill men, and the widows and children of the dead.

In stretching out a social safety net to catch the unfortunates and break their fall, company officers were themselves caught in a complex web of contradictory emotions and pressures. They mixed compassion with a concern for costs. Managers were confident when playing the role of employer, yet uncomfortable when playing their brothers' keeper. John Daniell, agent of the Osceola mine in 1878, expressed the ambiguity he felt when handling his company's casualties:

> A family left the location within a month. The man had not worked for a year. Sick. We have a case now. The man sick. We employ the boy for charity sake. We always find cases of this kind, and a person must be more than flinty-hearted, to be unaffected, but if in charge of the work cannot help feeling relieved when they are taken off his hands.[21]

The mining companies most definitely did not want their villages to fill with impoverished families, headed by broken men or widows. When a family was in need, managers often felt a humanitarian impulse to offer assistance. At the same time, however, they often felt compelled to restrain their largesse.

Mine officers rarely believed their company was responsible for a family's misfortune. If a man's family suffered because he was a "whiskey heart," or alcoholic, that surely was not the company's doing. If a mine accident maimed or killed, in most instances the agents believed the blame lay with the victim himself, or with his negligent co-workers. Since the companies were not responsible for most accidents, any more than they were responsible for workers falling to tuberculosis or pneumonia, it followed that they were not absolutely obliged to render assistance.

Many times, mine managers wanted to offer charitable aid, but were leery of the consequences. They feared that each act of charity, as it became known, might spawn a host of entreaties for similar aid. Personally, mine agents were not anxious to entertain pleas for assistance from needy families, or from their ministers or priests. As a matter of company policy, they feared setting precedents and raising expectations. If they gave to one, they might be asked to give to all. Agents were concerned that small, individual acts of charity not grow to become broad-based company welfare programs.[22]

In the early history of the copper district, neither the mining companies nor the local, state, or federal governments conducted any formal benefit programs. Aside from medical treatment, virtually no social safety net existed for men and their families who were jeopardized by unemployment, illness, accident, or death, save that which the workers raised themselves. As a buffer against catastrophe, by at least the mid-1850s mine employees began forming voluntary associations to collect regular monthly dues and mete out financial assistance in time of need. In 1857, the earliest known "Miners' Club," probably founded by employees at the Minnesota mine, had 216 members. The club provided $12 per month to men unable to work due to illness or injury, and in the event of a man's death, his widow collected $2 from each club member.[23]

Such mutual aid and benefit societies remained important at the Lake Superior mines well into the twentieth century. The members of a given society typically worked at the same mine, or at neighboring mines. They shared the same ethnic background, spoke the same language, and probably attended the same church. Between 1865 and 1910, at least 50 mutual aid and benefit societies formed in Houghton County alone.[24] The Calumet and Hecla Scandinavian Benevolent Society accepted members in and around Red Jacket who were Scandinavian by birth or descent; who worked as miners, mechanics, or laborers; who were under 40 years of age; and who were in good health and of good character. The Calumet Polish Roman Catholic Stanislaus Koslki Benevolent Society specified church membership as one of several eligibility requirements, as did the Slovenian and Croatian society formed at the Baltic mine. Members there had to be Roman Catholics who received communion at least once a year and attended the funerals of all deceased brothers. They also had to eschew membership in organizations condemned by the church, such as socialist clubs.

Virtually every ethnic group on the Keweenaw formed several localized aid and benefit societies: Cornishmen, Irishmen, Scots, Poles, Germans, Finns, Swedes, Italians, Croatians, Austrians, and Lithuanians. Most charged an initiation fee, plus a monthly fee of about 50 cents. Their monthly benefit payments, running from about $12 to $15, usually started after a specified period of incapacity, such as one week, and continued for a given duration, often one year. The benefit societies also paid lump-sum death or burial benefits, running from one to several hundred dollars.

Because workers formed their own benefit societies (and many men belonged to more than one), mine officials felt little pressure to organize and fund company programs. Some, such as Copper Range, assisted their men in managing a benefit society.[25] Copper Range collected and banked the monthly fees established by a miners' club at Champion, Trimountain, and Baltic, but contributed no company funds to the treasury. The mines, in fact, were a bit leery of the proliferation of benefit societies. If several different societies contributed to the welfare of an ill or injured worker, he might receive more while in bed than at work—and that jeopardized the work ethic and encouraged malingering.[26]

Only one company, Calumet and Hecla, operated a benefit society and matched the contributions paid in by the men. Started in 1877, the C&H aid fund paid $25 per month (for up to eight months) to men incapacitated by illness or accident; $300 for a disabling injury, such as the loss of a limb or of eyesight; and $500 upon accidental death while at work. From its founding through 1911, the C&H Aid Fund paid out $1,240,000 in benefits.[27]

In the early history of the fund, the benefits paid out were less than the monies put in by the company and the men. By 1894–95, the aid fund had about 3,200 members, who received a total of approximately $40,000 annually in death, sickness, and accident benefits. Meanwhile, the aid fund account showed a surplus of about $100,000. This surplus prompted a review of the policy of providing benefits for only eight months to seriously ill or injured workers. Alexander Agassiz, instrumental in establishing the aid fund, also oversaw its expenditures. He championed fiscal restraint and control, and in 1894 opposed any augmentation of benefits:

The aid fund never was meant to help men indefinitely. It was purposefully limited and the aid is not a question of what we can afford to do. It is a question of [a] contract between the men and the Co. The fact that we have a bal. [balance] has nothing to do [with it]. We have for ten years spent over $13,000 a year more than we ought to and I propose to run the aid fund in [accordance with] business and not sentimental principles.[28]

In accordance with Agassiz's business principles, much of the aid-fund surplus went into Calumet and Hecla stock shortly before the company entered a highly prosperous period. C&H paid a record $2.5 million in dividends in 1896, and the figures rose steeply from there, reaching a high of $10 million in 1899. Agassiz still thwarted any attempt to enhance the benefits paid out by the aid fund, but he did recognize that its treasury was flourishing. So while C&H profits ran exceptionally high, for three years the company maintained the aid fund while suspending the monthly contributions from its men.[29] At the same time, the company temporarily suspended workers' contributions to the medical/hospital fund. Over these three years, the payment suspensions spared a married worker a total of $54 of monthly fees—while Calumet and Hecla paid out $22 million in dividends. Indeed, the company could well afford to be generous in the late 1890s.

The Calumet and Hecla aid fund came with a few strings attached. The fund gave employees a measure of security. It gave the employer something, too: a source of information, and an occasional lever to work its will on a given family. Aid fund payments became part of a man's personnel record, and managers used this information to gauge his commitment to hard work. If an employee tapped his aid fund too often, he became suspect.

In March 1899, Frank Valsuano appealed to S. B. Whiting for reinstatement to his C&H job. After many years of company service as a timberman and then watchman, he had been fired for falling asleep on duty. Whiting had personnel officers run a check on Valsuano. They came back with information on his exact dates of employment, his various jobs, and his aid fund record.[30] Valsuano did not get his job back. One thing that hurt his case was the fact that in 13½ years he "was on aid about 200 days—an average of 15 days a year." Over three-fourths of his aid days were due to sickness, not accident. Frank Valsuano's record did not speak well of his dedication to work.

The aid fund's death benefit became an issue in negotiations between James MacNaughton and Christian Maier's widow, who occupied a company house. C&H knew that employee relations suffered if it did not treat widows kindly. The firm did not like to take direct actions against widows, such as evicting them, but managers still liked to free themselves of widow problems whenever possible. In the case of Mrs. Maier, her husband had worked 27 years for Calumet and Hecla before becoming seriously ill. He had had one operation and was scheduled for another. Unable to cope with his ailments, he fell victim to depression. Finally, one night he got up from his sick bed, took his gun upstairs in his company house, and "done the dreadful deed."

Mrs. Maier received $700 from two noncompany benefit societies to which her husband had belonged. The C&H aid fund, however, balked at awarding her its normal death benefit, because of the suicide. Mrs. Maier reasoned that since illness prompted her husband to take his own life, the benefit should be paid. James MacNaughton received her request and arrived at a solution that side-stepped religious or moral considerations, and made, he thought, very good business sense. He had his land agent inform Mrs. Maier that C&H would pay the death benefit, just as soon as she vacated her house and returned the lease and key to the company.[31]

While other mining companies lacked formal benefit or welfare programs, they undertook ad hoc charitable measures and espoused that "no person living on the company's land . . . will be allowed to be in dire necessity." This assistance took several forms. In the event of a fatal accident, the company might initiate or contribute to a special charity drive to aid the victim's dependents. In 1882, a rock car struck and killed John Harbeke, a recent immigrant who was sole supporter of an aged mother and blind brother, left behind in Germany. Upon learning of the family's predicament, the Franklin mine's agent, Johnson Vivian, started a subscription fund on their behalf.[32]

In general, however, companies did not believe themselves responsible for the perpetual care of a family just because the father once worked for them. Families were to support themselves through work and effort, and not throw themselves on the mercy of the company. Unfortunate families that begged for charity, or demanded aid as if it were their right to receive it, actually hurt their own cause. The companies preferred to help those who helped themselves.

In 1894 the Central mine shut down and threw its entire labor force out of work. Alexander Agassiz took note of the closing and wrote his general superintendent, "It's too bad about the Central and I am very sorry for the men. I wish times were better and we could take on some of their men because they must have a lot of excellent men with nothing to do."[33] Under Agassiz, Calumet and Hecla consistently held hard-working family men in high regard, and so did other companies. They honored work, seeing it as the source of a man's dignity. They honored the husband, seeing him as the rightful head of the household. So when illness or injury jeopardized a man's self-worth and his ability to serve as a provider, mine officials propped him up, assuming, of course, that he wanted to help himself as much as possible.

Calumet and Hecla reserved slots on its payroll for men who were blind, "decrepit," or crippled. They served as "scavengers," cleaning up the location; as keepers of the company's 15 or 20 dryhouses; and as flagmen, gatekeepers, or watchmen. C&H reserved slots for several blind ex-miners at a broom factory, which it set up in 1903–4 at a cost of $1,000. The shop produced about 500 dozen brooms per year, used throughout the company. As Agassiz noted, "Better have them work than pay them [charity] to support them."[34]

If a man could not handle even "a cripple's job," the company put one of his sons to work as soon as possible; the boy served as a surrogate for his father. After Michigan passed legislation against child labor, James MacNaughton proved generally strict about not hiring underage boys. Still, he made exceptions, particularly when a boy had to replace a breadwinner. Victor Meunier, while working for the C&H railroad, lost a leg and had to undergo surgery several times. He asked the railroad's superintendent, T. H. Soddy, to hire his boy, even though he was under 16. Soddy allowed the boy to be a locomotive wiper, as long as MacNaughton approved the hire. MacNaughton acquiesced: "if the son of Victor Meunier can fill the bill it would be a good thing to give him the job, notwithstanding the fact that he is not of age, and you have my approval for doing so."[35]

Similarly, C&H (and other companies) often employed the sons of deceased workers. By doing so, they preserved the dignity of the family, which did not have to rely on pure charity. The loss of a breadwinner became more complicated, however, if no sons were ready or willing to take a company job, because the mines almost never offered work to a deceased man's wife or daughters.

Well into the twentieth century, the industry's labor force remained almost exclusively male. In 1913, when it employed 4,337 men at its mines and mills, Calumet and Hecla employed just 28 women, who worked at its hospital, library, and in the women's department of the company bathhouse.[36] All physically demanding and skilled mechanical jobs went to men; less arduous jobs went to weaker men or to boys; even all clerical jobs went to men.

When a man died at a mine, his widow could not follow him onto the company payroll. She might leave the district, if fortunate enough to have savings or death benefits to cover the cost of travel. She might remarry, establishing a new family with a new breadwinner. Some widows found domestic or service jobs in local homes or shops. More commonly, perhaps, a widow shunted herself and any young children into one floor of a company-owned house, and boarded company employees on the other floor. It seems to have been common practice at many of the mines to allow these women to stay on, and often to reduce or eliminate their rent. Taking in boarders was an accepted way of supplementing the incomes of intact families; for many widows, it was absolutely essential. When widows keeping boarders still struggled to make ends meet, many companies offered additional assistance, such as free wood or coal for fuel and an enlarged garden plot.[37]

Under Agassiz, Calumet and Hecla had general guidelines for handling widows, but no hard-and-fast policies. The company finessed this social problem on a case-by-case basis. Lacking a "widow policy" that lower-level managers could administer, their cases often went all the way up to the superintendent, or even to Agassiz himself, before being resolved. As for the widows, they never knew exactly where they stood, or what they were entitled to.

Alexander Agassiz and James MacNaughton exhibited a mixture of compassion and cold calculation in August 1904, after three C&H employees died in industrial accidents. Agassiz noted that "fortunately" two of them were single, "but I notice that one man killed by the name of Pelland left eight children." Agassiz feared that a widow with eight children might prove a very sympathetic plaintiff in any court proceedings, so he instructed MacNaughton "to look up this case and see whether the Company is liable in any way."

MacNaughton replied that he doubted C&H could be held liable, because the man had been warned from time to time of the very hazard that killed him. Having dispensed with that worry, the general superintendent informed Agassiz that the aid fund had paid the widow $500, and that one of the sons, about 16 years of age, would be given work at the stamp mill "at the very first opportunity." "Even with all this," MacNaughton wrote, "I have no doubt that his widow will require some additional assistance." Agassiz concurred: "I think the only thing to do at present is . . . to keep an eye on his widow and see how she can be helped in the natural course of things."[38]

One of C&H's unwritten guidelines held that officers' widows merited special treatment. In 1898 a company physician, Dr. Messimer, felt run down, perhaps due to pernicious anemia. Agassiz told superintendent S. B. Whiting to arrange a sea voyage for the doctor at company expense and to continue his salary during his absence. But Messimer died, and in December Agassiz's concerns shifted to his widow: "I take it that Mrs. Messimer may want to remain in her house at Calumet at any rate till next spring and when I come up we can settle as to what should be done for her. In the meantime I hope you may be able to ascertain in what circumstances she's left. Let her anyway receive Messimer's salary for the present till I come to the mine in April."[39]

Similarly, in 1904 Agassiz and James MacNaughton deliberated over the proper consideration to be shown the widow of Captain John Duncan. Duncan had been with the company in Michigan since 1868, longer than any other employee. He had been "most faithful and devoted." On the afternoon of his funeral, C&H suspended its shop and surface work to accommodate the many men who wanted to pay their last respects. The company directors passed a resolution praising Duncan and voted to allow his widow to remain in her company house rent-free for an indeterminate time. They also allowed her a death benefit of one-fourth of her husband's $5,000 salary.[40]

Another guideline held that a widow not already living in a company-owned house would not be moved into one. The company sought to extricate itself from perpetual care cases, and not become more intangled in them. Rose Dellacqua's husband worked at C&H for 25 years. The family lived in Laurium. When her husband died, leaving her with five children, she asked MacNaughton to rent her a company house, one of the old log ones, which carried the cheapest rent. MacNaughton replied that, regretfully, he could not accommodate her request.[41]

Survivors living in an employee-owned house standing on leased company land sometimes did—and sometimes did not—receive help in maintaining their properties. On 22 April 1901 Mat Zagar took a job at C&H, and on 22 August 1902 he and his wife, Mary, purchased a house on mining company property. Two months later, Mat Zagar died in a rockfall, leaving Mary behind with a two-year-old son and a newborn. Mary received $500 from the C&H aid fund; $400 of this went toward paying off her house. From the end of 1902 until 1910, she received monthly welfare checks of $6, $7, or $8 from the county, plus free fuel from C&H. Mary supplemented the family income by moving herself and her two sons into two upstairs rooms, and by renting the three downstairs rooms.

By the time the Zagar boys were 10 and 8 years old, the family could not squeeze into the two upstairs rooms, so they moved downstairs. Mary lost much of her income from boarders, so she took a job cleaning the Blue Jacket School, while the older boy went to work at the Miscowaubik Club. Barely able to make ends meet, Mary Zagar had no money to put into her house, which after a decade of neglect was in need of serious repair.

Twice, Calumet and Hecla sent workers to Mary Zagar's house to make modest repairs, which they charged to the company's charity account. The widow, however, complained about both the quantity and quality of the work. She wanted the company, or so it thought, to put her house "in first class condition inside and out & painted inside & out & would like a stone foundation." Mary Zagar bickered with the land agent and boss carpenter for a year and then jumped over their heads and went right to James MacNaughton. "So knowing of the generousity of the C&H Mining Company towards its widows," she wrote, "I turn to you as the only recourse to save my house from going to ruins." MacNaughton, sensing that the widow could not be pleased at a price he was willing to pay, decided it was time for C&H to release itself from any obligation to Mary Zagar. He wrote that, "it is impossible for us to expend money on houses other than our own."[42]

When C&H owned a widow's residence, the company was clearly involved in her fate and the maintenance of her dwelling. Some widows lived in C&H houses for many years, particularly if they had sons who joined the company payroll. The difficult cases arose when a woman had no children at home, or had only very young children who might have to be supported for a decade or more before going to work. In such instances, Calumet and Hecla preferred that the widow leave, so it could reassign the house to an active employee.

Calumet and Hecla did not want to evict widows heartlessly, and yet sometimes it definitely wanted its house back. In 1905, Mrs. Szatkoska complained to Alexander Agassiz that MacNaughton was taking away her company house—which, in fact, he was. Agassiz questioned his general superintendent about the details of the case, and MacNaughton wrote back, "You may be assured the lady . . . is not being harshly dealt with . . . , and she will be required to move when her moving can be brought about without causing any hardship. . . . She must go from the house eventually, but she will not be harshly dealt with."[43]

C&H apparently allowed many widows to remain in their company houses rent-free for perhaps one year before moving them out—but at the same time the company cleverly arranged an enonomic incentive to encourage widows to leave sooner than that. In lieu of staying in their houses for up to a year, some widows instead took the cash value of a year's rent—$60—and surrendered their houses immediately. The rebate on the rent, coupled with any death benefits received, allowed a widow to relocate, and released the company of further involvement in her life. James MacNaughton acknowledged that "we have no hard and fast rule whereby we permit a widow to live in one of our houses after her husband dies, nor have we any hard and fast rule whereby we obligate ourselves to rebate rental if the house is vacated in less than a year."[44] Nevertheless, although C&H never guaranteed its widows such considerations, it often bestowed them at the discretion of management.

Managerial discretion, coupled with C&H's belief in patriarchy, also determined whether a widow received health care at the hands of company doctors. Here, the key issue was the nature of the human link between the woman and Calumet and Hecla. If she were linked to the company only through an employed son—and her *husband* had never worked for C&H—then the company generally excluded her from its aid program and from medical attention. However, if the woman's husband had worked for C&H, and she stayed in the area upon his death, company doctors often continued her medical treatments and dispensed medicine free of charge. At the same time, the doctors were never absolutely certain if these widows could be admitted to the C&H hospital. James MacNaughton denied that any "general ruling" covered the treatment or hospitalization of widows. He affirmed to the head physician that "each case must be decided on its merits," which meant that MacNaughton himself resolved the issue.[45]

Another important group regularly occupied a portion of James MacNaughton's time: aging, long-term employees. Calumet and Hecla's roots went back to the late 1860s, when a number of young men joined the company. Others came aboard in the 1870s and 1880s. Many had stuck for life, and by the turn of the century C&H had hundreds of men of advanced years and long service on the payroll. Because it valued loyalty, the company had not systematically weeded aging men out, had not discarded them when their performance had fallen off. Now, some of these men posed a problem for the company. C&H did not want to dismiss summarily its oldest and, at one time, best employees, particularly if forced retirement would cause them undue hardship. On the other hand, it did not want to carry men over 60 on the payroll as charity cases, until they one day obligingly quit or keeled over. To give itself some latitude in easing such men out, in 1904 C&H inaugurated the first, and for many years the only, pension plan at the mines.[46]

True to form, Calumet and Hecla saw to it that its pension plan did not open up a Pandora's box of benefits. It did not want to bind itself to making pension payments in perpetuity. It did not want to discourage family thrift and savings and encourage a dependency on company largesse. It certainly did not want to encourage older men to hang on, just because they hoped one day to finally obtain a pension. To avoid these pitfalls, nobody automatically qualified for a C&H pension just by meeting specified age and length of service criteria. These criteria—60 years of age and 20 years of service—defined those eligible for consideration. They did not define who actually got a pension. Managerial discretion—and James MacNaughton's recommendation in particular—remained the final arbiter. And just to insure that the demand for pensions was not too great, the company first ran its pension plan as a corporate secret.

Company directors in Boston drafted the pension plan and Alexander Agassiz communicated its sketchy details to MacNaughton, who was left to administer it. Pensioners received an annual amount, paid in monthly installments, as determined by a simple formula: for each year of service, a man received 1 percent of his average annual salary or wage over his final 10 years of work. The pension ran for a maximum of five years, and if a man died prior to that, the payments stopped and did not carry over to survivors. A pension came with strings attached. The pensioner had to vacate any company-owned house within six months and could not remain a member of the aid fund. The pension bought the company out of sheltering and doctoring a man and his wife until they died. Finally, Agassiz confided in MacNaughton that "We do not wish to give formal notice of this plan or to establish a fixed method of action but are willing quietly to try this experiment on the lines mentioned."[47]

Nominally, a seven-member advisory board ran the unpublicized pension plan, but MacNaughton fleshed out its details and made key decisions as to who benefited. He generally reserved the pensions for men in need. He was unlikely to give a pension to someone who appeared ready to retire voluntarily. He would not pension off a man, even if he met the age and length of service criteria, as long as he still performed at adequate levels. Indeed, the pension system "was gotton up for old employees who were supposed to be working but who could not do a day's work."[48] Another important factor that entered into MacNaughton's decision was the number of pensioners already served, and the monthly cost to the company. Individuals received pensions running from as low as $10 to as high as $50 per month. The general superintendent seemed bent on holding the program's total cost to about $1,000 per month. He put potential new pensioners on hold until old pensioners had used up their five years' of payments, or had died.[49]

MacNaughton and other C&H officials were not entirely satisfied with their pension plan, partly because after several years it was no longer a secret. At the start, they fully controlled it. They targeted the men they wanted to retire and made the offers. Later, workers targeted them for appeals. Pensioners with little savings requested that their five-year payments be extended. (These requests were denied). Men who had left the company voluntarily, and later learned of the pensions, asked to receive one retroactively. (These requests, too, were denied). The pension plan, because of the way that C&H designed and implemented it, became a source of contention between top management and some of the company's oldest employees. Also, the pension plan alone did not solve one of the company's basic problems: the aging of its labor force.

If the pension plan smacked of age discrimination, of a company trying to rid itself of men over 60, it was nevertheless true that mining was a hard industry, one largely borne on the backs of much younger men. When the pension plan failed to solve the problem at the top end of the age scale, C&H decided to practice age discrimination when hiring new workers. By about 1910, the company followed the general policy of not hiring any men over 40, "unless there is some special reason for it." MacNaughton turned down applicants over 40 because he believed his company already had "too many old men in our employ from whom we are not getting anywhere nearly the efficiency we should get."[50]

The social safety nets strung up by the companies offered workers some security but were far from perfect. The companies refused to set firm policies for their welfare programs. Hence, little was guaranteed. Mine officers personally decided who would get help in time of need. They might informally set limits on how many "make-work" jobs their companies could afford to keep for handicapped or older men, or how many houses they could afford to keep for widows. They might make considerable effort to aid the families of longtime, loyal employees, who had sons ready to join the ranks, and make little effort to aid single, unskilled laborers, whom they deemed expendable.

Certainly, the security of a man and his family was enhanced if he attached himself to a profitable company; if he was a diligent, reliable, and skilled worker; and if he participated in associational life by joining a church and one or more benefit societies. In the event of a calamity, his family might receive assistance from several different sources. On the other end of the spectrum, single men who spurned churches and perhaps jumped from employer to employer, risked losing everything to one incapacitating illness, disease, or accident.

By at least the early 1870s, widows, orphans, mentally and physically impaired persons, and unemployed men and their families had become acknowledged social problems on the Keweenaw. Of these groups, the mentally impaired merited the least respect and care from local society. The mining communities shipped off their "cranks" to the state "mad house" at Kalamazoo, glad to be done with them.[51] More compassion, in terms of public and private charity, was shown to the others.

The leading public charity in Houghton County after the early 1870s was its Poorhouse and Farm, operated by the Superintendents of the Poor.[52] This institution, which received its funds from the county board of supervisors, took in children and adult paupers of both sexes. Residents of the poorhouse helped work the farm, so they could learn industrious habits while producing modest revenues. (The farm's leading cash crop in 1873 was $186 worth of strawberries.)

Persons in need did not have to reside at the farm to receive aid from the Superintendents of the Poor. Their nonresident program consumed a greater share of their budget and served more individuals. From the late 1870s through the mid-1880s, the poorhouse typically lodged only 20 to 50 residents, while 450 to 600 individuals received outside aid.[53] In 1885, the recipients included "twenty-two needy families on the Peninsula mine, left on their own through that industry closing down." The Superintendents of the Poor provided housing subsidies, medicines and treatment, clothing, food, and furniture. In the early 1880s, the Superintendents usually expended some $3,000 to $4,000 annually on the poorhouse and farm, and another $8,000 to $10,000 for outside relief. These expenditures amounted to 10 to 15 percent of the county supervisors' total budget, and put Houghton in the top 10 percent of all Michigan counties in terms of welfare spending.[54]

Still, the welfare measures provided by the companies, churches, benefit societies, and the county were never enough. On 16 June 1881 the *Portage*

Lake Mining Gazette noted that "there are now in the district poor miners who are sightless and maimed, dependent upon cold charity. These unfortunates should appeal to the benevolent and arouse them to take some action in the direction of the establishment of a miners' hospital at some point in the district." The following year, on 17 August, it reported that three parties were ready to donate a suitable site for a "Disabled Miners' Home." This opened the way "for rich men connected with the copper mines of the district to come forward and say how much money they will donate to help erect buildings, etc." Despite the plea, no home for disabled miners was built. While the "rich men" failed to create such an institution, at least the "prominent women" of Houghton County staged charity balls for worthy purposes; a private Houghton County Charitable Association established soup kitchens and gave away clothing in winter months; and in 1899 local citizens chartered the private Good Will Farm to take care of needy children.[55]

The Good Will Farm operated solely on the basis of contributions. It served as an orphanage, a foster home program, and as an adoption service. Its records indicate that the social safety nets raised by company paternalism were indeed imperfect, that even the richest company produced unfortunates in need of outside assistance. Between 1899 and 1912, the Good Will Farm took in 573 children. Of these, 164 came from Calumet and Hecla.[56]

By the turn of the century, even as Copper Range built its new company towns south of Portage Lake, paternalism was faltering on the Keweenaw. The system had first arisen largely of necessity, when the territory was a frontier, broad-based social institutions nonexistent, the population small, and government nearly invisible. None of these conditions still held. Mine managers felt less compelled to cater to all the social needs of their workers, because the surrounding society, now mature, provided many services. The companies were by no means ready to abandon paternalism. It was a firmly entrenched institution, and they still counted it among their chief assets in controlling the cost of living in the area, and in providing stability, principally in the acquisition and maintenance of large numbers of men in skilled positions. At the same time, however, they recognized that the labor force was changing dramatically, and that the paternal bonds between employer and employee were growing weaker.

This was in part due to the increase in the scale of the industry. Managers and workers became strangers to one another as the mine organizations became larger, and as mine employment rose from about 6,000 to 14,000 men in the single 1890s decade. Between 1890 and 1900, for instance, total employment at the Quincy Mining Company jumped from 484 to 1,366.[57] While such growth augered well for production and profitability, it adversely affected labor relations, in terms of distancing the personal relationships between employer and employee.

The companies dealt with more men. They also dealt with different men, with greater numbers of Finns, Italians, and eastern Europeans. These men knew neither the English language nor the mining industry when they came—

and both "deficiencies" rankled the mine managers and created a widening breech of estrangement and alienation. The companies succeeded far better in maintaining paternal bonds with older employees than they did in establishing bonds of cooperation, trust, and loyalty with their newer men.

At Calumet and Hecla, early signs of trouble appeared in the late 1880s and early 1890s. The firm's report of 1890–91 stated that "relations to the men have, as in former years, continued excellent; though it is becoming more difficult, from the number of men who do not speak English, to deal directly with our employees." A few years later, a wildcat trammers' strike erupted, the most serious breech of labor-management relations at the mine in 20 years. The C&H report of 1893–94 noted, "It was a great disappointment to find that there were among the employees of the Company so many men ready to forget the friendly relations which had always existed between the men and the officers of the Company." In truth, however, this brief labor rebellion had not been waged by men who had forgotten longstanding friendly relations they had once shared with their employers. Instead, its instigators were men who had never experienced such friendly relations in the first place. When more and more men of this kind joined the ranks of labor between 1890 and 1910, it did not bode well for labor harmony at the mines. If paternalism failed to gain their loyalty and allegiance, Calumet and Hecla, and the other mining companies, would turn to sterner measures to get and keep them under control.

12

Social Control at Calumet and Hecla

> There is no middle ground; you are either for the mine owners or against them. If you take any view except their view, you are against them. If you do not accept as well founded their absolute right to control the district, you are opposed to them. That is their view of it.
>
> William J. MacDonald, U.S.
> congressman, Michigan's 12th
> Congressional District, 1914

Despite occasional strikes, from 1890 to 1910 the Copper Country remained relatively calm while a sea of troubles beset the new industrial America. During these decades the United States made difficult, painful adjustments to a modern social order (or *dis*order) that seemingly violated basic American values. In this democratic society, ordinary men and women were to be free to carve out a dignified, decent way of life. They were to live in harmony with fellow citizens, respectful of each other and cognizant of their mutual obligations. Americans were to share power and control across a broad base of egalitarian social institutions.

Harsh realities starkly contradicted such idealized values. Cities swelled with immigrants, herded into ghettos where the way in was easier to find than the way out. In overcrowded streets, crime, filth, and disease flourished. In new factories, men, women, and children labored long hours for too little money in unhealthy, often hazardous, environments. Giant corporations had arisen whose activities were little checked by government regulation or oversight. Big capitalists squeezed profits from their host society and gave back little in return.

Near the turn of the century, gross inequities spurred some Americans to embrace radical politics and issue a virtual call to arms. The evils of capitalistic America had to be confronted and overthrown. Power and wealth had to be stripped from the elite and redistributed amongst the masses. Others—the "Progressives"—balked at the idea of revolution, but nevertheless believed that serious reforms were necessary if hard-working Americans were to have a fair shot at a decent life. The disadvantaged had to be propped up and led out of squalor. Congested, dirty cities had to be made more habitable. The public health had to be protected and preserved. Industrial workplaces had to be transformed from sweatshops into clean, well-lighted places. Children had to be snatched away from the employers who exploited them, and put back into the schools where they belonged. Adults needed fairer wages and shorter work weeks. Scientists, engineers and other experts had to use their knowledge not just to turn a dollar but to build a better world. Governments—local, state, and federal—had to exert greater control over the nation's future in order to achieve social justice for all.[1]

Humanitarian impulses drove some social reformers. Others were motivated by their fear that American society was being ripped apart. Street violence and labor wars in numerous cities and industrial settings offered strong evidence that something was dreadfully wrong.

Throughout this troubled era, the American metal-mining industry provided more than its share of turmoil. In numerous western states, mine managers and workers waged war. When fledgling labor unions challenged entrenched companies over issues of wages, working conditions, and control over the workplace, all hell broke loose.

Meanwhile, the Copper Country rather quietly went about its business, out of the national limelight. Michigan's leadership role in the copper industry had been eclipsed by newer, larger producers. After 1890, activities at the Michigan mines fell into the category of "old news," even as they tremendously expanded their operations. Still, if the Michigan mine managers felt their achievements got scant attention, they certainly preferred neglect to notoriety. When western mining districts became battlefields and stole all the headlines, the Copper Country's principal achievement lay, perhaps, with the labor peace it sustained.

This peace, to be sure, was not absolute. The Copper Country was not immune to the troubles that rocked many mining communities across the United States. The copper mines suffered many social tremors, and a schism widened between management and labor, and between the mining companies' interests and the interests of the public at large. Nevertheless, through 1910 the mining companies, by exercising vigilance and by employing a variety of tactics, managed to quash small rebellions and avert large ones.

At Calumet and Hecla, Alexander Agassiz did not let outsiders—either radicals or reformers—tell him how to run his company. Nor did he idly stand by, watching labor unrest escalate. As a staunch autocrat, he was driven to control events and not let problems get out of hand. As C&H's labor force got larger, more polyglot, and potentially more fractious, Agassiz and his managers devised a hierarchy of tactics intended to keep their hegemony intact. These tactics ran from acts of corporate generosity and goodwill, to subterfuges and covert maneuvers, to outright displays of force. Employees had to do the company's bidding. They would do it out of loyalty, or because of subtle manipulations, or because of their exposed impotence in the face of exerted corporate power. But they would do it.

Alexander Agassiz believed that he and his managers knew what was best for Calumet and Hecla, and what was best for the company was best for everyone involved. Mutual self-interest bound together stockholders, managers, and workers. Profitability had to be preserved, and efficient order maintained. Authority had to be honored. These things brought success. And the more successful the corporation, the greater the benefits that trickled down to the mechanic, miner, timberman, and trammer. Since C&H had been so successful, it followed that its employees enjoyed an impressive array of paternal benefits. But these benefits were not freely given; they had to be earned. They were rightfully due the men, but only if they upheld their end of the bargain and worked for the greater good of the company.

Agassiz valued his company's reputation as a generous employer, yet C&H was not committed to using only benevolent paternalism to control labor-management relations. Agassiz's company was always prepared to shuck off paternalism in favor of coercion, covert manipulation, armed deputy sheriffs, or mass firings. Agassiz preferred to hold these sterner means of social control in reserve, but he would not apologize for using them when necessary. Calumet and Hecla was, indeed, eminently fair. Its men deserved whatever treatment they got—good or bad.

By 1901, the president of C&H believed that mining companies on the Lake had been too generous. They had coddled their men, had given them better treatment than they deserved. Some disloyal laborers openly expressed discontent, they performed their work poorly—yet they paid no price for such indiscretions. Agassiz felt the times called for an infusion of assertive management. In that year, S. B. Whiting resigned as C&H's superintendent, and Agassiz brought in "Big Jim" MacNaughton. MacNaughton, young, strong-willed, and ambitious, was the perfect replacement. He had been raised at Calumet and Hecla, so he knew the system. He had a college education and a refinement that squared well with Agassiz's tastes, and mentally prepared him to lead a nineteenth-century company into the twentieth century. He had practical experience as well, having served as superintendent of the Chapin iron mine. Also, MacNaughton was another autocrat seemingly barren of self-doubts and indecision. He could both give and take orders. He could get done whatever needed doing, using finesse or force. Agassiz would not be disappointed with MacNaughton's performance.

Calumet and Hecla had an advantage over all other mines in the Copper Country. It had the first pick of men from the available labor pool. C&H for years could weed out employees, cull out "the weak ones" and "keep only the best," because there was no shortage of good men who wanted C&H jobs.[2] But in the late nineteenth and early twentieth century, as immigration patterns changed and the mines faced greater competition in the nationwide market for workers, the company found it harder to fill jobs with men having desirable traits: married, quick to learn English, prior mining experience, industrious and reliable, young and strong, and temperate in all personal habits. Even the great C&H often settled for less than the best, and the company held many of its new men in low esteem.

C&H officers expressed dissatisfaction with the loyalty and quality of employees as early as the late 1880s and grew far more critical of the labor force over the next 20 years. The underground fires of 1887 and 1888 signaled an important breech of trust between management and labor. These conflagrations greatly embarrassed the company and forced lengthy shutdowns. The exact cause of the 1888 fire was never determined. Most mining experts attributed it to friction: a steel hoisting rope, traveling 1,500 feet per minute over a jammed wooden roller, caused the roller to burst into flames. Calumet and Hecla's leaders, however, had a different view of the fire's origin: it was arson. Suspecting that some disgruntled or deranged employee set the fire, they offered a $10,000 reward for his arrest.[3] The reward went unclaimed. Meanwhile, C&H's public allegation of arson clearly indicated that all was not well between managers and men.

More evidence of conflict came in the form of labor demands for higher wages or improved working conditions. C&H, after being hit by strikes in 1872 and 1874, largely escaped other labor troubles until 1890, when workers at its stamp mill—where a 12-hour shift was the norm—organized a short, unsuccessful strike to obtain a shorter working day. In 1893 a more serious outbreak occurred in May when trammers struck the company, attempted to keep miners from going to work, and briefly seized the Superior hoist-engine house. C&H summarily fired the strikers and called in the county sheriff to disperse them.[4]

After this display of corporate resolve and force, C&H entered another strike-free period beginning in the mid-1890s. C&H enjoyed record profits, and Alexander Agassiz opened the spigot to let more paternal benefits trickle down to workers. During this time the company suspended workers' monthly payments to the medical and benefit funds for three years. To short-circuit labor unrest, C&H also announced a new grievance policy. Any man with a complaint was to take it to his immediate supervisor. If that hearing gave no satisfaction, he was invited to go to his boss's boss. Indeed, the man could pitch his appeal all the way to the general superintendent or even the president, if he felt his wrong was not being righted.[5]

Such measures restored labor peace at C&H, but across the district as a whole labor unrest was definitely on the rise. Neighboring companies suffered periodic strikes from the mid-1890s through 1910.[6] These tended to be brief flare-ups at individual mines, often involving trammers. All the strikes taken together had little effect on the mines' production levels or profits, but they ushered in an era of greater mistrust between mine managers and workers.

C&H responded to labor agitation after 1890 in both public and private ways. Publicly, it still donned the mask of a highly paternalistic employer. Its schools, libraries, churches, and houses remained visible symbols of the company's benevolence. Privately, however, C&H adopted secret strategies and practices for quelling labor trouble. It cleverly manipulated wages and the timing of any raises or cuts. It became less imperious in dealing with other local mines, because labor unrest was a regional problem, and some day the various companies might have to meet it with a united front. In accordance with strong anti-union sentiments held by Alexander Agassiz and others,

C&H monitored all attempts to organize labor. It used industrial spies to report on agitators and organizers. It interfered subtly in labor's right to assemble in large groups. Finally, it practiced ethnic discrimination in hiring and firing, hoping to keep troublemakers off the payroll.

Alexander Agassiz was the continuous thread running through Calumet and Hecla's history, and as he crafted and refined C&H's labor policies, he drew upon his long corporate memory. Agassiz was in his late thirties, and C&H only a few years old, when serious strikes hit the mine in 1872 and 1874. Until his death at age 75 in 1910, these two strikes profoundly affected the course Agassiz set in any times of labor trouble.

In 1872, C&H managers were forewarned of a strike set to begin on 1 May. Superintendent R. J. Wood tried to head it off with an open letter sent to all employees on 27 April. Wood informed workers that when President Agassiz had visited the mine in February, several important decisions had been made in the interest of employees' welfare. Workers would receive a raise on 1 May. In addition, C&H would inaugurate a profit-sharing plan and operate a bank that would encourage thrift by paying interest on savings accounts. Wood's letter addressed the employees as "My Friends." His letter was conciliatory, not confrontational:

> The objects aimed at in these arrangements are to secure your cheerful co-operation in prosecuting the Company's enterprises, and to encourage all in habits of industry and thrift. We do not try to conceal the fact that the Company has its interests subserved in all this as well as yourselves. We have a vast amount of work to do—work that we want pushed along without trifling or delay—and we know that a man is useful to us just in proportion as he is careful and prudent in his own affairs. We feel sure that that large class of men who want permanent employment, homes for their families and education for their children, with the many other comforts that come of having settled plans in life, will appreciate and prize our scheme.[7]

Although signed by Wood, this letter clearly expressed Alexander Agassiz's early views on the friendly, mutual interests that should bind together a company and its men. The letter, however, did not dissuade the men from striking. Violence erupted as strikers confronted loyal employees and the Houghton County sheriff and his men. Sheriff Bartholomew Shea telegraphed Governor Henry Baldwin for reinforcements on 13 May, and the strike ended 10 days later, after a company of infantry arrived in Houghton.[8]

The men achieved a 10 percent wage increase, but at the cost of engendering in Alexander Agassiz a determination to be unbowing in any future labor disputes. He took the 1872 strike as a personal rebuke. When newly arrived Finns and Swedes instigated a short-lived, unsuccessful strike in 1874, no "My Friends" letter passed from management to labor. Agassiz gave clear instructions to James North Wright, now the company's superintendent:

> We cannot be dictated to by anyone. The mine must stop if it stays closed forever. . . . As I have written you before we have always treated our men fairly

and honestly, they have received higher wages than any other corporation. I have attempted formerly to try and get their good will by offering them a share of the profits. They spit in my face as it were and all we can do is to sit quietly and await results. Wages will be raised whenever we see fit and at no other time (if they don't like it they must go and get employment elsewhere . . .).[9]

In 1872 and 1874, Agassiz learned that when a strike was imminent or under way, that was no time to initiate a dialogue. Workers only interpreted that as a weakness, and it encouraged them to press their demands. C&H should never bow to discordant chants or to placards carried in the streets. The best way to handle labor demands—particularly those involving wages—was to anticipate them, and to defuse them before they exploded in public view.

Agassiz taught his mine superintendents to be wary of economic factors that could spark labor unrest. They monitored workers' apparent satisfaction with their wages. They watched how the supply of labor balanced against the demand, the prices being paid for copper, and the wages being paid by other local mines, by western copper mines, and by Great Lakes iron mines. If a superintendent expected that some change, such as an increase in copper prices, would likely cause agitation for a raise, he huddled with Agassiz to determine if a raise was affordable or desirable. In 1895, Agassiz wrote S. B. Whiting that C&H's policy was to effect raises "whenever *we* think" the time is right. Three years later he sounded the same note: "In the spring it may be necessary to raise wages and if there is any growling we want to do it before we are compelled to do so."[10]

C&H sometimes gave in—privately and quietly—to workers' demands for wage increases that had not even been voiced yet. Such raises enhanced the company's reputation as a liberal employer, while preempting strikes. Calumet and Hecla avoided the humiliation of making wage concessions to an angry band of men parading in the streets. By making concessions early and out of sight, C&H turned its "loss" into a public relations victory. A raise carried the message that employees could trust their employer to pay higher wages whenever it could afford them. Since C&H offered its raises freely and voluntarily, the men should also understand that whenever the company cut wages back again, it did so regretfully, and only because it had to.

Calumet and Hecla wanted to avoid strikes, but in certain situations it would not run away from one. Company leaders sometimes decided not to appease anticipated labor unrest. When they concluded in advance that a wage increase was unwise, they would not give one. If a strike erupted—such as the trammer revolt of 1893—so be it. A strike, after all, was not the worst thing that could happen. The worst thing would be a *successful* strike, and Agassiz determined that that must never happen. If the company felt it wise to accommodate labor in any way, it would do so on its own terms. But if the men walked out and voiced their demands, they would fall on deaf ears.

C&H successfully used various strategies to control its men after 1895, but keeping labor content became more difficult as strikes hit other mines on the Keweenaw. Dissent at one location tended to ripple outward, rocking the boat at other companies. Traditionally, C&H little concerned itself with the operations of other mines. But after the turn of the century, officers recognized that they had a stake in how other companies confronted labor problems. Thus, in 1905, Calumet and Hecla for the first time shared employment and wage information with other companies and consulted with them on the propriety of altering existing wage scales.[11] For labor, this did not bode well. When C&H decided to forgo corporate secrecy in favor of pooling information with other companies, this opened the door to the creation of a united front of mining companies, standing shoulder to shoulder against labor.

Calumet and Hecla started cooperating with other companies about one year after the Western Federation of Miners launched an organizational drive at the mines. Compared with other mining districts, the companies had seen few strikes, and most had been staged by men not affiliated with any union. Nevertheless, the mine managers keenly opposed unions. They saw union drives as conspiracies launched by outside agitators. C&H and the other companies could not tolerate any labor organization that threatened managerial control or preached the overthrow of capitalism.

Before the 1872 strike at C&H and the Portage Lake mines, the new and short-lived International Workingmen's Association had been active on Lake Superior, and this organization was seen as an instigator of the three-week-long labor rebellion.[12] This strike, which so profoundly affected Alexander Agassiz's view of how to handle labor disputes, also colored his attitude about labor organizations: they were to be kept out of Calumet and Hecla, one way or the other. He had an opportunity to put his views into practice in the mid- to late 1880s, when the Knights of Labor enrolled workers in the Copper Country.

The Knights of Labor began as a craft union formed by Philadelphia garment workers in 1869. In 1878 it went national; in 1881 it opened its membership to all working-class "producers"; and in 1884 its general assembly announced that "our order contemplates a radical change in the existing industrial system. . . . The attitude of our order to the existing industrial system is necessarily one of war." The Knights of Labor espoused a brotherhood of all workers; recruited broadly from different industries and occupations; and sought fundamental reforms of the capitalist system that would bring its members greater esteem, higher earnings, and better working and living conditions. Despite its confrontational rhetoric, under Terrance Powderly's leadership the Knights of Labor generally downplayed strikes and pushed for social reforms through education and arbitration. Powderly realistically assessed that in the 1880s poor workers stood little chance of defeating wealthy corporations on industrial battlefields.[13]

Nevertheless, Michigan mine owners deemed the Knights of Labor a sinister organization. Its membership peaked in 1886 when nearly 750,000 persons belonged. That year was also known as the "Great Upheaval." The U.S. Bureau of Labor Statistics recorded 1,432 strikes and 140 lockouts, involving over 600,000 American workers. Some of the worst confrontations occurred around the lower Great Lakes, which the Michigan mine managers felt was too close for comfort. Chicago was a center of labor radicalism, and in 1886 molders at McCormick Harvester unsuccessfully struck their company. In the aftermath of this strike, workers held a rally in Haymarket Square. Police charged the crowd, and a bomb exploded, killing police and protesters alike. That same month, immigrant iron workers in Bay View, Wisconsin—an industrial suburb of Milwaukee—massed large crowds to shut down the foundries. The strike was broken only after the state militia came in, used their weapons, and left 10 strikers dead.[14]

After the events of 1886, Calumet and Hecla's leaders became particularly leery of foundrymen's unions, and they avoided hiring unskilled workers off the streets of Chicago, fearing they might have been tainted by radical ideas. In general, C&H simply grew far more antagonistic toward all labor organizations. When confronted by problems, it became commonplace for Calumet and Hecla to jump to the conclusion that outside agitators or union men were behind them.

In 1886, serious labor problems had erupted across the nation. In 1888, a fire of suspicious origin burst out underground at C&H, and a strike broke out at its stamp mill in 1890. During those years, Knights of Labor organizers worked the Copper Country, and Agassiz—if he could not prove their culpability—nevertheless suspected them and wanted them removed. In 1887 and again in 1891 he told his superintendent not to hire men belonging to the Knights of Labor and to discharge members already on the payroll: "Every means possible should be taken to ascertain which of our men have joined and then to discharge these men as fast as any breech of our regulations or their contracts or duties gives the occasion. . . . If this is too slow then use express reason of joining K. of Ls."[15]

Due to company actions, and to a rapid decline of the Knights of Labor as a nationwide organization, its membership in the Copper Country fell to insignificant numbers by the mid-1890s. But just as the Knights of Labor were dying out, in May 1893, mine workers organized the Western Federation of Miners in Butte, Montana. A decade later, the WFM arrived in the Michigan copper district. The WFM's reputation preceded Joy Pollard, who came late in 1903 to begin an organizational drive. Michigan mine managers already knew the WFM as one of the more radical unions in the United States.

The Western Federation was born of the labor wars that erupted in several western mining districts in the early 1890s, and particularly of the conflict in Coeur d'Alene, Idaho, in 1892.[16] Coeur d'Alene had started as a gold mining camp, dominated by small enterprises engaged in placer mining; it had grown to become a silver and lead mining region, dominated by larger, more capital-intensive companies. Several local unions germinated in the late 1880s in response to a wage cut for trammers and muckers, and in 1889 these locals formed the Coeur d'Alene Executive Miners' Union. The union protested the failure of a few key companies to support a hospital for workers; objected to the increased use of machine rock drills, which displaced miners and forced them into lower paying jobs; and then pushed for a uniform $3.50 per day for underground workers, regardless of occupation. The companies met this demand for a uniform wage and then, uncomfortable with their capitulation, formed a Mine Owners Protective Association and hired Pinkerton detectives who infiltrated the union, serving as spies.

In January 1892, all the mining companies near Coeur d'Alene shut down operations for several months as a ploy to force railroads into reducing the freight rates charged on mine output. The shutdown, however, served the dual purpose of forcing hardship onto laborers in mid-winter. The companies hoped to make them less obstreperous and more thankful to receive whatever wages the mines wanted to pay. When they announced plans to reopen, the companies rescinded the $3.50 uniform wage and offered "car men and shovelers" $3 for a 10-hour day. Union men, uncowed, refused the wage cut. The companies brought in scabs protected by armed guards; the union men tried to turn them back. The uneasy standoff finally erupted into violence. A state of martial law was declared, and troops quashed the rebellion by arresting the union men and herding them into "bullpens." Organized labor lost the battle at Coeur d'Alene, but this loss sparked a long-term "war" between labor and management in western mines.

In 1893, representatives from small, local unions in Colorado, Montana, South Dakota, Utah, and Idaho met in Butte and formed the Western Federation of Miners. The inaugural meeting produced a constitution and bylaws spelling out a rather conservative agenda. The WFM sought fair compensation for men working in a hazardous industry, and safeguards against needless risks to life and limb. They vowed to seek these ends "through education, organization and legislation." Union members sought humane mining laws that would ban child labor, and ban management's use of armed guards and Pinkerton men to force their will on employees. The WFM would try to procure employment for union men over nonunion men and "imported pauper labor." The WFM pledged to use "all honorable means to maintain friendly relations" with employers. It would apply "mediation and arbitration in order to make strikes unnecessary."[17]

Western mine owners, who had fought bitterly against small, independent locals, strengthened their anti-union resolve when confronted by a larger, more powerful WFM. And after the WFM had entered into disputes over wages and working conditions—winning some and losing others—its rhetoric adopted a more radical cast. At the WFM convention in 1897, one leader called the delegates' attention to their constitutional right to bear arms:

> This you should comply with immediately. Every union should have a rifle club. I strongly advise you to provide every member with the latest improved rifle, which can be obtained from the factory at a nominal price. I entreat you to take action on this important question, so that in two years we can hear the inspiring music of the martial tread of 25,000 armed men in the ranks of labor.[18]

In the face of adamant opposition, the WFM became more militant and leftist. Many rank-and-file members sought only a fair shake from their employers, but WFM leaders embraced socialism and condemned capitalism as the real enemy. They threw out the cautious language of the first WFM constitution, and newer versions spoke of "wage slaves" engaged in a "class struggle." This struggle would continue until the worker, the real producer, "is recognized as the sole master of his product." The WFM asserted that "the working class, and it alone, can and must achieve its own emancipation."[19]

In 1903–4, when the WFM took its first steps to emancipate wage slaves in the Lake Superior cooper district, the organization claimed a membership of 30,000, or about one-fourth of all workers in the U.S. metal-mining industry. The union could not lavish money on its initial recruitment drive in the Copper Country, because it was then engaged in protracted labor wars in Cripple Creek and Telluride, Colorado.[20] Nevertheless, the WFM managed to establish local miners' unions at Laurium, Hancock, Painesdale, and Allouez; a smelterman's local at Lake Linden; and a district union office headquartered in Lake Linden.

The Michigan mine managers kept abreast of the WFM's situation in Colorado and watched the fledgling WFM locals in their own neighborhood. At the same time, they observed that the WFM was not alone in preaching the gospel of socialism in the Copper Country. In 1904, as the union locals formed, so did local socialist clubs. And a Finnish socialist newspaper, *Tyomies* ("Worker"), established a year earlier in Worcester, Massachusetts, moved its headquarters to Hancock.[21] The mine managers did not sound a public alarm about any of this. They did not want to call undue attention to the WFM or to the socialists, so they ignored them.

The mine managers' strategy seemed to work. Instead of growing into a powerful force, the WFM locals collapsed within a year or two. Nationally the WFM was in disarray. It was reeling from losses in Colorado, and its leaders' attentions were diverted toward the completion of a short-lived alliance with the "Wobblies," the International Workers of the World. Locally, the union drive suffered from a lack of membership and financial resources. To attract more men from the many ethnic groups found in the mines, the WFM would have to return with organizers who could speak to Finns, Italians, Austrians, Slovenians, and Poles in their own language.[22]

The demise of the early WFM locals did not signify that labor unrest disappeared in the Copper Country. Quincy faced brief walkouts in 1904 and 1905, and then a three-week strike in 1906. The last strike—over workers' demands for a 10 percent wage increase—was more acrimonious than earlier ones. Charles Lawton, Quincy's agent, offered only a 5 percent increase, which remained on the table, nonnegotiable, until workers finally accepted it. Lawton blamed the affair on a small group of dissident Finns and Italians, who had not learned American ways, and who were particularly susceptible to manipulation at the hands of unionists and socialists.[23]

In the same year as the three-week-long Quincy strike, Finnish trammers struck the Michigan mine in Ontonagon County. They sought a wage increase and the hiring of some Finns as trammer bosses. Strikers and sheriff's deputies engaged in a riot, and the workers got what was for them a rare taste of western-style violence. When the melee ended, two Finns lay dead and dozens were injured.[24] Because of such incidents, when the WFM returned to the Copper

Country in 1908–9, its organizers found the region somewhat more fertile ground for a union drive. They first reestablished locals at C&H, Quincy, and Copper Range. Later, they added others at Mass City in Ontonagon County and at Ahmeek in Keweenaw County.

At C&H, Alexander Agassiz and James MacNaughton initially took no extraordinary measures against WFM union men or socialists. In April 1904, James MacNaughton summarized the first WFM union drive in a letter to Agassiz. He told him that wage increases for miners and trammers and an eight-hour day seemed to be the rallying cries. A self-assured MacNaughton concluded, "I don't think there is anything in the entire situation to cause us worry at the present time." In reply, Agassiz agreed:

> It seems to me that the only thing we can do is to sit still and await developments and to try to keep our heads cool. . . . All that we can do at present is, as in the past, to deal with our operators as they ought to be dealt with and any other issue they choose to inject into the matter should occasion arise we must simply throw out.[25]

The two leaders of C&H did not adhere literally to Agassiz's words and just "sit still and await developments." They used covert means to discourage unionism as best they could. They still manipulated wages to head off labor unrest. They began sharing labor and wage information with other mines. They decided not to fire current employees for joining a union, but they protected themselves by not hiring any new men with union ties. In 1904, when the WFM was beaten in Colorado and its men were deported from that state, Agassiz told MacNaughton to "instruct the captains to be specially careful who they hire and not to take Colorado men on any account."[26]

In 1905, Agassiz and MacNaughton weighed the merits of erecting a company foundry, and the issue of unionism became key. They wanted no business with the Foundrymen's Union, which was "one of the strongest in the country." They did not want a foundry if it meant that "we have to be dictated to by union men." Agassiz did not want to "let loose a swarm of hornets" in C&H's own backyard. Agassiz and MacNaughton finally decided to go ahead with the foundry, but only after devising a plan to keep the union out. They would identify a small cadre of molders on whom they could rely to get things started. Then, instead of hiring outsiders from Milwaukee or Chicago to fill needed positions, they would "educate our own men as we have done in the machine shop, blacksmith shop, painting department, tin-smithing and plumbing department."[27]

In their thought and behavior, Agassiz and MacNaughton were often reactive. They were driven in certain directions by a labor force that only in theory had no power. Agassiz and MacNaughton wore the trappings of omnipotence, but had an underlying fear of rebellion. To calm that fear, they kept various labor factions apart. If any insurrections were to come, better they came from only a segment of the labor force, and not from all workers in combination.

Because of its fear of collective labor action, Calumet and Hecla was leery of sanctioning large assemblies of workers. In 1895, Alexander Agassiz questioned the wisdom of letting the Salvation Army occupy C&H property near the railroad depot for over a month. The intended purpose seemed worthy enough: the Salvation Army wanted to hold a religious revival. But as Agassiz wrote to S. B. Whiting, "I don't quite like the idea of holding a mass meeting for thirty-five days."[28] There was just no telling what the men might do after listening to zealous tent preachers for so long.

At the turn of the century, the villages surrounding Calumet and Hecla were lauded as fully modern communities. In his report for 1899, Michigan's commissioner of mineral statistics noted that the only thing missing was an electric streetcar line. It seemed odd that no public transit system tied together the mine, the mine housing, the residences in Red Jacket and Laurium, and their business districts. The 40,000 residents should have had the convenience of hopping trolleys to get around. That they did not, however, was no mere oversight. Alexander Agassiz opposed laying a network of trolley tracks down the steets near his mine. Streetcars could be carriers of trouble. In times of labor unrest, they would make it too easy for workers to leave their homes to attend mass rallies.[29]

In 1905, fire destroyed a C&H clubhouse equipped with bowling alleys and other entertainments for working men. When he contemplated rebuilding and upgrading this facility, Agassiz consulted with a Mr. Boal, who had erected men's clubs at several mines in the West. Calumet and Hecla's president wanted Boal to tell C&H how to use a clubhouse to keep the men content. He also needed assurances that a clubhouse would not have the opposite effect and encourage dissent.

Boal's advice mixed a liberal appreciation for the validity of working-class customs with a conservative appreciation for the company's desire for social control. A clubhouse, he said, should cater to the habits of the men. Mine workers were a hardy, spirited breed. If you wanted them to work hard on company time, let them play hard on their own time. Let them belly up to the bar. Build them a clubhouse at company expense, but make sure it's suited to their needs, and let them manage, staff, and police it themselves. The grateful workers would repay the company with their loyalty.

Agassiz thought that maybe C&H should have a place "for the men to go . . . in the evening and have a drink and play cards or billiards." And he rather liked Boal's idea of letting the men run such an establishment. But he worried about all the mingling and talking that might go on. Could such a clubhouse "be used as a focus for . . . labor unions or things of that sort?" Could it accommodate "a general meeting of the men without our knowing it?" One way to avoid that, Boal suggested, was to set up "a number of small clubhouses, say three or four, one for the Germans, one for the Swedes and Finns, one for the Italians and one for the English." Agassiz saw some wisdom in this. No one club would be "large enough to be able to foment trouble about wages or administration." Still, there was the chance that "the clubs might do individually, perhaps, as much harm as a large club."[30] After much thought, Agassiz decided he was not ready for the progressive idea of company-built, employee-operated barrooms or clubs. The men might talk, instead of just drink.

In the late nineteenth and early twentieth century mine managers (and industrialists of all sorts) had very willing allies in monitoring the affairs of labor: private detective agencies. For a fee, they planted spies in a labor organization or work force, and then shared what was learned with mine managers and owners. Since labor spies participated in covert operations that companies sought to conceal, the exact extent of their use in the copper mines is difficult to assess. But at least 23 operatives infiltrated Quincy's work force between 1904 and 1920, and Calumet and Hecla hired detective agencies to spy on the Knights of Labor in 1887, on the WFM in 1913–14, and on the American Federation of Labor in 1919.[31] It seems likely that C&H also had spies planted in its labor force between 1900 and 1910. Agassiz and MacNaughton wanted to monitor labor unrest, and a spy was a readily available tool for the job. Actually, James MacNaughton held spies in low esteem: "I have had to do with all kinds of detectives; the best of them are none too good, and all of them are nuisances."[32] Nevertheless, in periods of labor trouble, mine managers often paid for a spy's steady stream of wide-ranging observations and gossip.

Calumet and Hecla tried to keep its labor force under control in another covert manner: it practiced ethnic discrimination in hiring and firing. The most explicit evidence of this is found in 1912 and the first half of 1913. At the time, the local mines found themselves labor-short by 400 to 500 men, most of whom were needed as trammers and unskilled laborers. Calumet and Hecla, desperate for men, launched a special recruitment drive. Even so, James MacNaughton made it clear that certain men were not wanted.

MacNaughton did not want the "riffraff" from American cities, particularly Chicago. He believed that idle men in Chicago were unemployed "because they don't want to work." Besides, Chicago was "the worst labor center in the world and attracts the most undesirable people." MacNaughton also rejected immigrants who "have been in the country for some time and have learned all the bad features of American civilization and none of the good." Instead he preferred "men who had just arrived in the country, who . . . were wholly unacquainted with the customs and manners of the country." As MacNaughton put it, "We would rather make American citizens of these people in our own way than have anyone else do it."[33]

MacNaughton wanted fresh immigrants from specific locales that he believed provided men who were strong in stature, industrious, and compliant. He preferred Swedes, Germans, northern Italians, and Austrians. These were "just the people we want." He did not want "fruit peddlers and the like," men who were ill-suited for heavy manual labor. He rejected a shipment of Syrians, because "there wasn't one of them but was as broad at the ankles as he was at the shoulders; their hands are soft and they were altogether a hopeless lot." He rejected Greeks and southern Italians, and felt that C&H had "enough of these Poles." Of greatest significance, MacNaughton shunned the ethnic group the mines had relied upon most heavily since 1890 to sustain their labor force expansion. He informed the Commissioner of Immigration at Ellis Island and the U.S. Secretary of Commerce and Labor that "we do not want Finlanders."[34]

Mine managers had always found the Finns somewhat intractable: clannish, slow to learn English, and difficult to Americanize. Local society in general harbored suspicions about the Finns and their customs. In 1881, before Finns had become such a dominant ethnic group in the region, the *Mining Gazette* reported, "we have heard from what should be good authority that amongst a certain class of Finlanders it is not an uncommon occurrence to bury their dead without either coffin or box; but as this has now come to the ears of the authorities a stop will be put to the practice in the future, at any rate in the Company's burial ground."[35] In 1887, an early settler of the Keweenaw reflected on its various ethnic groups. The Americans preferred being the boss. The Cornishmen liked mining and deemed their vocation "among the most honorable, if not aristocratic." The Irish made good miners, but often left the underground because they also made good businessmen and politicians. The Germans, too, left the mines in favor of running a store, hotel, saloon, or brewery. The French Canadians were good citizens, but "could not be induced to become a miner." They wanted surface work. As for the Finns: "The old settlers looked down upon them with the same sort of aversion as the west coast people do on the heathen Chinee."[36]

The Finns who arrived in the Copper Country in the late nineteenth century were anything but heathens. Most came from two rural, northern provinces, Oulu and Vaasa, and they carried from their homeland a deep sense of religious piety rooted in Lutheranism. The Finns quickly established a number of local churches, usually affiliated with the Finnish Evangelical Lutheran Church in America, or with the Apostolic Lutheran Movement in America. These Finns were hardly radical troublemakers. In the 1890s, they did not flock to socialist clubs or unions. Instead, they formed temperance societies where they debated key social issues of the day, such as: Were dancing and card-playing wholesome amusements, or did they violate Christian morality and the ideals of temperance?[37]

But even in their piety, the Finns offered detractors evidence of their clannishness, their unwillingness to adopt American ways. In 1896, the Suomi Synod of the Evangelical Lutherans opened Suomi College and Seminary in Hancock. True, the school instructed students in American history and government, and its business department taught classes geared towards local employment opportunities. Nevertheless Suomi College looked to outsiders like a piece of Finland dropped into Hancock. Its very name, "Suomi," meant "Finland." Instructors and students were Finnish; the institution's major mission was to prepare pastors for Finnish churches; it instructed students in Finnish language and literature and seemed bent on passing down the folk culture of the homeland. Also, even in their religion the Finns seemed to display a fractious temperament and an inability to get along, even among themselves. Early in 1890, local Finnish congregations gathered to decide with which major Lutheran movement they would affiliate. In Calumet, Pastor J. W.

Eloheimo anticipated trouble, so he had his flock shepherded by armed representatives of the local police. When he announced his choice of affiliation, about 500 of the 700 Finns in attendance rose up to shout him down. Pastor Eloheimo excommunicated them on the spot, and they walked out to form a new congregation.[38]

In 1900, about 7,250 foreign-born Finns lived in Houghton County. By 1910, that figure had climbed to 11,500. The Finns who came after 1900 felt the sting of a more intense social discrimination than that which had greeted earlier arrivals. Their discrimination took on a harsher cast, as mine managers discovered traits in the new Finns that they distinctly disliked. These men tended to be less interested in Lutheranism and more interested in socialism; less interested in temperance societies and more interested in unions; less interested in the dream of saving up to buy a small farm, and more interested in raising hell.

In fact, distinctive differences existed between the Finns who arrived before and after 1900. Many of the later arrivals came from more urban provinces like Turu-Pori and Uusima. These were home to "free-thinkers" who had embraced unionism and socialism, in large part due to a political struggle between the Finns and Czarist Russia. The Finns, situated between the Swedes and the Russians, had long been dominated by one or the other. In the nineteenth century, Finland was not an independent nation, but a Grand Duchy of Russia. From 1809 to 1898, the Finns promised their allegiance to the Czar, and in return enjoyed considerable autonomy in their affairs. But in 1899, Czar Nicholas II curtailed this autonomy and implemented a Russification program. He named a new Governor General of Finland, who issued a series of edicts that censored Finnish papers, limited freedoms, and conscripted young males into the Russian army. These moves turned many Finnish workmen into leftists dedicated to fighting repression. Some gave up the fight in Finland, but took their new politics with them to the Copper Country and other Great Lakes mining districts.[39]

Between 1900 and 1910, mine managers across the upper Great Lakes increasingly blamed Finns for instigating labor troubles. Charles Lawton condemned their role in the 1906 strike at Quincy. A year later, Finns incurred the wrath of mine managers on Minnesota's Mesabi iron range. After helping to lead an unsuccessful strike there, the Finns found themselves locked out of employment. At Calumet and Hecla, James MacNaughton did not want this ethnic group to dominate the ranks of unskilled labor. Besides declaring in 1912–13 that C&H did not want to hire more Finlanders, it appears certain that between 1905 and 1910 MacNaughton weeded out many Finns already employed, and replaced them with other immigrants, principally Austrians. In 1910, Finns formed the largest foreign-born ethnic group in Houghton County. Nevertheless, by the middle of 1913, Finns (both foreign- and native-born) comprised only the fourth largest ethnic group working underground at C&H. Holding just 13 percent of the company's 2,200 underground positions, they ranked behind the Austrians (22 percent), Italians (17 percent), and Cornish (15 percent).[49]

By 1913, Calumet and Hecla surely had the lowest percentage of Finns working underground in the copper district. Other mines, though, seem to have followed the same discriminatory policy. In 1905, Finns represented the largest ethnic group by far at the Champion mine. Between 1905 and 1912, Champion added nearly 200 men to its underground roster. But while the number of Austrians rose from 181 to 282, the number of Finns fell from 322 to 266, and their percentage of total underground employment declined from 41 to 28.[41] In 1914, the *Engineering and Mining Journal* reported that the Copper Country and other Great Lakes mines seemed to be eliminating Finnish workers.[42] The journal was correct in its assessment, but six to eight years late in breaking the story.

The measures that C&H officials took to preserve their hegemony extended well beyond the workplace. Beyond the mine proper the company nurtured influential allies who could be called upon to support C&H in disputes with other parties, including its own employees. One such ally was the press.

On the state level, in 1907–8 Calumet and Hecla cemented the loyalties of H. A. Gilmartin, a respected Detroit *Free Press* political reporter and editorialist, by paying him over $4,000 to cover fees and expenses. Gilmartin, codenamed "Gibbons," used his press credentials to gather information for MacNaughton in Boston, Chicago, Grand Rapids, Detroit, and Lansing, and used his paper to advance C&H's interests. He served as MacNaughton's eyes and ears downstate, at a critical time when Michigan was holding a constitutional convention. For example, Gilmartin informed MacNaughton of the maneuvers of the "Osceola people," who were trying to make it illegal for C&H to purchase stock in other Michigan mines. He tracked down the source of an Associated Press article on labor agitation in the Copper Country. He squelched news items in the *Free Press* deemed detrimental to C&H, and wrote at least one editorial emphasizing the company's importance to the state's economy.[43]

On the Keweenaw, the mining companies had friends and foes in the local press. Newspapers published in English were staunch allies of the mine managers. The editor of Hancock's *Evening Journal* knew his work was appreciated, when he received money and a note of thanks from the Quincy Mining Company for running a particularly favorable story. Another important press ally was Houghton's *Mining Gazette,* which was a weekly from the 1860s to the turn of the century, and thereafter a daily. The *Mining Gazette* had a consistent editorial policy; it never met a union organizer, striker, or socialist that it liked. After the *Mining Gazette* became a daily, its publisher, W. G. Rice, wrote MacNaughton that "the idea of the organizers was to have a paper which would be devoted to the interests of the mining companies in the Copper Country."[44] As if to reinforce that stance, Rice stocked the *Daily Mining Gazette*'s board of directors with mining company men.

Foreign-language papers were more troublesome for MacNaughton and other mine managers. Early in the twentieth century, Swedes, Italians, Croatians, Slovenians, and Finns published their own newspapers.[45] Of these, the Finnish papers elicited the most concern. Although the Finns were slow to speak and read English, they were highly literate in their native language. Managers saw the Finnish press as a potentially dangerous carrier of subversive ideas. While MacNaughton and other superintendents could do little, directly, to keep the Finnish socialist paper, *Tyomies,* out of workers' hands, they covertly tried to counter its influence by assuring the economic viability of rival, more moderate Finnish newspapers.

The first scheme to garner some control over the Finnish press seems to have been launched by Rice at the *Daily Mining Gazette* in 1902. He wanted to purchase the daily *Paivalehti,* and the semi-weekly edition of this paper, *Amerikan Uutiset.* In part, Rice was motivated by a desire to profit from such an acquisition. The semi-weekly's circulation of 5,000 papers rivaled that of the *Mining Gazette* itself, and the daily had a circulation of about 1,400. As Rice wrote to MacNaughton, the asking price was low, "and for many other reasons which will suggest themselves to you, as they have to other mining men in the country, the move is important and will surely be prolific of good results."[46]

The *Daily Mining Gazette* failed to get control of *Paivalehti,* and an alternate plan to do so was under way two years later. From MacNaughton's point of view, this was just as well. Since the *Mining Gazette* was "looked upon as a mine paper," its ownership of any Finnish paper would have discredited that paper and blunted its influence within the Finnish community. MacNaughton preferred O. J. Larson's 1904 plan to take control of Red Jacket's Finnish daily. Larson, Houghton County's prosecuting attorney, was one of the area's most prominent Finns. He also became a front for the mining companies as they infiltrated the Finnish press.

A schism existed between the moderate and radical elements of the Finnish community, between the "church" Finns and the socialists. Tangible evidence of the split was best found in Hancock, where within a few blocks of one another one found Suomi College and Seminary, which preached piety, and the publisher of *Tyomies,* which preached rebellion. Local wags had it that the only thing the two groups of Finns had in common was a love of the sauna.

More moderate Finns, such as Larson, shared many of the mine managers' social objectives. They did not want radical Finns, by agitating labor unrest and spouting the rhetoric of class struggle, to make that struggle more difficult for others who wanted to get ahead in Copper Country society. Larson and MacNaughton both had a stake in seeing to it that a moderate Finnish paper remained influential within the region.

By 1904, *Paivalehti* had a daily circulation of about 2,500. The paper was short of capital and offered $3,000 worth of stock for sale. O. J. Larson solicitated monies to prop up *Paivalehti* from William Paine of Copper Range and the leaders of a few other companies. Using their money, Larson bought stock in his own name. Then he turned to MacNaughton and asked if C&H wanted to invest in the Finnish paper. MacNaughton, in consulting with Alexander Agassiz, acknowledged that he favored the plan because "no one knows nor would anyone know that the mining companies are interested in the paper." This secrecy would preserve its "independence" and credibility: "We have a large number of Finlanders in our employ and this Daily is practically the only Daily they get. I am told that the paper aims to advise them along the proper lines regarding labor agitators and labor organizations, and think if we were to put any money into a newspaper it would result more advantageously to us by putting it in this way than by having it known we were directly interested." MacNaughton asked Agassiz if C&H should take 50 shares at $10 each. Agassiz wrote back to take 100 shares, because "it was a very good plan to help them and much more likely to help us."[47]

The mining companies carefully hid their interests in a major Finnish paper, but made no attempt to hide their control over local politics. Every sizable mine had trusted employees—ranging from foremen right up through the agent or superintendent—sitting in important township, village, or county posts. In 1913, just as Houghton County entered its most tumultuous period ever, to be sure a few farmers and bartenders sat on the county's board of supervisors. But mining company interests overwhelmingly dominated that board. James MacNaughton chaired it. He was joined by the former chief clerk of the Atlantic mine; the superintendent of the Winona mine; the superintendents of the Osceola and Copper Range stamp mills; a past superintendent of the Copper Range Railroad; the superintendent of the Hancock Consolidated Mine; and by Charles Lawton, Quincy's agent.[48]

Collectively the mining companies held great sway over Houghton County at large. Still, their power was not absolute, a fact that became clearer at the turn of the century and beyond. Certain pockets of independence—and of anti-company sentiment—grew strong, particularly in Hancock and Red Jacket. There, the mining companies retained many allies among businessmen and professionals, but the immigrant populations distanced themselves from company control and followed their own social and political agendas. Thus, while C&H reigned supreme in Calumet Township, and Quincy ruled over Quincy Township, only a short distance away in Red Jacket or in Hancock, prominent socialists or union men—including such a national figure as Eugene Debs—proselytized the population.

Alexander Agassiz found it more and more difficult to work his will, unchecked, in the Copper Country. The company's interest and the public interest seemed more divergent than before, as a growing number of residents challenged C&H's right to rule. In 1899 and 1900, Agassiz engaged in a bitter political battle over streetcars, and lost. Agassiz had heretofore thwarted such a transit system, because he did not want trolleys to transport worker unrest. On top of that, he could not tolerate the idea of C&H rock trains, on route from mine to mill, being stopped dead in their tracks while waiting for trolleys filled with shoppers to pass a grade crossing.

Agassiz threw a presidential tantrum over the idea that some independent streetcar company might be allowed a right-of-way across C&H property at the mine or at the mill in Lake Linden. The people wanted a streetcar line; shopkeepers wanted one; and normally supportive newspapers "abused" Agassiz for fighting it. Even other C&H leaders thought it wrong to try to block the streetcars, but the president browbeat them into joining his opposition. Agassiz was fighting for a principle. He would not allow his company "to commit the folly of having the newspapers and outsiders dictate to us what we should do and how we should do it." He was "bound to see if the Lake properties are to be run by the mining co's.," or by other interests."[49] "I mean to fight this to the death," he wrote to his mine superintendent, S. B. Whiting

Agassiz had Whiting line up political opposition to the streetcar line. At Lake Linden, James Cooper, head of the C&H smelter, exercised his influence within Torch Lake Township. Within Calumet Township, Captain John Duncan, assistant superintendent at the mine, did the same. Agassiz himself lined up other companies, Osceola and Tamarack, in opposition to the streetcars. The key problem areas became the villages of Red Jacket and Laurium, one on either side of the mine. To dissuade those villages from granting a streetcar franchise a right-of-way over public streets, Agassiz instructed Whiting to pass along some thinly veiled threats to public officials. Agassiz wrote that "the Red Jacket people can do what they please in their baylywich, so can Laurium, but they must not expect any favors from us hereafter." The principal "favor" that Agassiz had in mind was water. C&H pumped and sold water to both Red Jacket and Laurium. Whiting was to hint that their public water supply might suddenly be turned off. Laurium also needed to run a new sewer across some C&H property. Agassiz allowed as how "they can have a sewer or R.R. [railroad] but not both."

Despite Agassiz's protests, Calumet and Hecla lost its battle with the Houghton County Traction Company, and with the various governmental units that welcomed the transit company and gave it rights-of-way. Construction of the line started in 1900, and within a few years the streetcars ran from Houghton across the bridge to Hancock, and then northward to Laurium and into Red Jacket. By 1910, the company's 27 miles of track also ran to Mohawk and over to Lake Linden and Hubbell.[50] Agassiz had been able to inconvenience the traction company somewhat. Its lines into Laurium and Red Jacket took rather circuitous routes, bypassing the heart of the C&H mine, and its tracks had to be elevated over those of railroads at all crossings. But he had not been able to stop it.

Between 1900 and his death in 1910, numerous events in the Copper Country did not sit well with Alexander Agassiz: the growing litigation over mine accidents, the loss of the streetcar battle, the sporadic labor revolts, the appearance of more radical Finns, and the arrival of the WFM and of socialist clubs. The chief architect of C&H paternalism sensed that his company's dominance was eroding and that company benevolence alone could not cement workers' loyalties and prevent trouble. Indeed, in emphasizing paternalism, perhaps C&H had erred by being too soft. Perhaps paternalism had encouraged more open dissent and disatisfaction on the part of workers who mistook paternalism as a weakness, as a lack of corporate resolve and strength. In 1901 he wrote, "The trouble at the Lake it seems to me has been fomented by our own officers and lack of force exhibited in the handling of discontented employees."[51] That same year, he secured the services of "Big Jim" MacNaughton as C&H's general superintendent.

For nearly a decade, MacNaughton took his managerial cues from Agassiz. Maintain peace with labor as much as possible, but always on your own terms. Use a variety of tools and approaches: a bit of paternalism here, a bit of guile, surveillence, and subterfuge there. Stay ahead of trouble by getting rid of potential troublemakers. Never reveal weakness or doubt. Act decisively. Do not negotiate. Never let workers dictate to managers. Lay down the law and then enforce it.

MacNaughton performed up to Agassiz's standards. When Agassiz died, Quincy A. Shaw, Jr., succeeded him as president in Boston, but MacNaughton became the key man at Calumet and Hecla. Since "Jim" and "Quince" were friends, MacNaughton ran the mine pretty much as he saw fit, without fear of being second-guessed out in Boston. That gave him considerable confidence in dealing with C&H's problems: in trying to maintain its leadership role despite the declining copper values of the Calumet conglomerate lode; in pressing for new technologies, such as the one-man drill; in overseeing the other companies in which Calumet and Hecla had purchased controlling stock; and in monitoring and controlling the labor situation.

MacNaughton needed all the confidence he could muster. In truth, Calumet and Hecla was not the company it once had been. Its most glorious days had passed. Agassiz had enjoyed luxuries that MacNaughton never would. He had worked the conglomerate lode at its best, when its superlative riches put C&H in a class unto itself among all the mines. MacNaughton worked the lode in its decline. Also, Agassiz, the proponent of imperial, cool leadership, had enjoyed the advantage of being in Boston, out of the day-to-day fray. As a man of science as well as of industry, he had a life beyond copper mining. He escaped into his specimen collections at Harvard, and periodically sailed on far-flung expeditions around the world. MacNaughton lived at the mine, where there was no escape from work and its problems, save for occasional short holidays. He was the resident man in charge. While he enjoyed his local status and prestige, it came with a price.

MacNaughton served in a taxing position, within a managerial structure that could hardly be called modern. Largely because of the way his company practiced paternalism, too many small problems and requests reached his desk. It's therefore not surprising that he occasionally became abrupt or testy in dealing with them. He grew particularly weary of trying to finesse the myriad labor problems which showed no signs of going away. He continually patched things up in one way or another, yet there was no escaping the fact that he was saddled with a great many men whom he deemed inferior, and who were not respectful of his authority. If nothing came along to heat up the conflict between managers and men, MacNaughton would muddle through. But if some inflammatory acts did prompt a showdown, it would not be wholly unwelcome. MacNaughton could then truly assert himself. As his mentor had taught him, he could exhibit force—rather than mere finesse—in handling the discontented.

13

Showdown: The Strike of 1913–14

> An empty cupboard, a disheartened woman, crying children, disconsolate man. Picture of distress. A desolate home. In a district where only peace and plenty have reigned for fifty years. NO! God grant that it may not come to this in the copper country. God grant that the workers will come to appreciate who their real friends are. Parades, the sound of martial music, inflammatory speeches by irresponsible agitators, who ride to meetings while the workers walk, will not fill that cupboard. BUT WORK WILL!
>
> *Daily Mining Gazette,* 27 August 1913

In Denver, the executive board of the Western Federation of Miners grew worried in mid-July 1913 over the prospect of a strike in the Lake Superior copper district. The WFM had organized the copper miners in Butte, Montana, several years before, and the union had long wanted to win a confrontation with mine owners on the Great Lakes. Nevertheless, as its locals seemed bent on engaging the Michigan copper companies in battle, the WFM's executive board knew full well that the long-entrenched firms would be hard to beat.

The WFM did not allow its locals to take a strike vote unless such a referendum had the prior approval of the executive board. This arrangement kept locals on a leash, but out in Michigan the locals, much encouraged by recent, rapid growth, virtually refused to heel. They strained ahead, eager to strike, pulling the executive board along behind them. The executive board did not want to quash this spirit of rebellion, but feared a strike in the summer of 1913 would be premature. The locals needed more seasoning, and the WFM treasury needed enrichment, before tackling the likes of Calumet and Hecla, Quincy, and Copper Range.

In November 1908, workers had reestablished a WFM local in Calumet. The union foothold secured by the local was tenuous; a mere 27 men belonged at the start of 1909. Still, its founding rekindled interest in the WFM, and in February and May 1909, workers formed small locals in Hancock and then

South Range.¹ Since a small minority of copper workers on Lake Superior had demonstrated a renewed commitment to the WFM, the union channeled funds into an organizational drive. By autumn 1909, no fewer than six paid WFM organizers worked in Michigan, traveling back and forth between the copper and iron ranges.

Half a dozen WFM organizers worked in Michigan each month during 1910, and from mid-1911 to mid-1913, between 8 and 12 regularly worked the mines. Their ethnic diversity mirrored the region's polyglot labor force, because to recruit as many men as possible, the WFM organizers needed to communicate in several languages. So organizers Guy Miller, William Tracy, S. G. Chadbourne, and Harry Brown were joined by the likes of Ciagne, Belletii, DeMeio, and Bartalini; by Aaltonen, Moleta, and Valimaki; by Judda, Judesh, Strizich, Verbos, and Urbanac; and by Arseneault and Koepeke.[2]

The organizers told the Michigan copper workers that because of the WFM, underground workers in Butte worked only an eight-hour day. They worked for a minimum daily wage of $3, which was substantially higher than the average pay of Michigan workers. The locals attracted members who felt cheated by such unfavorable comparisons with Montana, and who believed they worked too long for too little reward. The locals attracted men from the later-arriving ethnic groups who had been discriminated against, both in the allocation of good jobs and in the parceling-out of paternal benefits, such as housing. The roster of WFM members included workers who felt the underground had grown too uncomfortable and dangerous, and who believed that arbitrary managers were pushing them harder than before. The WFM roster also included some socialists who wanted to topple the mining companies, dismantle capitalism, and restructure the entire economic system.

Different workers joined the union for different reasons, but from 1909 through the end of 1912, the Copper Country locals in fact grew rather slowly. Over these four years, total membership increased from about 300 to 1,000 names.[3] The WFM organizers slowly built a solid foundation for the locals, but their recruitment efforts lacked a rallying cry or a single intense issue that could quickly pull in men by the thousands.

Mine managers thought the union more meddlesome than menacing, as long as it had too few members to pose a real threat. The situation radically changed, however, in the first half of 1913. By early July, the three WFM locals had grown to five, and total membership had increased sevenfold to 7,000 men. A new Mass City local, in less developed Ontonagon County, rose from 137 to 411 members, while the Keweenaw County local, not formed until 27 May, swelled to 1,596 members in less than two months. The three Houghton County locals also showed impressive gains. The South Range local grew from 30 to 1,580 members; the Hancock local from 363 to 1,450; and the Calumet local from 440 to 2,048 men.[4]

The WFM became far more popular because the companies gave the union a cause that appealed to a broad spectrum of underground workers. Over the winter of 1912–13, the companies made clear their full commitment to the one-man drill, and this machine made union men out of thousands of miners and trammers. The drill destroyed the tradition of miners working together on teams. It promised harder work, and more dangerous work, because the miner now had no partner to look after him. The machine also threatened half of all miners with unemployment, or with displacement into lower paying jobs. As for trammers, they did not want to compete with demoted miners for work. They also opposed the one-man drill because it limited their chances of moving up into miners' jobs.

In the Copper Country, miners and trammers had viewed each other as two different groups, set apart by ethnic background, skill level, income—and by the amount of favor bestowed or withheld by the mining companies. So it was quite remarkable and surprising when miners and trammers came together fast in the first half of 1913 to oppose the new technology. The recognized that the companies would have their way with the one-man machine, unless the men combined under the banner of the WFM. They also knew that striking in the midst of a Lake Superior winter was impractical. As a consequence, the locals had the spring of 1913 to enroll as many members as possible, and the summer to wage their battle with the mining companies.

On 25 March, Charles Moyer, president of the WFM, had urged caution on Thomas Strizich, one of the union's paid organizers working out of Calumet:

> I was much pleased to hear of the progress being made in the way of organizing in Michigan and sincerely trust that the men there will realize the importance, in fact the absolute necessity, of deferring action that may precipitate a conflict with the employers until they have practically a thorough organization.[5]

But contrary to Moyer's wishes, the Copper Country's five WFM locals, combined under the umbrella of District Union No. 16, took actions in late June and early July to bring a strike about. While the union's executive board fretted over the foolhardiness of starting a major strike with less than $23,000 in the treasury, the membership in Michigan fretted over the passage of too many summer days. They were running out of time to have their strike in good weather. They had to speed things up. Goaded on by the Hancock local, home to a number of socialist Finns, the district union held a referendum between 1 and 12 July.[6] The ballot carried two yes/no questions:

> Shall the miners' unions, acting through the district union, ask for a conference with the employers to adjust wages, hours, and working conditions in the copper district of Michigan?

> Shall the executive board of the copper district union, acting in conjunction with the executive board of the Western Federation of Miners, declare a strike if the mine operators refuse to grant a conference or concessions?

While this referendum was under way, WFM president Moyer was out of the country. Vice-president C. E. Mahoney urged District Union No. 16 to defer making demands of the mine operators until after 14 July, when the national board would convene in Denver and thoroughly discuss the situation in Michigan. When the executive board met, it tried to figure out how to slow things down out on Lake Superior, where a reported 98 percent of the ballots had supported a possible strike. The executive board drafted a polite letter for District Union No. 16 to send to the copper companies. The letter asked for a meeting with mine officers on or before 28 July and made no mention of a possible strike. Then the executive board learned to its dismay that the district union had already sent its own letter to the mine managers.

The letter was not nearly as polite as the one drafted in Denver. It stressed that the men wanted to bargain collectively with the mining firms, because they realized that "as individuals they would not have sufficient strength" to correct the "evils" or lessen the "burdens" placed upon them. "Friendly relations" could continue between employer and employee in the Copper Country, but only if the mining companies agreed, by 21 July, to met with WFM representatives. Should the companies "follow the example given by some of the most stupid and unfair mine owners in the past," and refuse to discuss grievances, then they would throw away the opportunity "to have the matters settled peacefully," and they could expect the WFM to call a strike.[7]

The district union's letter sent a greater shock wave through the WFM's executive board than it did through the ranks of the mine managers. The board scurried to get some of the union's top brass out to Lake Superior. They needed to appear as if they were leading this oncoming strike, even if in fact they were being dragged into it by their rank-and-file members. As for the mine managers, they tendered no response to the district union's letter. They did not—and would not—recognize the WFM as the bargaining agent for workers. As a consequence, they would not answer a letter from that organization.

Failing to get a response from the companies, on 23 July 1913 the WFM called a strike. The main target, from the start, was Calumet and Hecla; no strike could succeed unless that company capitulated. But at C&H, James MacNaughton never gave a moment's thought to capitulation. MacNaughton believed it essential for C&H to adopt new technologies such as the one-man drill, and to enforce a more disciplined work ethic underground. Recently, in the face of labor unrest, he had slowed the pace of change, hoping to avert a strike. But now that a strike had come, MacNaughton almost welcomed it. He felt "more or less relieved." After a particularly trying and turbulent first week of the strike, when his family had to abandon their house for safety's sake, and when MacNaughton himself slept in a different bed every night, protected by an armed guard, he was still pleased that this festering labor dispute had come down to a real fight. He wrote to his friend and company president, Quincy Shaw:

> There is nothing in the situation to worry about. So long as yourself and the other directors stand behind our organization we can handle this situation to a finish. I have not felt blue or depressed for one minute; I know there is only one outcome and that we can bring it about. I haven't worried, have been in perfectly good health, haven't lost an hour's sleep nor a meal by reason of all this

trouble. Indeed I rather surprised myself. . . . I never felt more fit or more like fighting than I have during the whole procedure.[8]

Quincy Shaw remained in Boston, a safe haven far removed from the early street violence engendered by the strike. Yet despite his remote location, C&H's president seemed to appreciate, better than MacNaughton, the gravity of the situation facing his company and its surrounding community. Shaw, a day after the strike began, anticipated its long-term effects:

> The worst part of all this is that it is going to undoubtedly drive away a great many of our better men who don't want trouble but quiet places to live in and I sort of feel as if, no matter what the outcome, it will take a long time, and perhaps never, to get back to what the community was before.[9]

Almost as soon as the strike began, Quincy Shaw sensed a loss. Not a literal loss to the Western Federation of Miners—he was sure his company would win that battle. Instead, he feared a loss of tradition, a loss of an old order. As keeper of a great, paternal company, he keenly sensed that this strike was more serious than earlier labor disputes, and that it represented a dividing line in the company's history. Life at the mine would never be the same.

On the strike's first day, James MacNaughton expected trouble. He opened the mine as usual, counting on many underground workers to appear. He also expected nearly all shop and railroad employees to report, because they were "loyal and anxious to safeguard the interests of the Company." The WFM had attracted few surface workers into its fold. Thus, MacNaughton depended on carpenters, machinists, hoist operators, and boiler tenders to protect C&H from the WFM.

The WFM claimed as many as 9,000 members on the Keweenaw, out of a total work force of about 15,000 at the mines, mills, and smelters. The companies insisted that only 15 to 20 percent of their workers had joined the union.[10] At C&H, MacNaughton believed that for each WFM striker, he could secure two loyal men to serve as deputy sheriffs. In fact, Houghton County Sheriff James Cruse immediately deputized at least 200 C&H employees. MacNaughton placed one important restriction on these men: he refused them guns and permitted them only billy clubs. To lessen the chances that loyalists might be out-gunned by strikers, MacNaughton warned Calumet's National Guard unit to safeguard its rifles and ammunition. He also had all dynamite removed from company magazines and shipped back to the powder mill at Senter location.

Late in the afternoon of 23 July, strikers met in Laurium and then marched over to C&H. The mining companies had ignored the WFM's requests for meetings. They refused to acknowledge the union's existence. They refused to validate the idea that workers had any legitimate grievances. Confronted by majestically arrogant companies that would never willingly negotiate, the WFM had little recourse but to try to force its opponents to the bargaining table. The union had to capture wavering workers and scare off staunch loyalists. The strikers had to shut down Calumet and Hecla and the other mines and inflict severe financial losses. Lacking that, the companies would break the union and reaffirm their absolute right to rule the Copper Country.

Strikers confronted the C&H deputies, assaulted workers finishing the day shift, and chased off men coming to work the night shift. They hurled bottles, rocks, and pieces of steel; they clubbed and pummeled workers and deputies cornered at different parts of the mine. C&H's deputies were "hammered wherever they showed up and it was useless . . . to continue," so MacNaughton called off the night shift. Convinced that "professional gunmen" led the strikers, he thought C&H might have to "arm at least five hundred picked men" to reestablish order and control. The alternative, as MacNaughton discussed late at night with Sheriff Cruse, was to request Michigan's governor to send the National Guard to Houghton County.[11]

MacNaughton described Sheriff Cruse as a man "willing to do anything we tell him," but a man lacking "initiative and force." After the strike's initial melee, Cruse admitted he was unable to handle the situation, so MacNaughton consented to the sheriff's request of troops from Governor Woodbridge Ferris. Cruse telegraphed Ferris at 2 a.m. on 24 July, and later that day the governor ordered up the entire Michigan National Guard. By the strike's second day, the WFM had routed the men who wanted to work; the state's governor had detailed 2,765 men north to keep the peace; and Calumet and Hecla, Quincy, Copper Range, Tamarack, Osceola, and the other struck companies had suspended operations. By shutting the mines down, the strikers had achieved a primary objective. But it was James MacNaughton who was elated: "If we had planned the whole affair beforehand we could not have played into our own hands any better than the strikers did. The mob violence practiced by them put us in the best possible position. Outside of the ranks of the strikers themselves there is absolutely no sympathy for them anywhere."[12]

The general superintendent of C&H did not want too much blood to flow in the streets. After the first day's battles, he tried to avoid clashes between loyal employees and strikers, and he determined not to hire "strike breakers or gun men of any kind" unless "absolutely necessary." He believed, "It will be much better if we . . . get through with this thing without loss of life."[13] Still, MacNaughton appreciated the public relations value of some battered skulls. As long as the strikers did the battering, they cast themselves as villains and the companies as victims.

The various mine managers, much chummier during the strike than either before or after it, were not terribly distraught over the shutdown of operations, because they could turn it into an object lesson. Surely the companies were going to suffer production losses, and they would incur considerable expense in keeping loyal surface workers on the payroll, even though no mining was being done. (At C&H, with 1,570 surface workers, this cost amounted to $100,000 per month.) But the costs would be worth it, if the shutdown taught strikers—as well as nonstrikers who were locked out of work—that rebellion always came with a price: the loss of income.

Managers wanted underground workers to miss paydays, feel hunger, and have to tighten their belts. Quincy Shaw believed that although "it may seem hard on the better men, I believe that they and the Companies in the long run will profit most by a reasonably long shutdown." MacNaughton concurred, suggesting it might be a good idea to let the company's lesser mining operations fill up with water. Stopping the pumps, he thought, might convince workers that C&H would sooner quit mining than deal with the WFM.[14]

The Lake mining companies had a tradition of keeping considerable cash on reserve, and they had added to their reserves when a strike loomed on the horizon. Thus they could well afford to shut down long enough to force hardship on workers' families. At the start of the strike, when the WFM treasury held less than $23,000, C&H had $1 million cash on hand; another $1.27 million due it in bills receivable; 16 million pounds of copper ingot to sell; and 12 to 13 million pounds of mineral that would yield an additional 8 million pounds of copper when smelted. Quincy Shaw told James MacNaughton that when it came to the strike's cost, "don't let it worry you for a minute."[15]

With the companies content to close down, and with National Guardsmen camped out at the mines, strike violence generally subsided by the end of July and the first part of August. Labor and management alike reserved the two or three weeks after the Guardsmen arrived for public posturing, proselytizing, and propaganda. The strikers staged parades and rallies and listened to nationally known labor figures, such as Mother Jones, who praised their courage and urged them to stay away from guns and drink. On 9 August, the union began publishing *Miners' Bulletin,* a paper that promised "to Tell the Truth Regarding the Strike of Copper Miners." In its first issue, the *Bulletin* provided a succinct statement of the union's objectives: "What are our demands? Recognition of the union, eight-hour work day, minimum wage of three dollars for underground workers, 35 cents flat increase in wage of surface workers, and two men on all [rock-drilling] machines."

The mining companies always denied practicing collusion, but throughout the strike their superintendents met regularly, planned common strategies, and memorized the party line on all issues. In determining that party line, James MacNaughton took the lead, the other mine managers followed, and the companies spoke with a unified voice. Only in response to the WFM's demand for an eight-hour day would the managers ever demonstrate any softness. They rejected the granting of uniform raises or the setting of a minimum wage. The different companies, they said, could not all afford such increases. As for the new drilling technology: "The conditions of competition, the low grade of our rock . . . , the increasing expense with depth, and other conditions have made the use of the one man drill imperative for the continuance of operations."[16]

Laying aside dollars-and-cents issues, company officers simply could not stomach the idea of recognizing a union, especially the WFM. They saw the WFM as "a foreign body preaching socialistic doctrines, and inciting class hatred and disloyalty." The WFM had "a notorious record for crime, disorder and brutality." It practiced intimidation, coercion, and murder. Its constitution was immoral and un-American. The key issue driving the strike for the mine managers was not wages, or the need for new, more productive machines. From start to finish, the managers had to wreck this union to preserve their own hegemony. At the start, MacNaughton reported to Quincy Shaw that the managers were "all of one opinion, namely, that the Union must be killed at all costs." Near the end, MacNaughton had learned no mercy: "If we want to be insured against a repetition of this thing within the next 15 or 20 years we have got to rub it into them now that we have them down."[17]

The mine managers tried to emasculate union members by treating them as invisible men, by refusing to communicate with them, and by denying their complaints had legitimacy. Initially, the mine managers' smug arrogance infuriated WFM members and drove them to violence in the strike's opening days. Then the union retreated from violent tactics, and WFM vice president Mahoney and an attorney met with Governor Ferris to urge him to help mediate the strike.[18] The Democratic politician tried to oblige. He was chagrined by the fact that when he ordered up the National Guard, he had exposed himself to charges of strike-breaking and union-busting.

The governor first had General Abbey, head of the National Guard force, approach the mine managers about a mediated settlement. Then he sent Alfred Murphy, a circuit court judge from Detroit, to investigate the strike and try to bring the mine bosses and strikers together. But mine officials only reluctantly agreed to meet with such emissaries.[19] They never welcomed the participation of outsiders from the state or the federal government. They never agreed to arbitration or negotiation. They planned to wage and win this strike on their own terms. When James MacNaughton testified at a congressional investigation into the strike, he was questioned about letting some "high authority," some "disinterested power" arbitrate the dispute. Would he not have confidence in Michigan's governor, or even in the President of the United States? At this point the imperious, cool MacNaughton apparently brandished his wallet at the congressman and replied: "This is my pocketbook. I won't arbitrate with you as to whose pocketbook this is. It is mine. Now, it would be foolish to arbitrate that question. I have decided it in my own mind."[20]

The companies' refusal to talk with the WFM, and their rebuff of the governor's peace-making attempts, opened them up to criticism. They were often condemned in the press—but not in the English-language press of Houghton County. The editors of Houghton's *Daily Mining Gazette* and the *Calumet News* never took the mine bosses to task. As MacNaughton acknowledged to Quincy Shaw, those papers were "with us in this dirty fight from the very first. They have done everything in their power for us."[21] These papers tried to convince workingmen that the strike was misguided folly, destined to failure. Less than two full weeks into the strike, that message, coupled with the companies' refusal to budge, seemed to bear fruit.

By 4 and 5 August, working-class solidarity on the Keweenaw began unraveling. A large crowd convened in the Red Jacket Town Hall on the fourth to launch a back to work movement independent of the WFM. On the fifth the group sent representatives to meet with MacNaughton—and he received them, which was in stark contrast to his treatment of all union representatives. MacNaughton opened the door for men to begin sidestepping the WFM, and he counted on old loyalties and traditions, plus some new economic hardships, to drive a wedge between select men and the union.

The back-to-work men did not go begging to MacNaughton. They presented grievances, and in some ways made greater demands on MacNaughton than the WFM. A $3 minimum daily wage for trammers and timbermen was all right; but they wanted $3.50 for miners. They wanted miners to be paid for six full shifts per week, even if they only worked five. They wanted two men on each drill, or at least one man and one boy. They did not demand, however, company recognition of the WFM.

MacNaughton told the men he could not yield to their demands, because the WFM would surely claim credit for any concessions. Similar back-to-work movements took place at Quincy and Copper Range, and managers Lawton and Denton, following MacNaughton's lead, promised nothing. Nevertheless, these encounters paved the way for the companies to take back many underground employees and resume underground operations by mid-August.[22]

Popular support for the WFM began to erode quite rapidly. Many men could not afford to stay off work, and the union's promised strike benefits had not yet come. Some quickly figured that the companies were never going to give an inch, so it was foolish to prolong the agony of the strike. Others backed off because of the opening days' violence, or because of the socialistic, radical leanings they saw in certain union leaders.

Ethnic rivalries led to divisions in the ranks. Many Cornishmen felt uncomfortable in the midst of newer arrivals to America—just as many pious Lutheran Finns felt uncomfortable being lumped together with socialist Finns. The companies counted on preachers steering many good Christians back to work, and they also counted on many miners drawing away from the WFM. The companies had always treated miners better than trammers; they had housed them better and paid them more. Many Copper Country miners cooled to the WFM, when the egalitarian organization continued to emphasize that all underground workers were equal, and that miners and trammers should receive equal pay.

Finally the relationship between worker and union—only recently formed—was often less secure than the older tie between worker and company. Most C&H employees could measure their WFM membership in months or even weeks. By contrast, when the strike started, 1,660 men had worked between 15 and 45 years for the company—and 1,352 of the company's 4,300 employees were the sons of fathers who also had worked for C&H.[23] Many of these men stayed out of the union altogether, and of those who joined, many soon retreated from their experiment with organized labor.

In mid-August, the state of Michigan, which paid for quartering all the National Guard troops at the mines, ordered half the original force home. As Guardsmen withdrew, more of the burden of protecting mine property fell to about 50 guards brought in from the Waddell-Mahon Corporation in New York City, and to 600 local deputies they had trained. After earlier denouncing the WFM for bringing in thugs, the companies exposed themselves to the same charge by securing the services of the Waddell "gunmen." Just as the Waddell men became more visible and important, the companies increased the likelihood of violence by preparing to reopen their mines and take back men wanting to work.

In this emotionally charged atmosphere, on 14 August the WFM garnered its first martyrs, men gunned down at a Croatian boarding house at the Champion mine's Seeberville location.[24] While returning to Seeberville from South Range, strikers John Kalan and John Stimac walked across a patrolled part of the Champion mine declared off-limits to union men. When told to stop, Kalan and Stimac ignored the guards and walked on through. The guards then informed a Waddell man, Thomas Raleigh, of the trespass. Raleigh marched six deputies, including four Waddell men, over to Seeberville to arrest Kalan and Stimac.

Spotted outside his dwelling, Kalan refused arrest, and a fight ensued with the armed deputies. When a hurled bowling pin struck a Waddell man on the back of the head, he whirled around and fired, shooting Steve Putrich. The guards then surrounded the house, firing a fusillade through its windows. Inside, bullets struck Alois Tijan, Stanko Stepic, and John Stimac. Tijan died instantly; Putrich died the next day.

Earlier, the WFM had been the headbashing organization. Now the mining companies and the county sheriff bore the stigma of violence. They were the ones who imported armed "goons" and turned them loose. For many of the 5,000 individuals who turned out for the funerals of the Seeberville victims, the deaths inspired greater solidarity and resolution. Yet at the same time, others decided it was time to go back to work. The workers' deaths and funerals coincided with the reopening of the Champion mine on 14 August, and of the Calumet and Hecla branches of C&H on 15 and 16 August. Quincy restarted a few days later. Almost all Houghton County mines hoisted some rock again by 1 September.[25]

The companies used the resumption of mining as a tool to break the union. Any man taken on had to swear he was "not now a member of or in any way connected with the Western Federation of Miners or any branch or local thereof." He had to promise that in consideration for his employment, he would not become a member of the WFM. Finally, if he had been a member of the union, he had to surrender his personal membership book.[26]

The companies knew who the union members were, because loyalists had recorded the names of men seen at WFM parades and rallies. Generally the companies forgave union membership, as long as a man had since quit the organization, but they did blacklist especially active WFM members. At C&H, MacNaughton wrote President Shaw on 2 and 3 September that he thought the strike was all but broken because nearly half the underground force was back to work. Still, "there are at least 250 or 300 of the Calumet and Hecla employees that we will never take back. We wouldn't feel safe a minute having them in the mine, nor will we leave even a nucleus of an organization underground."

As soon as they reopened, the companies all thought the strike was won. They were considerably shy of their pre-strike levels of production and employment, but they believed it was only a matter of time before more men returned and the WFM collapsed. They sensed the strike was narrowing down to one of trammers and unskilled laborers, because many miners were coming back. The strike was breaking down along ethnic lines as well. By early October, C&H reported that 98 percent of its Cornishmen were back to work, as were 80 to 90 percent of all Scots, Irishmen, and Scandinavians; 60 to 65 percent of the southern Slavs; and half of the Italians. Its holdouts were predominantly Finns, Hungarians, and Croatians, who had returned at the rates of 35, 10, and 7 percent, respectively.[27]

The mine managers expected one of two things to deliver a killing blow to the WFM locals: the first hint of winter, which would send men scurrying back to work, or insufficient funds. The WFM had promised its members benefits, which were to start a month after the strike began. These ranged from $3 per week for single men, to $7 for married men with five or more children, and then up to a maximum of $9 in cases of emergency.[28] These benefits were key to continuing the strike; without them, the men could not afford to stay out.

James MacNaughton used influential friends to keep track of the financial condition of the Federation in the local banks, and R. R. Seeber, superintendent of the Winona mine, did the same. (Seeber shared his information about union bank balances and creditors with MacNaughton; not lacking a sense of humor, he signed one letter to MacNaughton as "Travelling Auditor, WFM.")[29] Based on their knowledge that the union had collected a paltry $8,000 for the strike through mid-August, and on their estimate that strike benefits for 4,000 workers would cost $100,000 per month, the mine managers believed the WFM incapable of sustaining the struggle.[30] The Western Federation, however, had a major surprise in store for the mine managers.

On 30 August, WFM president Charles Moyer arrived in the Copper Country for the first time since the strike began. He announced that the WFM had started soliciting financial assistance from unions all across the country, and it had the support, among others, of Samuel Gompers, president of the American Federation of Labor. The reopening of the mines had cast a pall over the strike, but Moyer invested the strikers with renewed vigor and hope. The WFM, he said, had the tactics and resources to win.

To cement strikers' loyalties and get them money to live on, the WFM levied assessments of up to $2 per month on all members outside Michigan. It borrowed $100,000 from the Illinois District of the United Mine Workers of America, and $25,000 from the United Brewery Workmen of America. The WFM set up a special Michigan Defense Fund to receive contributions. The Defense Fund received only $1,876 in August, but took in ten times that amount in September, and then about $30,000 per month in October through December. Locals of the AF of L and the United Mine Workers made substantial contributions, as did diverse union organizations, particularly from Chicago and Butte. Croatian, Slavic, and Finnish socialist societies also sent funds regularly.[31]

As part of its strategy, the WFM wanted to portray the companies as oppressive belligerants, while securing for itself the high moral ground of fairness and reasonableness. In September, Charles Moyer and the renowned attorney Clarence Darrow approached Governor Ferris twice to request that he help set up arbitration. The sympathetic governor used "moral persuasion" and the symbolic power of his office to try to soften the companies' anti-union stance. Unfortunately, just two years before, Michigan had repealed a law that provided for state investigations of serious labor disputes where the parties refused arbitration. Lacking a legal lever to move the mining companies, Ferris was never sanguine about his chances for effecting arbitration: "When James MacNaughton says that he will let grass grow in the streets before he will ever treat with the Western Federation of Miners or its representatives, I believe what he says."[32]

The WFM anticipated that the mining companies would be little moved by the governor's public accusations that they were unfair and arrogant, so they tried to budge them by taking to the streets. They stepped up the pace of parades and began picketing the mines. Pickets sought to stop any more defectors from returning to work. They hoped to intercept and to intimidate men who had already gone back, causing them to come out of the mines again. Finally, they wanted to turn away any new men, scabs, who had been imported into the district on guarded trains to help break the strike.

In the first 20 days of September, daily confrontations between strikers and workers, deputies, and National Guardsmen became routine. A woman and a girl provided the strikers with their newest brave victims of the abuse of power in the Copper Country. "Big Annie" Clemenc, marching while draped in an American flag, was raked across the wrist by a Guardsman's bayonet. Near Kearsarge, Guardsmen confronted 200 marchers who were protesting the use of scabs. Marchers hurled rocks, the troops discharged their weapons, and 14-year-old Margaret Fazekas fell with a serious gunshot wound to the head.[33]

Women indeed "manned" the front lines during September. A striker said that workers' wives were the "heart and soul in the cause. They urged us to strike and are urging us not to give in." A National Guardsman reported, "The women are the real active ones in the district. The men are more inclined to sit back and let the women do the fighting, for it is more difficult [for Guardsmen and deputies] to deal with them than with the men." James MacNaughton informed Quincy Shaw that after one morning's "rioting" in Calumet, 15 of the 20 persons arrested were women. He also painted a vivid picture of how the women supporting the strike violated his sense of decency and decorum:

> I cannot tell you to what length the Finnish and Croatian women are going in this matter. . . . At the Trimountain mine last week, the Finnish women dipped up a pail of human excrement from an outside water closet. They put a long stick

through the handle and carried the pail down the street followed by five other women with brooms, the intention being to smear any non-Union man they could find. Indeed one non-Union man was caught by the women who watched him go into a water closet, they plastered him thoroughly over the head and face with human excrement. Here at Calumet they have not been quite so bad in their acts but the language they have used is beyond description, and a "S-O-B of a scab" is a common expression among the politest women of the strikers.[34]

In the midst of this turmoil, the mining companies rebuffed a settlement overture made by the newly created U.S. Department of Labor. Its Secretary, William B. Wilson, had assigned Bureau of Immigration personnel to investigate and assess the copper strike. On 15 and 17 September, federal officials met first with MacNaughton, and then with the other mine managers. All steadfastly refused the Department of Labor's offer to set up a workers' committee to negotiate a settlement.[35] The only way to settle this, the managers said, was to have the WFM get out of the district, and to have all the men renounce the union and go back to work.

Instead of seeking peace through negotiation, the 17 struck mining companies sought peace through legal injunction and turned to the law firm of Rees, Robinson & Petermann (Calumet and Hecla's regular attorneys) for assistance. The mining companies and the lawyers had a secret agreement whereby mutual and combined legal action was channeled through this law firm. Rees, Robinson & Petermann handled the companies' correspondence with the U.S. Department of Labor, Governor Ferris, and the National Guard. They filed complaints and warrants that led to the arrest of strikers. They arranged for lawyers to defend the Waddell men and deputies charged with the Seeberville murders.[36] Now they applied for an injunction to stop the strikers' picketing and their harassment of men who had gone back to work.

The companies' request for an injunction went to an old legal foe, Patrick O'Brien, who once had specialized in filing personal injury lawsuits against the mining firms. Now, as an elected judge, he was caught in the middle of this strike. His personal sympathies clearly fell with the strikers, but he wanted to preserve law and order. He wanted a strike settlement to come about through negotiation, and not through threats or intimidation. Reflecting his personal ambivalance, O'Brien's legal actions vacillated. He granted the companies their injunction on 20 September, but just nine days later he lifted it, emphasizing that the strikers had rights of assembly and speech. When the mines' attorneys applied for a second injunction, O'Brien rejected it, which prompted Allen Rees to carry an appeal to the Michigan Supreme Court.

On 8 October, the state supreme court reinstated the injunction. Strikers could parade peacefully, but they were not to intimidate men who wanted to work. A few weeks later street violence picked up again, and Judge O'Brien issued orders to enforce the injunction. This led to some mass arrests of strikers: 141 at Allouez and another 68 at Mohawk. But when those individuals came before O'Brien, he became tolerant and lenient. He reminded them to obey the law, but then noted that they had been pushed into violating it. The strikers were not bad men: "The men on strike are most interested in showing they are law-abiding citizens and willing to do their duty as citizens and inhabitants of this county."[37] Instead of having them prosecuted then and there, O'Brien continued their hearings and released them on their own recognizance.

The mining companies were especially keen on having an effective injunction in place, because in September and October they launched major drives to import scabs, both to break the strike and to get up to normal operating levels. Near the first of October, Calumet and Hecla had about 80 percent of its men back, but Quincy and Copper Range had seen only about a third of their workers return. Some of the absent men were active strikers. Others stayed out because they were afraid to go back in. Still others, from 1,500 to 2,500 men, had left the district since the strike began.[38]

To get additional men, Quincy turned to a licensed employment agency in New York City, while C&H sent out its own recruiters. C&H rounded up men not only for the Calumet and Hecla mine proper but for other mines in which the company held an interest: Ahmeek, Allouez, Centennial, Isle Royale, Osceola, Kearsarge, and Superior. By mid-December, C&H had brought in 1,200 new men, and another 500 arrived over the next two months.[39]

Once the new hires boarded a train coach bound for the mines, they often found armed guards at the doors. The WFM claimed the guards brought the scabs in under duress; the companies claimed the guards protected the men from the WFM. The potential for violence increased as the men traveled closer and closer to the mines. At Copper Range, superintendent Denton proposed bringing the imported men in on special trains, with a car of men sandwiched between two cabooses; "These cabooses being supplied with cupola lookouts command all sides of the train, it being possible to seat two men on each side with a window for each man to look out through, and to shoot through if necessary."[40]

Out at the mine locations, while many imported men were put up in hastily constructed bunkhouses located near the shafts, many strikers still occupied company-owned houses, even though they had not paid any rent for months. The companies had agreed among themselves to forego the forced eviction of strikers. The mine managers realized that mass evictions, in this northern climate, would constitute a public relations disaster. To pressure the strikers, and to teach them a lesson in corporate control and worker dependency, some companies sent out eviction notices—but then stopped short of enforcing them.[41] Upon receipt of a notice, some strikers moved themselves out. But perhaps as few as two or three families were actually put out into the street, along with all their possessions, by the mining companies. They flashed the weapon of eviction, without really using it.

By the start of November, Moyer, Mahoney, and other WFM leaders knew they were beaten. In September, they had intensified the struggle; now they looked for a face-saving way to end it. On 2 November, Judge O'Brien summoned James MacNaughton to a meeting, which the mine superintendent later described in a letter to Quincy Shaw:

He told me he had talked to Union officials and they said they were whipped. They realized they could not get recognized in this camp and were ready to pull out provided we would take the men back without requiring them to withdraw from the Union. This I positively refused to do.[42]

For MacNaughton, this was no time to back off. A final victory was in sight that could be won wholly on the companies' terms. A week after MacNaughton met privately with Judge O'Brien, he applauded the arrival of a new organization dedicated to ridding the Keweenaw of the Western Federation. Self-styled as a popular movement, the "Citizens' Alliance" shrouded its origins in mystery. Nobody publicly stepped forward as its creator. Yet there was no secret as to whose side the Alliance was on. It was named after a similar group that had once routed the WFM from Colorado. Members espoused that the union's "poisonous propaganda of destructive socialism, violence, intimidation, and disregard of law and order" had to end, and because it was "a menace to the future" of the Copper Country, "the Western Federation of Miners must go."[43]

In the three weeks prior to the birth of the Citizens' Alliance, more than 400 strike-related arrests had been made. The National Guard force was down to fewer than 200 troops, but Sheriff Cruse had deputized 1,200 men, and Calumet and Hecla alone employed nearly 400 "Deputies and Hotel Men" to protect mine property and scabs.[43] Copper Country residents had grown weary of all this tension and turmoil. Local sentiment hardened against the WFM as residents longed for an end to parades and arrests. Many merchants, once supportive of the strike, now opposed it because the Western Federation had opened its own stores in the region. Now it paid strike benefits not in cash but in coupons redeemable only at the union stores. So in one sense, the Citizens' Alliance did signal a growing, grassroots opposition to the WFM. But the mine managers and their lawyers almost certainly created the Citizens' Alliance themselves, and they definitely organized and financed many of its activities.

On 27 November, the Citizens' Alliance published the first issue of *Truth,* its propagandistic rebuttal to the *Miners' Bulletin.* The publication laid out the organization's anti-union stance, and touted the fact that the Alliance already claimed 5,236 members. The Citizens' Alliance distributed *Truth* free of charge throughout the district, thanks to the covert largesse of the mining companies, who bore its printing costs.[45] One day after secretly helping to unleash popular opposition to the WFM, the mining companies openly undercut the union in another way: they announced that a shorter workday and new grievance procedures would go into effect on 1 December.

Two years before the strike, Calumet and Hecla and Quincy fought against a bill in the Michigan legislature to establish an eight-hour day. Company presidents Shaw and William R. Todd agreed the bill was "most objectionable," and it failed to pass. Shortly after the strike began, the mining companies said they had been considering the merits of the eight-hour day for some time, and they might implement it in the future. In truth, the companies opposed the shorter workday, but they no doubt reckoned that this was a Progressive Era reform that would eventually come to pass. That being the case, the mine managers reserved the eight-hour day as a bone they could toss to their underground men when the time was right.[46]

The companies also promised to pay more heed to workers' on-the-job problems. At C&H, this meant that MacNaughton set aside "a fixed day every week for hearing complaints and grievances." The general manager of Calumet and Hecla never sympathized with hard-core socialists or WFM members, but the strike did teach him that some good workers had legitimate complaints about how they were treated by their immediate bosses. MacNaughton, before the strike ended, admitted that many petty bosses had been "dictatorial." They had been "overbearing, arbitrary and short" when dealing with employees.[47] MacNaughton said he would fix all that.

The companies did not want the eight-hour day and new grievances procedures to appear like concessions to the WFM, so they delayed them until the mines had been reopened for several months. The companies expected their "voluntary" moves to constitute a public relations coup that would stimulate more anti-union feeling. Also, they thought their actions perfectly timed to cause men to bolt from the WFM at the last possible instant before hard winter. Surely the strike would now collapse.

But the month of December, instead of witnessing the winding-down of the strike, saw a string of highly divisive, emotional, and tragic events. It was almost as if the pressure of the long, hostile strike could not merely dissipate; it needed, finally, to explode.

On 5 December, Judge Patrick O'Brien held the final hearing for 139 of the persons arrested late in October for interfering with returning workers at the Allouez streetcar station. O'Brien found them guilty of contempt of court for violating the injunction, but then suspended all sentences. The judge expressed his view that the strikers acted "through enthusiasm for their cause." They were not criminals, but men "engaged in a heroic struggle for the mere right to retain their membership in a labor organization." The mining companies had provoked the strikers by doing everything possible to "increase their bitterness and hostility."[48]

On 6 December, the Citizens' Alliance condemned Judge O'Brien's remarks as an invitation to WFM violence and lawlessness. But at the same time, the mining companies' strategies seemed to be working, as more men returned to the mines. On this day, Arthur and Harry Jane, two brothers in their early twenties, arrived back in the Copper Country by train. They had left just before the strike. Now, thinking the disagreeable conflict was all but over, they came back to the Champion mine and to their Cornish friends who boarded with Thomas Dally.

At about 2 a.m. on 7 December, John Huhta, once recording secretary of the WFM's South Range local, hunkered down outside the Dally house. Beside him were two Finnish union men, Hjalmer Jallonen and John Juuntunen, and Nick Verbanac, a paid WFM organizer. Armed with rifles and bent on scaring off scabs, the men fired numerous rounds into nearby houses.

In the Dally residence, the Jane brothers died almost instantly. Thomas Dally took a bullet in the head and died before morning. Next door, a rifle shot hit a 13-year-old girl.[49]

The killers' identities were not known at the time, but Citizens' Alliance members harbored no doubts: the gunmen were Western Federation men. Coming on the heels of Judge O'Brien's liberal treatment of the strikers, the Painesdale murders propelled many residents to conclude that the Copper Country was on the brink of anarchy. Fear and outrage led to near-vigilantism. The Citizens' Alliance quickly assembled a rally at the Calumet Armory on 7 December and called for Sheriff Cruse to press for "absolute enforcement of the law" and "to rid this community of these murder inciting mercenaries."[50] The Citizens' Alliance announced plans for mass meetings of sympathetic residents to be held at Calumet and Houghton on 10 December to protest the Painesdale murders.

On 8 December, Houghton's *Daily Mining Gazette* announced that "FOREIGN AGITATORS MUST BE DRIVEN FROM THE DISTRICT AT ONCE." As the Citizens' Alliance and the local English-language press beat the drums for union banishment, their efforts drew strength from strong undercurrents of ethnic discrimination and hatred, which now surfaced. The strike was seen as headed by Finnish radicals, backed by southern and eastern Europeans. When men from the bottom rungs of society started killing Cornishmen, it was time to bring this "reign of terror" to an end.

On 10 December, the mining companies gave their employees the afternoon off—with pay—so they could attend one of the Citizens' Alliance meetings. The law firm of Rees, Robinson & Petermann orchestrated the events.[51] They hired the halls, arranged for special trains to deliver workers free of charge to Houghton or Calumet, and booked the bands. (Later, the attorneys billed their client mining companies for all these services.) The rallies were well attended. A reported throng of 10,000 marched in Houghton, where 6,000 persons later jammed the Amphidrome to hear civic leaders condemn the Western Federation. At Calumet, the street crowd supposedly reached 40,000 and 4,000 later squeezed into the Armory, and 8,000 into the Colosseum, to hear leaders deplore the erosion of law and order, attack Judge O'Brien's lenient rulings, condemn the WFM for its violent disturbance of local tranquility, and commend the mining companies. It is interesting that the companies' attorneys, Rees and Petermann, spoke at these rallies. The job of the Citizens' Alliance, they said, was to rid the community of the WFM's "poisonous slime."[52]

In the week or 10 days after the Painesdale murders, Houghton County won low marks for protecting citizen's civil rights. After the inflammatory orations of 7 and 10 December, deputies and Citizens' Alliance members broadly harassed striking WFM members. Some serious, but nonfatal, shootings occurred, and deputies raided many WFM enclaves up and down the Keweenaw, confiscating guns and other weapons. Despite the turmoil and pressure, and despite the discontent generated within union ranks by the murders and the Citizens' Alliance "counter-offensive," the WFM locals did not collapse.

In a 17 December letter to Quincy Shaw, James MacNaughton begrudgingly acknowledged the strikers' tenacity: "the leaders are holding them together very well and but few of them are returning to work." But the next day, he and other mine managers tried to drive a final wedge between the rank-and-file strikers and their leaders. The companies jointly announced that all men who failed to return to work immediately would be replaced. Their jobs would go to new men imported into the district.

At this point, some local businessmen interceded on the strikers' behalf, probably in hopes of salvaging something good from what had been a disastrous lead-in to the Christmas season. The businessmen appealed to MacNaughton to delay the return-to-work deadline until after the first of the year. MacNaughton agreed. After all, the strike had worn him down, too, and he needed a respite from his troubles over Christmas. In July MacNaughton had "never felt more fit or more like fighting," but my mid-December he had changed his tune: "I get terribly blue and depressed at times, and as time goes on it requires more frequent short vacations to pull together."[53]

Strikers and their families certainly needed relief from the strain and hardship they had lived with for five months. They were now a distinct minority, surrounded by an ever more hostile and short-tempered majority. Christmas at least promised some diversion and cheer. To make the best of it for strikers' children, the WFM planned a party for the afternoon of Christmas eve, to be held upstairs at Calumet's Italian Hall, owned by the Societa Mutua Beneficenza Italiana. Italian Hall was an attractive, five-year-old masonry building, two stories tall, located a bit more than a block from the village's town hall and fire station. The Italian benefit society leased the first floor to a bar and to a Great Atlantic & Pacific Tea Co. store. On one end of the building, double doors set under a Romanesque arch opened to a straight flight of stairs leading to the social rooms above. Diverse working-class organizations often booked these rooms for parties and meetings.

Children and parents began arriving at the party early in the afternoon of 24 December. Western Federation people scrutinized all who entered, making sure that only staunch unionists partook of the festivities. By two o'clock, over 175 adults and about 500 children crowded the hall. They sang carols and listened to some speeches. Mothers and fathers shepherded their children into a long line that wound ahead to Santa Claus, who had treats for each child. After the children got their modest gifts, some families stayed in the hall, while others bundled up, bid their good-byes, and left. Then, suddenly, at about 4:30, the remaining crowd bolted toward the exit.

Some fortunate ones at the head of the charge made it safely down the stairway. They pushed open the double doors and left the building. But behind them, somebody tripped and fell, causing a tragic chain reaction. Dozens of individuals near the bottom of the stairs crumpled into a tangled mass. A human dam formed at the base of the stairs, and it held back the dozens more who kept pouring into the stairwell from the top. Screams subsided as victims lost all breath. People started dying from the bottom up, killed under the crush of their friends and neighbors, their brothers and sisters.

In 70 years of mining Lake copper, the worst fatal accident had taken place not far from Italian Hall, at the Osceola mine. There, 30 men and boys had died, most from asphyxiation. In the face of a real fire, they had stayed put, instead of scrambling to safety. At Italian Hall, the situation was tragically reversed. The bodies in the stairwell got there while frantically fleeing from a nonexistent fire. Apparently, a man's cry of "Fire!" triggered the mass hysteria, but when rescuers arrived they found no fire—only 73 contorted corpses on the stairs, and a seventy-fourth victim who would die the next day.[54]

The victims so tightly packed the bottom of the stairs that rescuers had to peel off the bodies, one by one, from the top. They carried them over to a temporary morgue set up in the village hall and laid them out. The dead included about 3 Italians, 20 Croatians or Slovenians, and nearly 50 Finns. Sixty of the victims were children, two to 16 years old. A coroner or an assistant stripped the bodies of six young girls, then placed them side by side, with some sheets covering their middles. With their eyes closed and their heads almost touching, with their bare shoulders and dirty feet sticking out from the covers, the girls looked like urchins who had finally collapsed at a slumber party, except they were dead. A photographer took their haunting portrait, and the battle for the bodies was started. Who could rightfully claim these victims, who was allowed to mourn?

Italian Hall was no private tragedy felt only by the immediate families and close friends of the dead. It was a public, communal disaster, shocking in its scope and particularly wrenching because it came on Christmas Eve. But it occurred in a place where all sense of community, of oneness, had been ripped apart in the last half-year. Did the victims belong to everyone, or just to the Western Federation of Miners?

Residents along the Keweenaw Peninsula grieved for the dead and immediately set up a relief fund for the suffering families. By noon of 26 December, they had raised $25,000. Contributions came in from mine officers and merchants, from men and women who had always opposed the WFM. Compassion, when it came to this event, knew no bounds. But then the president of the WFM, who was in the Copper Country over Christmas, delivered a hurtful snub to all these sympathetic souls. Charles Moyer disallowed their grief, saying that the union would bury its own dead and tend to the needs of survivors: "No aid will be accepted from any of these citizens who a short time ago denounced these people as undesirable citizens."[55]

Months before, the mine managers had enraged many workers by treating them as invisible men. When they asked to talk, the companies refused. Now, Moyer turned the tables on the mine managers and all their allies; he made them the invisible, no-account ones. When they offered aid, he cut them off, refusing to validate their sympathy. The rejection stunned and angered the good people. Then the union leaders and their allies, such as the Finnish socialist paper, *Tyomies*, went one giant step further: they laid the blame for the Italian Hall tragedy at the doorstep of the Citizen's Alliance.[56]

What exactly happened at Italian Hall—what caused the rush to the stairs— was not known in the hours and days immediately following the tragedy, and it is not known now. The same fear and panic that killed seventy-four also muddled and distorted the recollections of survivors. People enjoying themselves in different ways on different parts of the second floor suddenly found themselves trapped in a complex, chaotic scene. When it was over, all their reports taken together painted a picture of contradiction and confusion, not of consensus. But a number swore that they heard, at the start, a man yell "Fire!" A smaller number swore that they saw the man who shouted it. And an even smaller number, perhaps half a dozen, swore that he wore on his coat the membership button of the hated Citizens' Alliance.

The union, rightly or wrongly, blamed the catastrophe on an unidentified Citizens' Alliance intruder, who came up the stairs, yelled "Fire!" at the doorway of the party, gestured for everyone to race to safety, and then somehow escaped getting trampled himself. Many union members imputed even greater guilt to the Citizens' Alliance: after one member had yelled "Fire!," others at the foot of the stairs had barred the way, causing the pile-up of bodies. In the emotionally charged hours following the tragedy, the WFM leadership attempted to turn what was probably a deadly accident into a murderous conspiracy. This strategy resulted in an ugly, illiberal backlash.

For many, the WFM had gone too far. First it refused the compassionate help of ordinary citizens; then it charged the Citizens' Alliance with murdering 60 children and 14 adults. The charge demanded a swift response. On the evening of 26 December, a band of angry, self-righteous men snatched Charles Moyer out of his Hancock hotel room and passed him to a mob outside that dragged him over the Portage Lake bridge to Houghton. There, they loaded Moyer onto a train bound for Chicago. The fresh gunshot wound in his back was a reminder of how unwelcome he and his union were in Houghton County.[57]

The Italian Hall disaster and the frenzied deportation of the WFM president called forth a rash of investigations and much criticism from around the nation of the mine managers and the Citizens' Alliance. Governor Ferris finally traveled to the Copper Country to assess the situation first-hand. The U.S. House of Representatives authorized its Committee on Mines and Mining to investigate the strike and working and living conditions in the copper district.[58] On 12 January, the *New York World* editorialized, "Houghton County . . . is not a community of self-governing American citizens. It is one chiefly of aliens brought thither to serve the monopoly, and ruled from Boston in defiance of law and in despite of democratic institutions.[59]

In untold theatres, motion picture-goers watched newsreel footage shot in Calumet. They saw Italian Hall, the front of a Finnish church spewing forth casket after casket, and the long funeral parade leading to a mass burial at Lake View Cemetery. The film adopted the union's side of the story: a man wearing a Citizens' Alliance button had triggered the castastrophe. At the Empire Theatre in Chicago, the WFM used the film as a fundraiser for victims' families. The marquee read, "Pictures of Calumet Disaster," and beneath it a touter beckoned to pedestrians. Inside, a lecturer praised the union's struggle, spoke of the need for relief funds, and condemned the mining companies.⁶⁰ Before running the film, he projected slides, ending with views of the dead bodies in the temporary morgue at the village hall. One of the most riveting ones showed the dead little girls, all lined up in a row.

In January 1914, while Chicagoans watched the slides and film of Italian Hall, up along Lake Superior the mine managers and their friends shrugged off the condemnation leveled at them by outsiders. They never retreated from their mission for a moment—which was to deal the union the worst defeat possible. The champions of the status quo, of the old-time rules that had governed this closed and controlled society so well in the nineteenth century, were most pleased when, on 15 January, the Houghton County Grand Jury indicted Charles Moyer and three dozen other WFM members for conspiring to keep men from going to work. Then, on 24 January, the Grand Jury completely exonerated 17 men named in a bill of indictment for Moyer's assault and deportation. Many outsiders felt that things were going crazy in Houghton County. But powerful insiders believed that things were now coming together very nicely. The total rout of the WFM was on.⁶¹

As the winter of 1913–14 wound down, many diehard strikers from the locals wanted to continue the struggle, but Moyer and other WFM leaders knew it was hopeless. They needed to cut their losses in Michigan and run, before the strike did any more damage to their organization. On 4 April, the WFM's executive board took a measure calculated to encourage the locals to quit. They did not pull the WFM out of the fray, but they trimmed the strikers' benefit payments. In response to this move, on 12 April some 2,500 still-active WFM members cast strike referendum ballots, and nearly 80 percent voted to call it off.⁶² After nearly nine months, the strike officially ended with a whimper on Easter Sunday, when its staunchest supporters finally abandoned the fight.

The Western Federation had lost hard-fought battles before, but this one proved particularly devastating. The strike left the union buried in debt. It had spent $800,000 in its vain attempt to organize the Michigan copper workers, including $125,000 borrowed from other labor organizations; $275,000 that had come in as voluntary contributions; and nearly $400,000 of mandatory assessments paid in by WFM members.⁶³

The strike called by the WFM was warranted, but unwinnable. Despite the legitimacy of its cause, the WFM reaped a field of troubles it had largely sown for itself. After encouraging labor dissent in the Copper Country for several years, the unprepared and underfinanced WFM suddenly found itself being pulled into a long and difficult strike by overzealous, immature locals. Having engaged powerful, entrenched foes, the union failed to discover an honorable way of disengaging itself from the struggle, even after it effectively was lost. By failing to call the strike off earlier, the WFM paid too great a price for losing. It not only lost in Michigan—it lost in Butte and other mining districts, where it had assessed its members too much for too long, all the while promising them a victory. Later, many members felt betrayed and abused, and the WFM, crumbling from within, had to seek a new lease on life in 1916 by reorganizing itself as the International Union of Mine, Mill and Smelter Workers.[64] The new union was never as vigorous, as active, or as feared as the old one.

The strike produced numerous individual and institutional losers, besides the WFM. Some 2,500 workers uprooted themselves and left the district because of the turmoil. The thousands of working-class men who stayed on the Keweenaw and stuck to their union suffered almost nine months of economic deprivation, all the while living with tension and social ostracism. In a strike that allowed no middle ground, where everyone had to choose a side, the imported scabs suffered, as did the wife of Judge Patrick O'Brien. They felt the hatred of the men they replaced; she felt the scorn of Houghton County's polite society, which condemned her husband as a union sympathizer.[65]

Of local institutions, perhaps the churches suffered most. In Kearsarge, the Hungarian Magyar Reformed Church lost four-fifths of its members, who just picked up and left. Finnish churches were particularly susceptible to divisiveness, because while many Finns were on the front lines of the strike, many others adopted a conservative, law-and-order stance. A month after the strike ended, A Finnish pastor reported on conditions south of Portage Lake: "The strike period has been disastrous for the congregations. Trimountain and Painesdale congregations have suffered the most and are almost in ruins."[66]

The strike produced some unexpected losers, as well. The law firm of Rees, Robinson & Petermann worked hard to help the mining companies break the strike, and they paid many expenses out of their own pocket. When it was over, they sought fair compensation from their clients. Their reward, however, was slow in coming. Copper Range discharged the firm as its counsel, because the strike had made it clear that the attorneys always put Calumet and Hecla's interests ahead of Copper Range's. And Quincy, at least through October 1915, failed to pay much of its legal bill.[67] The lawyers simply had to write this off as a loss. After all, the mining companies had always alleged that there was no collusion involved in fighting the WFM, there was no organization of mining companies, and there was no direct connection between the companies and the Citizens' Alliance. The attorneys, then, could hardly take Quincy to court and solicit public testimony "to the arrangements and the various agreements which were made from time to time" between the companies. Those agreements, made in secret, needed to be kept secret.

As for the mining companies, their decisive victory over the WFM did not come for free. From 1910 through 1912, altogether they produced an average of 220 million pounds of copper per year. In 1913 and 1914, total production ran at only 140 and then 165 million pounds. In 1912, the companies had paid out a total of $9 million in dividends. They paid out $7 million in 1913 (when they no doubt dipped into cash reserves to give stockholders their dividends), and only $1.6 million in 1914. Not all of the strike's costs, however, were short term financial ones.

The strike drove workers away from the Copper Country just when cleaner, safer, and better paying industries were trying to lure them away. In the midst of the strike, precisely when the Italian Hall disaster cast a pall over the mining communities, in downstate Michigan Henry Ford announced his $5 per day wage and bonus plan for auto workers building the Model T. He doubled the earnings of his industrial workers, while the copper companies fought against a $3 minimum wage for theirs. Before the strike ended, many of the 2,500 workers who left the Keweenaw ended up in Detroit, as noted by the Copper Country's U.S. congressman, William MacDonald: "The automobile factories of Detroit are full of them. I went into one factory at the noon hour when the employees were at lunch, and I recognized and spoke to 25 or 30 young men from the copper country."[68] As a result of the exodus of experienced mine workers and the importation of green men to take their places, the mining companies ended the strike with a labor force distinctly inferior to the one with which it had started.

The mining companies figured their victory was well worth the price paid. Save for the adoption of the eight-hour day, and of modestly improved grievance procedures, the companies had given not an inch. By refusing to budge, they had killed off the WFM and reaffirmed their hegemony, including their control over the technologies to be used underground.

The companies switched more and more men over to the one-man drill during the course of the strike, and while doing so they typically raised miners' wages from about $2.50 or $2.80 per day to $3.13. By the summer of 1914 the transition from two- to one-man machines was nearly complete. As it turned out, miners had been right in fearing the effects of the one-man drill on employment. At Calumet and Hecla, miners lost jobs. The company employed 980 to 1,095 men in shaft-sinking, drifting, and stoping from 1910 through 1912. During the 1913–14 strike years, that figure fell into the 700s. Between 1915 and 1918, the number of miners declined each year, from 634 down to 485. The company effectively halved its force of miners, while increasing its production of stamp rock from 2.8 million tons per year in 1910–12 to 3.1 million tons in 1915–18.

As managers had hoped, the one-man drill allowed for substantial productivity increases. From 1910 through 1912, each stoping miner at C&H produced an average of 3.5 thousand tons of stamp rock per year; by 1915 to 1918, each stoper broke 6.1 thousand tons annually. The Quincy mine saw similar gains. Each miner broke 7.5 tons of copper rock per shift before the strike, and 13.8 tons daily after the strike, even though work shifts had been shortened from nine to eight hours.[69]

Still, the one-man drill was no panacea for all the mines' troubles. It did not provide a miraculous technological fix to the problems of increasing depth, diminishing copper yields, and increasing competition. The machine did not give the Lake mines a competitive advantage over other districts, because the mines did not have a monopoly on the new technology. Indeed, Ingersoll-Rand informed James MacNaughton, early in 1914, that it had already provided some 15,300 small drills to the American mining industry. About 10,000 had gone to western mines, where over 90 percent of them were being run by a single miner.[70] At the end of the strike, the Michigan mines had not pulled ahead with their new technology. More modestly, they had succeeded in not falling any further behind.

That fact was driven home shortly after the strike ended. For months the companies had struggled to reopen, to get their old workers back on the job, and to bring in needed new men. But right on the heels of the WFM's surrender, war broke out in Europe. The start of the First World War upset international trade and briefly depressed the copper market. Copper sold at 15.7 cents per pound in 1913; at 11.3 cents in November 1914. Just after the Michigan mines had built themselves back up, they were forced to scale down again. Half a year after concluding the strike, Calumet and Hecla went on a three-quarter time schedule and cut wages back to 1912 levels; Quincy laid off men and reduced wages by 12.5 percent; the Copper Range mines trimmed wages and operated half-time; and the Winona mine closed down altogether.[71]

For all the power they could still exercise in Houghton County, the companies found themselves less and less in control of their economic futures. During the course of the First World War, the copper market rebounded to great heights. Copper sold for 28 and 29 cents per pound in 1916 and 1917. In response to the price increase, several Lake mines reopened and the larger ones went on a production binge. Perhaps knowing that such good times would never come again, the companies gorged themselves on profit taking. In 1916 and 1917 they produced about 270 million pounds of copper per year, while paying dividends totaling $18.7 and then $23.9 million. By comparison, before the war the mines' highest annual production had been 233 million pounds; their largest dividend payments, $13 million.

But this tremendous performance was only temporary. At the war's close, some 2.2 billion pounds of surplus copper choked warehouses at home and abroad.[72] The price of copper plummeted. The Michigan mines crashed along with the market—and they never fully recovered. Winning the strike meant little in terms of granting the companies long-term security.

Winning the strike had, however, meant a return to the companies' paternal ways. Having exorcised the demon WFM, the companies again looked after their own in their own way. At C&H, James MacNaughton made sure that Andrew Hill got a "notice to quit" the small farm he leased from the company. Andrew Hill said he had never supported the strike, but MacNaughton knew better. His security chief, August Beck, swore that Hill and his sons had been WFM members. They had marched in every big parade. Furthermore, Hill "allowed strikers to cut their winter wood free of charge" off this leased company ground! So Andrew Hill had to remove himself. At the same time, MacNaughton offered special assistance to the loyalists—particularly the deputies: "We have taken care of almost every Deputy that struck with us during the Strike, and there are very few who did not get something to do if they wanted to work."[73]

Late in the summer of 1914, the mine managers sounded almost as if the recent strike had never happened. Because of the dip in copper prices, they had been forced to cut workers and wages, but they tried to soften the blow. At C&H, MacNaughton reverted to old form in first giving work to men with large families. At Copper Range, Superintendent F. W. Denton followed a similar approach. He figured he would see some turnover of personnel, and when jobs opened up, he wanted to fill them with needy married men:

> I am having a census taken of all the families living in our mining locations and not working . . . and have asked one of the citizens of South Range for a similar census to be made of their village. I think that we shall have enough changes in our force by Sept. 1st to give openings to all able bodied residents in our district who have families to support.[74]

At the Winona mine, Superintendent R. R. Seeber had helped fight the WFM by keeping MacNaughton abreast of the union's bank deposits and withdrawals. In August 1914, Seeber had the unpleasant task of shutting his mine down, of throwing everyone out of work. He wrote the assistant general manager of Copper Range's Champion mine: "I am enclosing list of underground men whose families are now on our location and with very few exceptions need work to keep their families from starving. They will be glad to do anything in the shape of work."[75]

Seeber's "List of Idled Married Men" contained 62 names. Seeber wanted to help these men, but he also wanted to be fair to the Champion mine, so he appended a brief evaluation of each man on the list. Only on this level, the level of individual comments made by Seeber, could one discern that the new paternalism was not identical with the old. The strike had indeed happened, it had introduced some new terms to the district, just as it had put some new twists on old loyalties. Besides rating a man's ability as "fair" or "good," Seeber carefully remarked as to whether he was a "striker," a "nonunion" man, or a "scab." Since the post-strike paternalistic era was still in its infancy, Seeber felt it necessary to make one additional comment: "Scab is a mark of honor."

14

Shutdown: Death of an Industry

> Besides closing down the mine, I have had to turn my attention to farming. We are plowing up all of our available farm land. Each family of 5 or less is supplied with a 50 × 100 foot lot and 2 bushels of seed potatoes and other seeds, such as rutabaga, beet, onion, carrot, cabbage, pea, bean, etc., and told to go to it so that now instead of having a lot of idle miners on my hands, we have a lot of busy farmers. They are as busy as the devil and the outdoor work will do them a lot of good.
>
> Telegram, William Schacht, president of
Copper Range, to Ward Paine, treasurer,
20 May 1932

The boom enjoyed by the Michigan mines during the First World War hardly lasted until the armistice. In September 1917, the Price Fixing Committee of the War Industries Board started setting the price of copper to be paid by the American government and manufacturers. The price of copper declined, and the Lake mines cut back their production. Inflation and increased labor costs ate into their profitability. To encourage men not to leave for the army or for better paying employers, the mines regularly increased wages, until they were 60 percent higher in 1919 than in 1915. Still, the companies found themselves labor-short in a competitive market for young men. Wage increases and a labor shortage eliminated marginal companies; about one-third of the mines operating in 1915 dropped out of production by 1920. Mine employment peaked at 15,000 in 1915–18. It fell to 11,000 in 1919, and to only 8,000 in 1920. By 1920, the mines produced just 160 million pounds of copper and paid a modest $2 million in dividends.[1]

Then, in 1921, the bottom truly fell out. Stockpiled copper was dumped onto the market, its price plummeting to only 12½ cents per pound. Michigan's production followed suit, falling to 90 million pounds, its lowest level since 1889. Many producers, including Calumet and Hecla and the large western mines, shut down altogether for up to a year. And for the first time since the Cliff mine paid its initial dividends in 1849, in 1921 the Lake mines paid out no dividends whatsoever.

On the whole, the American economy benefited from the First World War. Employment, production, and export levels rose. The United States became a creditor nation, and a more important figure on the world scene. After the armistice, the nation stumbled into an unplanned demobilization, and the economy pitched into a deep recession. Most businesses and industries recovered rather fast, but not so the American copper industry, whose postwar economic problems proved neither minor nor of short duration.

The United States market for new copper had declined, and American mines found themselves with excess production capacity. Compounding the problem of an oversupply of domestic copper, new mines in Canada, Africa, and Chile offered additional competition. To prevent a total collapse of the market, western American copper companies led major producers to join legal cartels, sanctioned by federal legislation. These cartels initially orchestrated production cutbacks to deal with the over-supply problem, and then set official international prices for copper. Through most of the 1920s, the price of copper lay flat, staying under 15 cents per pound—but it could have been even worse. Production and price agreements bought some time for the copper companies to regroup and adjust to their economic problems.[2]

The major copper producers in Arizona, Montana, and Utah took about three years to come to terms with postwar conditions and the 1921 market collapse. The Michigan mines needed five or six years to regain some semblance of equilibrium. By the time that occurred, Michigan's share of American copper production had fallen to just 10 percent, only half its share in 1910. Calumet and Hecla still made some pretense of being a major force in the industry, but in truth the major copper companies in the west—Anaconda, Kennecott, and Phelps-Dodge—clearly dominated American production.

In the 1920s, the Michigan mines worked at maximum depths of 8,000 to 9,000 feet. By contrast, miners took Butte copper from a maximum depth of 3,800 feet; Arizona's vein mines bottomed out at 2,915 feet; Arizona's underground porphyry mines, exploited by the less-expensive block-caving technique, extended only 1,000 feet beneath the surface; and in both Arizona and Utah, steam shovels scooped up thick, horizontal deposits of porphyry copper in open-pit mines (a technology first applied to copper in 1907). Mining techniques, geology, and production cost data all favored the western producers over Michigan.[3]

By the mid-1920s, the western mines caught up to and then surpassed their peak wartime production levels. Meanwhile, the Lake mines—which would never produce as much copper as they once had—pursued less lofty goals. They set about the task of mopping up their industry, of finishing it out in the most profitable manner they could. In doing so, they struggled to maintain a labor force of adequate size and skill, while they pitched some final organizational and technological changes against their production costs, trying to drive them down.

Since the 1913–14 strike, the copper companies had fared poorly in attracting and keeping men. Turnover ran high, and workers left the district at a faster rate than jobs were being lost due to economic decline. The Quincy Mining Company employed 1,801 men at the end of 1919. Of these, fewer than 300 men remained from before the strike. Short-timers were commonplace: 575 men had worked less than six months for Quincy, and another 302 had worked only six months to a year. Given such a labor force, it was small surprise that Quincy's captains had "become a little discouraged in trying to train green men for miners."[4]

Early in the 1920s, Copper Range recognized that "the day of the Cornish as miners is over. . . . Ford evidently is keen on Cornishmen and most of them go to Ford." In 1922, the superintendent of Copper Range reported to eastern officers: "The exodus from the Copper Country to the automobile cities seems to continue quite strong so far this month. Over 350 tickets have been sold up to yesterday. Two extra coaches were put on South Shore Saturday afternoon's train leaving for lower Michigan."[5]

The managers at Copper Range were not hopeful about slowing this outmigration. Wages were half again as high as those paid before the war, but still the men left. Contract miners receiving bonuses "have gone from us." Trammer bosses and surface men "drawing good salaries have left us." Managers understood that "the movement toward the large industrial cities is general throughout the country." They doubted that still higher wages—that the company could ill afford in any case—would hold their work force: "Unless the increase in pay brings wages nearly up to that in cities, men will continue to voice their discontent and threaten to leave—and will leave."[6]

To partly offset the loss of departing workers, in the 1920s Calumet and Hecla, Copper Range, and Quincy imported workers from Canada and Mexico, and from Cornwall and Germany. The federal government put restrictions on immigration in 1921, but allowed the mining companies to import workers due to their demonstrable need for labor. The importation programs, however, did not work very well. Some men "ran" on the companies, leaving their employ before paying back the costs of their transportation. Quincy's new workers from Mexico City, angered by the absence of a promised raise, picketed the office building for several days before their consulate in Chicago arranged for some Pullman cars to take them home. Other problems arose because the companies had promised miners' jobs to many Cornishmen and Germans, but had then put them to work as trammers. The companies were eager to hide this little deceit from the U.S. Department of Labor.[7]

The inward flow of newcomers paled in comparison with the outward flow of men and families. Houghton County had 88,000 residents in 1910. By 1920, the population was down by nearly 16,000, and by 1930 it fell another 19,000, leaving only 53,000 residents. In 20 years, the county lost nearly 40 percent of its population, and the townships immediately adjacent the C&H and Quincy mines lost well over half their residents.

Many individuals and families did not want to leave the Keweenaw, but felt compelled to move to Detroit or elsewhere to earn a decent living. Often, the exodus split families along generational lines. The parents stayed, while the sons and daughters moved away. The population decline played havoc with local businesses, with social institutions like schools and churches, with property values and the stability of neighborhoods. Yet the exodus was positive and necessary. It cleansed out the surplus population that a declining industry could no longer support. The American economy offered opportunities elsewhere, and people benefited by moving. Local communities shrank, but that was better than having them fill with thousands of unemployed. Outmigration kept poverty and unemployment in check, even while a half to almost two-thirds of all jobs in the mining, milling, and smelting of copper were lost between 1915 and the mid-1920s.[8]

In the 1920s, the companies adjusted some time-honored work practices, in an attempt to keep men more content while increasing their productivity. Confronted by the labor shortage near the end of the First World War, Copper Range did away with the tradition of weekly rotating underground workers from the day shift to the night, and vice versa. One boss, who had worked such shifts, allowed as how "the men never get used to it." The change "went very smoothly" and caused "much favorable comment," so Copper Range retained the new practice in the 1920s.[9]

The companies maintained the charade of using "contract miners," but hoping to boost productivity, they switched to a wage-plus-bonus system of pay. The contract specified a miner's minimum daily wage. But he could better that by drilling holes to a greater depth than a set standard. For C&H workers on the Osceola lode, the standard was 50 feet of holes per shift; Quincy's standard for the Pewabic lode was about 42 feet. A boss measured the cumulative footage drilled by each man each day. If a miner exceeded the standard, he received a bonus of about 10 cents per foot.

Besides measuring the depths of shot holes, bosses carefully inspected where they were put. To lower production costs, companies needed to increase the yield of ingot copper per ton of rock broken underground. They mined more selectively and gave their less skilled miners more explicit directions. At the Isle Royale mine, "the placement of holes shall meet with the approval of the shift boss." At Quincy, drill holes had "to be properly placed and approved by the Bosses," or the men would not be credited for drilling them.[10]

Matters of discipline and efficiency became paramount. Superintendents pushed captains, who pushed shift captains, who pushed level bosses, who pushed the men. At Quincy, in July 1926, superintendent Lawton wrote to president W. Parsons Todd, "We have stirred up the shaft captains and level bosses to greater endeavor." These men, he wrote, "must give a better account of themselves in the future," and to prod them on he demoted some captains who had "been doing too much along the lines of least resistance." The drive for efficiency produced some painful casualties, but it could not be helped:

> Captain Jerry O'Neill has been working for the Quincy Mining Company for thirty-five years, and has been an extra good man, but we have started out to change many of the ways and methods in the operation of [his] No. 6 shaft . . . ,

and we will not stop. It will go right forward. I do not know what we will do with Captain Jerry. His pugnaciousness would tend to eliminate any consideration for his future, and yet I believe we owe him something after thirty-five years of hard work in the Quincy mine, even though he errs now.[11]

Rather than stand pat during a time of economic difficulty, the companies invested in new tools and equipment.[12] They purchased a variety of one-man drills manufactured by the Ingersoll-Rand, Chicago Pneumatic, and Denver rock drill companies. They extended their use of power scrapers and conveyors to muck out rock, and they tried, albeit with little success, to put power shovels (like small steam shovels) to the task of mucking and car-filling. They invested in additional electric haulage locomotives, now powered by storage batteries.

In 1920, on the surface at its deepest shaft, Quincy added a nearly final monument to the glorious age of steam power: it erected the largest direct-acting steam hoist in the world, built by the Nordberg Manufacturing Company. This four-cylinder, cross-compound behemoth was designed to conserve on fuel—and its 30-foot drum could carry as much as 12,000 feet of hoisting rope in a single layer. Elsewhere, the companies advanced more into the age of electricity, often by generating their own. They invested in low-pressure turbine generators, powered by the steam exhausted from their various reciprocating engines, particularly steam stamps.

After about 1915, the technological changes of greatest economic impact were found at the lakeside stamp mills, not at the mines proper. Many of these concentrating plants had operated for decades, and massive deposits of tailings fanned out from these facilities, redrawing shorelines and filling lake bottoms. C&H's tailings, 120 feet deep in parts of Torch Lake, covered over 152 acres. Although once discarded as waste, the stamp sands in fact contained much copper: in the case of C&H's conglomerate lode tailings, more than 400 million pounds of it. Early milling technologies had captured only three-fourths of the copper brought to the surface, letting the last fourth escape. Now, new technologies that could recover fine copper turned tailings into valuable new "mines."[13]

Metallurgists and assayers had always known that considerable copper washed out of the mills. Their stamps crushed mine rock to about three-sixteenths of an inch in diameter, and these coarse sands contained flecks of copper. For decades, mill bosses had resisted trying to recover the smallest copper particles still bound up in the sands. If they had crushed the rock more finely to liberate more metal, it would have come to nought. The last of the liberated copper would have been too small and light to capture by means of gravity separation. The water coursing through the mill would have carried the copper particles straight past the concentrating machines and out into the lake.

At the turn of the century, the mills started acquiring the technologies that would liberate and then capture fine copper. As a first step, they imported a key piece of concentrating equipment developed by Arthur R. Wilfley in

Colorado in 1896. The linoleum-topped Wilfley table carried low riffles aligned in rows over part of its inclined deck. Water carried miniscule particles of copper and rock onto the tabletop. The rock flowed right over the riffles and tailed off the table, but the heavier copper collected behind the riffles. Back and forth vibrations imparted to the table moved the copper along the riffles and onto a smooth end of the deck, where it flowed off and was collected.

Calumet and Hecla acquired the Lake's first Wilfley tables in 1898. With this technology in hand, the mills, again with C&H in the lead, next obtained fine grinding machinery. At first the industry adopted Chilean mills, which ground rock under heavy wheels about six feet in diameter. Set upright on their edges, the rotating wheels broke and abraded the rock against a die at the base of the machine. By about 1914, companies turned to ball or pebble mills. A batch of sand was put into a closed, rotating container, along with a number of steel balls or flint pebbles. As the conically shaped mill turned around and around, the balls or pebbles wore away at the copper-laden sands.

Grinding equipment and Wilfley tables advanced the art of milling and concentrating native copper. But machines that used gravity alone to separate rock and copper had reached their peak limits of metal recovery (about 90 percent) by 1914–15. To obtain higher yields, companies added chemical leaching and flotation to their mechanical/gravity means of separation. Here again, Calumet and Hecla was in the vanguard of change. C. Harry Benedict, a Cornell graduate and C&H metallurgist, dedicated his career to wringing the last ounce of copper from every ton of rock.

In the leaching process, an ammonium solution washed over copper-bearing sands contained in large, enclosed tanks. The solution dissolved only one mineral in the sands, the copper, and the copper-rich solution was then piped into a separate, heated vessel. There, steam pipes caused the ammonium to boil off as a vapor. As the ammonium was driven off, the dissolved copper combined with oxygen and precipitated out as a black solid, copper oxide. Meanwhile, the ammonium vapor condensed back into a liquid, and was collected to be recycled through the process. This technique, first employed on C&H's sands in 1916 and later used on Tamarack's, ultimately accounted for the recovery of some 400 million pounds of copper.

The flotation process turned the traditional gravity separation process on its head. Gravity separation worked on the principle that the heavier copper would settle out from the lighter rock. Flotation worked in reverse. In a tankful of "slimes" (extremely fine particles of copper and rock suspended in water), the heavier copper could be made to rise to the top of the tank, while the lighter rock sank to the bottom. The key to overturning the natural order of things was to introduce the right oils and frothing agents into the tank.

Select fatty oils, such as pyridines and xanthates, when agitated, created a rising froth whose bubbles had an affinity for copper particles but not for rock. While the rock sank, the froth latched on to the copper and carried it to the top of the tank, where it was skimmed off. C&H built a complete flotation plant in 1920, and thereafter the process accounted for somewhat more than 10 percent of C&H's entire copper production. Following Calumet and Hecla's lead, other operating mills added flotation equipment during the 1920s.

The technologies for capturing fine copper often did double duty. That is, companies applied them to the copper rock coming fresh from the mines, and Copper Range and Quincy ultimately joined C&H in applying them to their old mill tailings, which they reprocessed in reclamation plants. C&H opened the region's first reclamation plant in 1915. A floating dredge with an iron snout dipping 115 feet into Torch Lake sucked up stamp sands, which were pumped to the reclamation plant on the shore. At first, C&H reprocessed the tailings using just regrinding machinery and Wilfley tables, but by 1920 it added leaching and flotation.

Calumet and Hecla's reclamation plant was a major copper producer and a very important contributor to the company's well-being. In 1925, C&H's "mine" on the bottom of Torch Lake surrendered 20 million pounds of copper—more than the Champion or Quincy mines produced that year—and by the end of its first decade it accounted for 121 million pounds of ingot copper. In the 80 years prior to 1925, only a dozen Keweenaw mines had ever produced more. Before closing in 1952, the plant reworked 38 million tons of tailings and reclaimed 423 million pounds of copper.[14]

In the troubled 1920s, reclamation proved a boon for Calumet and Hecla, because it produced copper at half the cost of deep-shaft mining, or even less. C&H could dredge up copper from Torch Lake, reclaim, and smelt it—for about about 5 to 7 cents per pound. Meanwhile, copper sold for 12 to 14 cents. To take greater advantage of the wide profit margin, in 1925 C&H erected a second reclamation plant on Torch Lake, which reworked the conglomerate lode tailings deposited by the Tamarack Mining Company. (Tamarack had sold all its properties to Calumet and Hecla in 1917.) In the 1920s and 1930s, these profitable reclamation plants accounted for nearly 60 percent of the dividends C&H paid out over that time, and for 35 percent of all dividends paid out by the surviving companies on the Lake.[15]

The companies also pursued technological changes at their smelters. The traditional smelter building had four small reverberatory furnaces inside, each used for both melting and refining, and each tended and tapped by hand. In the 1920s, companies replace the four small furnaces with two larger ones, the first for melting, the second for refining. After the mineral and mass melted in the first furnace, men skimmed off the molten slag, then tapped the copper into the second furnace, where it was rabbled and poled. Instead of hand ladling the contents of the refining furnace into molds, a furnaceman periodically tapped its contents into a mechanically controlled ladle, which poured the molten copper into ingot, bar, or cake molds held in a revolving casting machine. A single operator controlled the tip of the ladle and the rotation of the molds. Under this new system, smelting became a more continuous process, one that required fewer laborers and the fueling and maintaining of fewer furnaces.[16]

Between 1917 and the mid-1920s, more than a dozen smaller mines in Houghton, Ontonagon, and Keweenaw counties closed because they could not operate profitably. Only the strongest companies—Quincy, Copper Range, and Calumet and Hecla—could afford to fight decline by investing in new technologies. These companies followed other strategies as well. Quincy became a trimmer, more efficient operation. Instead of hoisting rock from four or five shafts, it confined operations to only two. At Torch Lake, it improved one stamp mill, but closed another. Copper Range similarly reduced operations at its Champion, Trimountain, and Baltic mines.

Calumet and Hecla did the most to adjust to postwar economic conditions through organizational changes. These changes followed logically from many important moves C&H had made since the turn of the century. When copper values had fallen off dramatically in its conglomerate lode, C&H tried to stave off decline by locating additional sources of copper on its own property. It moved 800 feet east of the conglomerate to establish works on the Osceola amygdaloid lode, and then went 2,200 feet further east to sink shafts into the Kearsarge amygdaloid lode. Then, from 1906 to 1910, C&H bought up substantial shares of numerous companies owning large tracts of mineral lands. C&H did so "with the purpose of assuring the continuance of the life of this company and the profitable use of its very valuable plant after the exhaustion of its own mineral deposits."[17]

C&H first pursued "friendly cooperation" between these companies as a road to greater profitability. But in 1911 it switched to the tactic of consolidating with nine other firms to form one new mining company. This newly minted, grander version of Calumet and Hecla would own contiguous lands from south of Calumet all the way northward into Keweenaw County. James MacNaughton and Quincy Shaw believed that with consolidation, "greater economies could be secured . . . , by abandoning intervening boundary lines, making common use of the more favorably located plants, operating fewer shafts, centralizing management, accounting, purchasing, milling, smelting, and many other minor operations."[18]

Legal challenges blocked the proposed consolidation in 1911. But in 1923, hoping to achieve the same cost-saving and life-prolonging benefits as before, Calumet and Hecla again pursued consolidation, and this time it went through. The consolidation eliminated the independent Ahmeek, Allouez, Centennial, and Osceola mining companies. These mines, plus any other mines that they had owned (such as the North and South Kearsarge) now merged operations and produced copper under the Calumet and Hecla Consolidated Mining Company label.

The consolidation made C&H the king of Lake Superior copper once again. Because the Calumet conglomerate lode had faltered, while other companies' operations on other lodes had expanded, C&H's share of Michigan's copper production had fallen from 60 percent in the late 1890s to only 30 percent by 1915 and thereafter. But the consolidation—particularly due to the absorption of the Ahmeek mine on the Kearsarge amygdaloid lode—bolstered C&H's share. From 1923 through the Depression and on to the end of the Second World War, the company always produced at least half of all Lake copper annually, and in some years it accounted for as much as 70 or 80 percent.[19]

By 1925, due to closings and to C&H's consolidation, only six companies still operated: Calumet and Hecla, Copper Range and its group of mines, Quincy, Isle Royale, Mohawk, and Seneca. The industry had sharply contracted in terms of employment and production, but on the positive side, productivity had increased, as had the yield of copper per ton of rock broken underground. The companies cut production costs, and most had resumed paying dividends. By 1929, the economic picture for the Lake mines looked brighter than it had for a decade, especially because the price of copper jumped that year from its usual 12 to 14 cents all the way up to 18 cents per pound. With their production costs then running about 10.25 cents per pound, the Lake mines expected to profit handsomely.[20]

But it did not happen, at least not for long. Just as the mines were poised to enjoy their best times since the war, the Great Depression beat them down again. Despite the efforts of the copper producers' cartel, copper prices fell to 13 cents in 1930, to 8 cents in 1931, and to only 5.6 cents in 1932, when the cartel disintegrated. Copper remained under a dime a pound until 1937—and from then through the end of the Second World War it ranged from 10 cents upward to only 13.5 cents per pound.[21]

The early 1930s, in particular, devastated the Lake mines. In the face of a terribly depressed market, the companies sought new and novel uses for their copper.[22] They supported an unsuccessful drive to have Michigan fabricate automobile license plates out of copper, instead of steel. Similarly, companies backed the idea of marketing ultrathin, letter-sized sheets of copper as a replacement for carbon paper. The copper sheet would receive an embedded, typed message on its surface, and then sit forever in file cabinets, preserving the information. But the companies mustered no new markets, technologies, or organizational forms that miraculously turned large operating losses into even modest gains. Unable to beat the Depression, they tried just to survive it.

Meanwhile, local communities found they had virtually no buffer against economic calamity. Other American communities generally had a more diversified economic base, and variety and breadth offered a measure of protection. Some industries were less hard hit by the Depression than others; some recovered faster than others. But the Keweenaw had the one dominant employer: copper mining. Its payrolls fueled the local economy. Copper money passed around—or was supposed to pass around—to the store clerks, barbers, gas station attendants, and church collection plates. When the copper money stopped, everyone was in trouble. Much of that money had stopped in the 1920s, when the total annual copper mining payroll declined from $14.6 million in 1919 to $9.8 million in 1929. By 1939 it had dropped even further: to only $3.3 million.[23]

The mining companies had benefited from their domination over local society. They had never encouraged other industries to locate on the Keweenaw. Other industries might have competed with them for labor, might have altered wage scales in the region—and might have carelessly let in unionism. For the mining companies it had been best that the Copper Country remained populated with one-industry towns, and the communities themselves had not objected to this state of affairs—until they started to see that one industry die. At that point, the communities were at a loss as to how to stave off precipitous decline. Their remote location, their notorious winters, and the Depression itself, draped like a pall over the national economy—all thwarted the hope that some significant new industries would locate on the Keweenaw.

The price of copper did not tumble immediately with the stock market crash of 1929, and the mines maintained considerable production throughout that year and the next. The trouble was that few people bought copper at the prevailing price. Instead of going to customers, much of it went straight into company stockpiles—and the costs of producing this unsold copper ate away at the companies' cash reserves. Then the price of copper really dropped in 1931 and 1932, and the local mines crashed to a near total halt, taking the surrounding communities with them.

On 22 September 1931, Charles Lawton, who had been Quincy's general manager for a hard quarter-century, wrote his company's president, W. Parsons Todd: "The day opens very bright and clear for the morning of the suspension of operations. It has been cold and rainy during the past few days. Everybody in the immediate vicinity naturally is very much depressed, and we are doing everything we can to maintain the proper spirits, and to look forward with interest to the future."[24] After mining the Pewabic lode almost continuously for three-quarters of a century, Quincy laid off all but a skeletal staff; mothballed its mine, mill, and smelter; and barely escaped bankruptcy in the process.

On the hill above Houghton, just across Portage Lake from Quincy, another survivor of the troubled twenties—the Isle Royale mine—succumbed to the thirties, suspending mine and mill operations in April 1932. It kept the mine unwatered for awhile, but then in December stopped its pumps. Over the border into Keweenaw County, the Mohawk mine mopped up production from two shafts in 1930 and 1931, and then stopped mining on 12 September 1932. Shortly thereafter, the company liquidated its considerable properties for just $25,000.[25]

In the 1920s, Copper Range had generated profits at its Champion mine and had endured steady losses at its Baltic and Trimountain mines. To cut these losses, Copper Range suspended operations at Trimountain in May 1930. "To give employment as long as possible," the company kept a small force of Baltic miners busy robbing shaft pillars of their copper, until it closed that mine, too, on the last day of 1931. In March 1932, mining stopped on the northern end of the Champion mine.[26] Champion continued hoisting only at shafts 3 and 4—and this on a very abbreviated schedule.

Conditions were equally bleak at the Calumet and Hecla Consolidated Copper Company. At the start of the Depression, C&H had sacrificed cash reserves to keep mining. It built up a large inventory of unsold copper, fell into debt, and borrowed heavily against its stockpile. By 1932 the company faced a real conundrum. It could afford neither to operate nor to shut down. It lacked the funds to cover the losses that continued mining would bring. On the other hand, if it closed down its many works, it had no funds to keep them unwatered and in repair. And if C&H lets its mines fill with water, it might never find the funds to pump them out and recondition them.

By the end of 1932, C&H closed its reclamation plants and halted all operations on the Osceola amygdaloid and at the Allouez, Centennial, and North Kearsarge mines. Only two parts of the proud, old company remained in production: its original works on the conglomerate lode, and its newer works at the Ahmeek mine. To escape total shutdown and liquidation, C&H quite literally decided to rob its past at Calumet in order to buy itself a future at Ahmeek.[27]

In 1933, C&H started to close the books on its original mine beneath Calumet by shutting down its pumps and letting it fill with water. Meanwhile, miners working from the bottom of the mine up, and keeping just ahead of the rising water, robbed the mine of its rich shaft pillars and last copper values. This work could be done very economically and achieve extremely high copper yields—but at the same time it rendered the mine irrecoverable in the future. For C&H, the future of mining lay to the north, where the Ahmeek mine on the Kearsarge amygdaloid lode was closed, but kept unwatered and in good repair.

While the companies undertook full or partial shutdowns, they remained paternalistic. Copper Range explained its course of action in its 1931 annual report:

> In order to cause the least hardship, work is given men in accordance with their needs. Those having dependents work three or more days per week, and others two days or less per week. All are rotated so as to spread employment to the greatest possible number. Where wages do not cover family needs, relief is administered through the local relief organization, whose funds are supplemented to some extent by contributions from the Company.

Calumet and Hecla's annual report for the same year sounded a similar note—take care of as many as possible, and take greater care of family men:

> To relieve, as far as possible, the distress of limited employment . . . , all replaceable single men having no dependents were dismissed from the Company's service and by a system of rotating all men retained, together with a plan giving preference to men in proportion to their family obligations, as determined by thorough investigation, every effort is being made to contribute to the welfare of the Company's employees and their families.[28]

During the Depression, Keweenaw residents retreated from consumerism and the modern world's cash economy. They adopted mid-nineteenth-century ways, when agriculture and gardening had been important activities at the mine villages, and when residents had done for themselves, or had done without. Copper Range laid out the plots, distributed the seeds, and encouraged idle mine workers to till the soil. Quincy did likewise. Upon shutting down, it let former employees without incomes remain in company houses rent free. It plowed up ground for gardens and allowed residents of Quincy Hill to cut firewood on company timberlands at no cost.[29]

From the start of the First World War through 1929, the Lake mines lost more than half their jobs. This left about 7,500 men employed in the mines, mills, and smelters at the start of the Depression. By the end of 1933, another two-thirds of the jobs were lost, leaving only 2,000 to 2,500 employees.[30] Salaried officers and managers had taken substantial pay cuts, and laborers usually worked part-time for reduced wages. Welfare needs ran high, as unemployed men choked Houghton and Keweenaw counties, and even the family men fortunate enough to have part-time jobs often needed financial assistance to make ends meet.

Before the major federal relief programs started, the local county governments were ill-prepared to cope with the Depression's economic devastation. Due to the decline of mining, the counties' tax bases had fallen off precipitously since about 1910. In Houghton County the assessed evaluation of all personal and real property had peaked in 1907 at $119 million. By 1929 it stood at about $40 million, and by 1933 at only $20 million.[31] Just when it needed tax revenues most, the county had less of an industrial base to carry the tax load. At the same time, Houghton County found it had to tend to more residents.

During the contraction of the copper industry in the 1920s, out-migration had swept the county free of almost all unemployed men. But in the early 1930s, no similar exodus occurred. Out-of-work men chose to stay in rural Houghton County, instead of moving to an industrial, urban area also plagued by unemployment. In fact, many persons who had once left the county to find greener pastures elsewhere now came back, where they believed they could better weather the Depression and live more cheaply. They could live with family or friends, or otherwise obtain extremely cheap housing, because there was no shortage of vacant dwellings on the Keweenaw. They could do a bit of farming, fishing, and hunting to help themselves get over the hump, and take their winter fuel out of the woods. Between 1930 and 1934, Houghton County's population rose by 4,000 persons, or 8 percent.[32]

The ranks of the unemployed in Houghton County swelled to an estimated 8,800 by the middle of 1933. The county Superintendents of the Poor could not cope with such numbers. Private charities and church organizations helped many who received no public assistance, and Copper Range and Calumet and Hecla—the only two companies that kept open even a part of their operations—also provided relief.[33] C&H eliminated 3,600 jobs in Houghton and Keweenaw counties, but then turned around and privately contributed relief to 1,600 welfare cases through the end of 1932. The next year, the state and federal governments came in to assume the greater share of the Copper Country's relief burden.

Through the mid-1930s, the copper companies managed but a few bold steps to try to break out of the grip of decline. Copper Range kept an eye to the future and took some measures deemed of benefit over the long run. In 1931 it invested in a hydroelectric power plant at Victoria in Ontonagon

County, which transmitted power northward to Houghton County. Once this power was available, Copper Range began substituting electricity for more expensive coal-fired boilers and furnaces. It replaced the steam stamps at its Champion mill with motorized crushers, and somewhat later it planned to erect "Lectromelt" furnaces at its smelter. More important, in 1931 Copper Range extended its life by diversifying and obtaining controlling interest in C. G. Hussey & Co. of Pittsburgh. For years, Hussey had been a major buyer of Copper Range copper. Now Copper Range itself got into the business of fabricating copper nails, rivets, ferrules, and building materials.[34]

Such investments remained rare, however, until an upturn in copper prices finally stimulated greater activity at the Lake mines in 1937. In April, the directors of the reorganized Isle Royale company approved a plan to unwater the mine and put it back into production. Quincy, another company idled for half a dozen years, resumed mining at two shafts. Copper Range still confined its mining operations to a portion of the Champion mine, but it initiated a reclamation plant to rework amygdaloid tailings. Calumet and Hecla reopened its Ahmeek and North Kearsarge mines and returned the Ahmeek mill to service, along with the Calumet reclamation plant.[35]

By the late 1930s, four companies again produced copper, but a few mine and plant reopenings did not reverse the general trend of decline. Now that the national economy had improved since the darkest days of the Depression, men and families once again left the Copper Country for better jobs elsewhere. Houghton County's population stood at 57,000 in 1934, but fell to 47,600 by 1940. Between 1930 and 1940, the number of copper jobs in this one county declined from 6,750 to 3,000, while annual production dropped from 169 to 90 million pounds. By 1940 the Lake mines supplied only one out of every 20 pounds of new copper produced in the United States.[36]

This old mining region hardly needed yet another strong signal of the fact that its best days were long behind it, but it received one in October 1939, when C&H permanently sealed off No. 12, its last operating shaft on the Calumet conglomerate. The company ended the working life of one of the world's greatest lodes, the one that had made C&H one of the world's richest mines. The company's annual report for 1939 at least paid some homage to this event:

> Active mining began on this lode in 1866 and continued almost uninterruptedly for a period of 73 years, during which time it produced upwards of 3,275,000,000 pounds of copper.[37]

For nearly three-quarters of a century, a two-mile-long strip of land atop the Calumet conglomerate had been filled with activity; with what had been hailed as the most sophisticated mining plant in the world; with hoists, manengines, boilerhouses, compressors, rockhouses, dryhouses, railyards, shops, warehouses, and stockpiled fuel and timbers. Now, although some shops remained active, the physical plant along Mine Street lay depressingly still, waiting for the wrecking ball and the salvage crews. The machines that had brought up copper for so many years now yielded up their own metal, their own iron and steel, to the scrap drive initiated at the start of the Second World War. By the middle of 1943, C&H, plus other mines in Houghton and Keweenaw counties, contributed 40,000 tons of scrap to the war effort.[38] Back into the furnaces went some of the finest examples of steam-powered technology ever built in America, such as C&H's Leavitt-designed hoisting engines.

Copper was a sought-after metal just before and during the Second World War, and the federal government encouraged its production by creating a premium price plan. The marginal Michigan producers (all active mines, except C&H) received a special price subsidy of their own. Before the United States ever entered the war, the Metals Reserve Company, a federal agency, agreed to buy Copper Range, Quincy, and Isle Royale copper for 15 cents per pound. Other American producers were expected to hold their price to about 12 cents. The Michigan mines required a higher price, it was thought, so they could afford to raise wages. If the mines did not pay more, their production would surely fall off due to labor shortages, because their men would flock to other, higher-paying war industries.[39]

Despite price subsidies and other financial assistance (such as the Metals Reserve Company's loan to Quincy to fund the erection of a reclamation plant on Torch Lake), the Second World War was not the boon to the mines that the First World War had been. Between 1939 and 1945, they paid out a modest $12.8 million in dividends, which was just a million and a half more than they had paid out before the Depression in the single year of 1929. Annual wartime production peaked at 93.5 million pounds, only half of 1929's level, and wartime employment, too, reached only half the 1929 level.[40]

In the years surrounding the war, the mines did not enjoy a great revival, but they did undergo considerable change on a variety of fronts. Key leadership posts turned over due to death or retirement. James MacNaughton, who had returned to his boyhood home in 1901 to serve as general manager of C&H, and who had led the rout of the WFM in 1913-14, retired from his company's presidency in 1941. C&H's annual report acknowledged that MacNaughton had been "bound up with the very life story" of the company: "His name is associated with those of Quincy A. Shaw and Alexander Agassiz as the powerful forces which made Calumet and Hecla one of the historical copper producers of the world."[41] In 1944, engineer William Schact died after 37 years of service with Copper Range, including the last 14 as president. At Quincy, Charles Lawton died in 1946, after serving as the mine's superintendent for 41 years.[42] The old guard, who had arrived early enough to see the Lake mines peak, who had led the 1913-14 assault against organized labor, and who had then shepherded their firms through two extremely difficult decades, finally were gone.

During the Great Depression, organized labor had benefited from liberal, New Deal legislation such as the Wagner Act, which demonstrated the federal government's support of collective bargaining. As labor became an ever more important part of Franklin Roosevelt's political coalition, his administration backed policies that encouraged a rapid growth in union membership. Shortly before the Second World War, and a full quarter-century after the WFM strike had been lost, organized labor finally gained a foothold in the Lake mines. In 1939 the National Labor Relations Board ordered Copper Range to dismantle a three-year-old company union, and employees then elected to be represented by the International Union of Mine, Mill and Smelter Workers of the C.I.O. By early 1941, Quincy and Isle Royale also operated under union contracts, and in 1942 the union came to Calumet and Hecla.[43]

It was fitting that C&H, with its long history of paternalism and antiunionism, was the last company to be unionized. Once collective bargaining was initiated (first by the International Union of Mine, Mill and Smelter Workers and then later by the United Steel Workers of America), contract negotiations almost always proved difficult. Workers pressed for wages approximating those paid in other districts, and the companies rejected such competitive wage levels, arguing that as marginal producers, they could not afford them.

During the Second World War era, Calumet and Hecla finally diversified operations to secure a longer future. It had closed its works on the conglomerate lode; had nearly exhausted its stockpile of tailings to reclaim; and could foresee the end of mining at Ahmeek. To compensate for these losses, during the war C&H unwatered old shafts on its Kearsarge, Centennial, and Seneca properties and put them back into production, and it sank new shafts into the Iroquois and Houghton lodes. Still, C&H recognized that it could no longer rely so heavily on the mining of native copper to carry the company, so during the war it moved into the scrap copper business, using its leaching technology to reclaim copper from sources such as brass shell casings. C&H further broadened its base by moving into manufacturing. In 1942 it acquired the Detroit-based Wolverine Tube Co., a manufacturer of nonferrous, seamless tubing, which it operated as a new C&H division. Three years later, in partnership with another firm, C&H formed the Lake Chemical Company, which produced copper-based chemicals for industrial and agricultural users. The company also devised plans for using some of its idled shops and personnel in Calumet. Before the 1940s were through, C&H's foundry produced iron castings for sale to outside firms, and its machine shop manufactured drill bits for the mining industry. The company set up another small firm to produce buffing compounds that used stamp mill tailings as a base.[44]

The Second World War concluded with the formal Japanese surrender on 2 September 1945. One day earlier, the Metals Reserve Company's special price supports for Michigan copper had ended, meaning that another wrenching contraction of the industry soon took place. In 1942, a wartime peak of 1,517 men worked underground in the mines of Houghton County; from 1946 until the cessation of mining there in the late 1960s, the county's mines employed only 250 to 500 underground workers.[45]

Calumet and Hecla operated as many as seven shafts at various Houghton and Keweenaw county locales after the war, but turned much of its attention away from local underground mining and toward its new scrap copper operation and manufacture of tubing, drill bits, chemicals, and castings, and toward fledgling mining ventures in other states, such as Illinois, Wisconsin, and Nevada. C&H also generated income from its surface resources in Michigan, which amounted to more than 240,000 acres. It sold the timber off its lands, and leased and sold lakefront property to vacationers.

Less than a month after the war ended, Copper Range, unable to settle a wage dispute with its union, shut down mining operations at Champion and closed its reclamation plant and smelter. While conducting some exploratory mining at White Pine in Ontonagon County, the company lived off the proceeds of its Hussey manufacturing division, plus its railroad, a motor-bus line, sales of surplus land and timber, and the operation of a water-supply plant that drew water from an old Champion shaft to serve Houghton and Hancock. Copper Range, after scrapping out much of its mine and mill equipment, resumed modest production at Champion in 1949, because its Hussey manufacturing division needed the mine as a sure source of copper.[46]

By eschewing development work and stoping out only opened up ground, the Isle Royale mine stayed in production for a few years after the war. Quincy did not even fare that well. Its mine closed the very day its last wartime contract with the Metals Reserve Company ended. For Quincy, the shutdown was a particularly hard blow, because it had been the oldest active mine on the Lake. Although the company long harbored hopes of reopening, and although it conducted some trials and explorations in later decades, its works on the Pewabic lode were closed for good, and some 150 miles of Quincy shafts and drifts filled with water. Despite closing its mine, the company still produced copper until 1967—by using its wartime reclamation plant to rework the tailings it had deposited in Torch Lake for over a half a century.[47]

By 1950, the United States had become a net importer of copper, and as the Korean War was starting the federal government took measures to boost domestic production of this needed metal. On the Keweenaw, both companies still mining native copper, Copper Range and Calumet and Hecla, took advantage of wartime incentives to launch new initiatives. For Copper Range, this meant opening up the White Pine mine, a move it had been considering for some time.

Located nearly 75 miles southwest of the heart of the traditional mineral range, the White Pine mine would be an entirely different type of Michigan copper producer. The name given to its mineral deposit bespoke of its uniqueness; it was the "Nonesuch Lode," because there was no other like it in the region. Its rock was a shale that overlay the conglomerate and amygdaloid rock strata historically associated with native copper. This buried, 300-million-ton ore body did not dip steeply into the ground, but lay almost flat. Finally, although it contained some native copper, about 90 percent of its copper existed as chalcocite, or copper sulphide, "in grains so small as to be practically invisible."[48]

A White Pine Copper Company incorporated in 1909 had earlier mined this ore body's native copper. During the First World War era, White Pine produced 18 million pounds of copper. But the company idled its mine when the postwar recession set in, and in 1929 Copper Range acquired White Pine and other extensive mineral rights in Ontonagon County for $165,000.

This investment brought no return for over a quarter of a century. Copper Range evaluated White Pine's potential between 1940 and 1945, but backed away from the project. The ore body was low-grade (on average, about 21 pounds of copper per ton). The sulphide ore presented novel mining, milling, and refining problems that had to be resolved. Also, putting a new mine into operation required more capital than Copper Range had, or was willing to risk. This last consideration changed, however, with the onset of the Korean War. As part of its program to increase domestic copper production, the federal government provided Copper Range with a $57 million construction loan to help open the mine, and the Defense Materials Procurement Agency contracted to purchase 243,750 tons of White Pine copper. With these inducements in hand, Copper Range incorporated the White Pine Copper Company as a wholly owned subsidiary, invested $13 million of additional capital in the project, and brought the new mine on-line in 1955.[49]

At White Pine, large cavernous rooms with flat floors constituted the underground, and men moved broken ore toward the surface not in tramcars or skips but in diesel-powered dump trucks. From the start, White Pine produced far more copper than the surviving native copper mines to the north. In 1955, its first year, White Pine produced nearly 60 million pounds of copper, while all surviving producers in Houghton and Keweenaw counties mustered a total of only 35 million pounds.

With the opening of White Pine, Copper Range's old Champion mine became an even more marginal operation. For a while in the 1950s, Champion had its own government contracts for copper. But when these ran out, Champion was only modestly profitable in good times, and at other times Copper Range occasionally closed the mine to cut its losses. It also shut the mine when faced with labor strikes. Champion shut down for at least part of each year in 1958, 1961, and 1964.

In 1963, when Copper Range produced 122 million pounds of copper, only 5.8 million pounds came from Champion, the remainder from White Pine. Shortly thereafter, the company pushed ahead with some last-gasp exploratory work, to see if any new, workable deposits could be found near Painesdale. Finding none, the company proceeded with mop-up operations. Early in 1967 it decided to close Champion later that year:

> The decision to close was difficult for us to reach due to the human element involved. This property has materially contributed to the economic life of several small adjacent communities. Prior to making a final determination, we discussed the problem with representatives of our employees in an effort to minimize the effects of any dislocation upon this loyal group, who have served us faithfully for so many years.[50]

Copper Range went ahead with its plans and closed the Champion mine for good, along with its associated stamp mill out at Freda on Lake Superior. This left Calumet and Hecla as the only company mining native copper, and its days, too, were numbered.

After the Second World War, Calumet and Hecla engaged in a variety of pursuits. Indicative of its more diversified operations, when the firm reincorporated in 1952 it shortened its name to "Calumet and Hecla, Inc."—dropping any specific reference to being a mining company. On the Keweenaw, C&H continued its chemical and foundry operations and still leased land and dealt in forest products. Off the Keweenaw, in 1950 it opened a new Wolverine Tube Division plant in Decatur, Alabama, and in 1951 it put a zinc-lead mine into operation in Wisconsin. In 1955, C&H broke with 90 years of tradition and moved its corporate headquarters from Boston to the Chicago area. In 1957, in tune with the new "high tech" age of nuclear energy, the company started work on a uranium mine in New Mexico, which went into production the next year. Paralleling this move, Wolverine Tube expanded beyond its old markets and moved into new fields, such as the tubing needs of the nuclear reactor industry.[51]

In the 1950s and 1960s C&H closed some operations in Houghton and Keweenaw counties, while opening (or reopening) others.[52] Buoyed by the rise of copper prices occasioned by the Korean War, and by a three-year government contract to buy up much of its copper, in 1952 C&H decided to unwater and reopen its works on the Osceola lode. Osceola, C&H thought, would provide at least another decade's worth of mining. The company never surrendered its hope of finding a new, rich lode somewhere amidst its vast landholdings on the Keweenaw. It kept some marginal works open, it said, just to keep skilled miners in the area, so they would be available when a new mine was launched. In the early to mid-1960s, C&H briefly thought that its new Kingston mine (on a conglomerate lode of the same name) would prove an important life-extender, but it did not work out that way.

While developing the Kingston, C&H also developed specialty coppers, or alloys targeted at select industrial markets. For example, it aimed "Lideox," a lithium deoxidized copper, at the electronics field. As far as common copper went, C&H had to buy it from other producers, because it no longer made enough to fulfill the needs of its own manufacturing divisions. The firm's 1961 annual report noted that "Copper, the 'grand-father' of the family, is still mined, but is not really a Company product since much more is used in manufacturing than is mined."[53]

As C&H expanded outward, locating new offices and plants in several states and Canada, its operations in and about Calumet declined in significance. C&H treated its native copper mines as high-cost, marginal producers, and was always on the lookout for cost reductions. This habit exacerbated labor-management problems after the Second World War. It became commonplace for the expiration of each three-year union contract to lead directly to a round of hostile negotiations, followed by a strike and a lengthy suspension of mining.[54]

In 1949, C&H closed its mines in May and did not reopen them until September, when workers accepted a wage cut. In 1952, the United Steel

Workers engaged Calumet and Hecla in a 63-day-long strike. In 1955, the men went out on 2 May and did not return until 22 August. This time, the company coerced the men to return to work by announcing a permanent halt to mining and a sale of its properties. Bowing to that threat, union members accepted the company's terms, and C&H stopped liquidation proceedings.

The threatened closure in 1955 seemingly chilled union members for some time; they worked through the next few contracts without strikes. But a 10-week strike broke out in 1965, and in 1968 labor and management failed to reach agreement again. Their dispute centered on issues of wages and select work rules, such as the honoring of the seniority system. The 1968 strike rang down the curtain on C&H, and on the long history of the Lake Superior native copper district, because the company "settled" it by carrying out the threat leveled 13 years earlier. It shut down operations and held a giant liquidation sale.

Actually, the old, independent Calumet and Hecla company did not make the decision to terminate all mining and manufacturing operations on the Keweenaw, thus eliminating approximately 1,200 jobs in Houghton and Keweenaw counties. This decision was made by Universal Oil Products (UOP), a conglomerate that recently had bought up Calumet and Hecla. UOP, a sprawling company, did a wide variety of things. It did not do everything from soup to nuts, but, quite literally, it did do the soup; its fragrances and flavors division produced sauces, gravies, and soup bases. UOP's largest source of revenue was the petroleum and petrochemical processing market. In addition, it sold itself as an advanced scientific and engineering research company, with broad expertise in materials science, minerals processing, catalytic converters, and plastics. As a metal fabricator company, it also manufactured truck suspensions, airplane seats, and airline food service equipment.[55]

Most of UOP's growth had come via acquisitions. It entered new fields by buying up companies already active in those fields. By the late 1960s, UOP was concerned that it might become the target of a takeover itself, so it sought further expansion in order to make such a takeover more expensive and difficult. UOP went in search of a sizable, yet affordable, company with which to merge. C&H attracted its attention because of its size and location. UOP was headquartered in Des Plaines, Illinois; C&H in Evanston. Also, C&H's profitable tube works and its markets in the electrical power industry squared will with UOP's interests in energy and metals fabrication. On 30 April 1968, C&H merged into UOP; each C&H share was exchanged for 0.6 shares of UOP common stock. The UOP stock traded to acquire C&H was worth about $100 million.[56] UOP then incorporated a new Calumet and Hecla, Corp., which it operated as a wholly owned subsidiary.

In August 1968, C&H's last contract with the United Steel Workers expired, and a strike ensued that halted the production of copper, copper alloys, castings, and chemicals. The union found itself in a standoff with a company whose leadership was less committed to the continuation of operations on Lake Superior than the old C&H, which had a century of history on the Keweenaw. The new UOP-owned C&H had a history measured in months.

The deadlock and shutdown dragged on into the spring of 1969. A brief flurry of renewed negotiations brought no settlement, and UOP announced it was closing out the Calumet division altogether. On 9 April, UOP informed the strikers that their employment had ended. They had no jobs to return to.[57]

With the announced closure, church bells tolled in Calumet, sounding what many thought was the "death knell of an industry and perhaps a community as well."[58] The loss of over 1,000 jobs and of a $9 million annual payroll would have a devasting effect on a set of small villages that had already endured four and a half decades of economic decline. The civic entity of Calumet had always been dominated by the corporate entity of Calumet and Hecla. Calumet had always heated its high school with company steam from C&H's Superior boilerhouse. What would it do now? After a century of paternalism, the corporate father disowned the community, and that community had to struggle with its newfound independence under the most trying circumstances.

Civic leaders tried to avert the closing by bringing labor and management together on some eleventh-hour agreement. But no agreement was reached.[59] UOP was not disposed to make a contract offer liberal enough to garner the acceptance of the membership of Local 4312. And among the rank and file, some were committed never again to cave in to management. Some thought the announced shutdown was just another bluff, like the closing C&H had announced in 1955. And some thought the announced closing was just fine— and even overdue. After years of marginal operations and of working at jobs that never paid as well as they should, it was time to let C&H and its last mines finally die.

Last-ditch negotiations failed, as did some attempts to find other investors or companies to take over C&H. Some hope prevailed, as long as UOP continued pumping out the Centennial and Kingston shafts. These had been the ones to receive the last infusion of C&H capital investment, the ones thought to have the best future. But late in 1970, UOP let them fill with water. The end was at hand. UOP scrapped out or auctioned off the Calumet division's equipment.[60] Eighty miles southwest of Calumet, the White Pine Copper Company, working its unique sulphide deposit, employed about 2,650 workers and produced some 130 to 150 million pounds of copper per year. But the traditional native copper industry, the one that had taken ten and a half billion pounds of copper from the spine of the Keweenaw Peninsula since the mid-1840s, was finished.

In 1882, a writer for *Harper's New Monthly Magazine* penned an article about "The Upper Peninsula of Michigan."[61] He began by noting that mining regions "are proverbially barren and rocky, and the upper peninsula of Michigan—at least that portion of it which is so productive of . . . copper—forms no exception to this rule. . . . The face of the country is rugged and seamed and worn. Were it not for its mineral wealth it would remain permanently a wilderness." The land was "generally valueless from the farmer's point of view," which ruled out agricultural communities. Lumber companies probably would have invaded the area and robbed it of its forests—but then they would have left their camps for other forests in other places. Save for mining, much of the Upper Peninsula would have become "an eddy where the stream of Western migration had left a few Indians and woodsmen to subsist by the methods of primitive life."

The successful founding of the Lake Superior copper industry meant that the Keweenaw Peninsula became far more than just a temporary destination on America's western frontier. The *Harper's* author, when visiting the area about 40 years after geologist Douglass Houghton had written of its copper deposits, had more than just trees, rocks, water, and wildlife to look at. He had mining communities to explore, especially Calumet. Calumet, with its massive steam engines and its deep shafts. Calumet, with its schools and hospitals, with its sage corporate leaders and with its mixture of nationalities, which formed the labor force. After taking it all in, the author praised the community for its "manly civilization."

The author no doubt selected the word "manly" because it seemed appropriate on a variety of levels. In this society, men outnumbered women. The society was founded on and wholly supported by an industry whose workers were, without exception, male. And if one looked at the physical requirements of underground work, it took strength, power, and endurance on the part of laborers. Because it was hazardous, it also called for risk-taking.

But in embracing the term "manly," the author meant to imply more than brawn, bulk, and courage. He meant that in Calumet, the workers were men; they were not something less than men. They were not degraded underlings, beaten down by their employer. They had not been stripped of their dignity. They were not pitiful or pathetic. They were not unhealthy. They did not drown their sorrows continuously in alcohol. They did not sulk or threaten. Instead, they were confident men of "moral tone." They were not unlike "the old-time rural population of New England." They said "hello!" to a stranger in the street "cheerily and heartily."

What gave the workers their confident, proud demeanor, what made them "the most efficient body of working-men in the world," was "the absolute certainty felt by the miner" that his employer, Calumet and Hecla, was "governed by stern principles of rectitude." Calumet stood for wealth and power. But it stood for an intangible, too. It gave "the strongest testimony to the immense silent power of character." It was an industrial society that valued a man, and did not just use him up.

The *Harper's* author somewhat confused the real and the ideal. He was not a particularly discerning judge of the "character" he so extolled at Calumet. For workingmen there (and indeed at other Lake copper mines) labored under a paternalistic system that empowered the bosses at the expense of the workers. The men were not nearly as independent as free men should be. The corporations made many decisions for them, and in fact little tolerated workers having an independent cast of mind. When the managers and the miners did concur, when they both bought into the system, then their paternalistic communities appeared remarkably humane and harmonious for their times. But the mining companies' paternalism was like a friendly mask that officers could don or drop at will. Paternalism was not only a means of social welfare, but a means of social control, and the companies had no intentions whatsoever of sharing control with their men.

This paternalistic, industrial society dedicated to copper mining worked best during the region's long period of economic growth over the second half of the nineteenth century. Over this time, not every endeavor was successful, and in fact the failures outnumbered the profit-makers. Mines opened; mines closed. While part of the Lake Superior copper industry built itself up, another part shut down. Still, the overall trend was for growth and expansion, due to the development of one giant, Calumet and Hecla, and a supporting cast of major mines, such as Quincy, Osceola, Tamarack, and the late-arriving Copper Range—companies that operated over several decades and regularly enlarged their production and employment.

The Michigan mines overwhelmingly dominated American copper production for much of this period. They helped usher in the modern era of industrial mining, with its new, larger organizations and its new technologies, such as powerful hoists, rock drills, and dynamite. The industry supported thousands and thousands of new jobs, and local communities kept pace by erecting new churches, stores, schools, and housing.

While riding the upward curve of growth, it was easier to sustain optimism and harmony. But growth was not an unmitigated benefit; it came with some costs or problems that inevitably had to be addressed. In a sense, the industry was always in the process of hastening its demise, because it profited only by consuming its own future. Every pound of inexpensive copper taken out today from near the surface was a pound that would not be there tomorrow. Every company dug its own grave as it went deeper and deeper for copper. At some point, that copper could no longer be had at a profit.

The expansion that created jobs also created increasing numbers of fatal accidents underground. The more men who worked in the mines, the more men who died, and the more men who were maimed for life. The ever deeper underground, which drove up production costs, also seemed more hazardous and less hospitable to men—and the men would begin to complain.

The expansion that created jobs also created a more diverse and fragmented society, and paved the way for increasingly strained labor-management relations. The increase in the size of the mining companies alone tended to distance managers and workers and make them less well known and understood by one another. Also, mining communities that had first been stocked largely with Cornish, Irish, German, and French Canadian workers were compelled to admit other groups as immigration patterns changed. The mining companies as they expanded, *needed* men from Finland, Italy, and Hungary—but they never really *wanted* these men. They leveled much discrimination at Finns and southern and eastern Europeans, because they found them strange in customs, and unskilled and undisciplined in work. The mining companies, once hailed as firms the men could trust, came to breed much enmity and mistrust among their own men.

Between 1900 and 1920, a tumultuous era, the companies and their communities rode a roller-coaster of good and bad times. The firms dealt with problems of growth and then, in short order, with the problem of being high-cost producers in a very competitive copper market. They sought to maintain firm social control over their communities at a time when labor became more assertive of its rights and interests. The mining firms, especially with the one-man drill, politicized technological change and made it an ever more divisive issue. The workers politicized the issue of mine safety and demanded more comfortable and secure working conditions. The companies stirred controversy by increasing the role of efficiency experts and instituting new work rules. The men wanted more say in how they spent their workday, in what tools they used—and they wanted higher pay. Two major events were hanging fire at this time. The first was the showdown between management and labor, the confrontation that both sides hoped would settle many long-festering disputes. That finally came in the strike of 1913–14. The second was the industry's passing of its peak, the onset of irreversible contraction. That finally came just after the First World War.

For decades, mine superintendents had coped with managing growth. During the First World War, for instance, they pushed their old mines to their highest peaks ever of production and profitability. But just after the war, the same superintendents confronted a steep and permanent decline. In 1921, a once robust and mature industry became like a man felled but not killed by a massive stroke. It grew uncertain of the future and far more conscious of its own mortality. The companies, now marginal producers of copper, would never again be in full control of their own destinies. By mustering some final technological and organizational changes, they could hope to survive, which they did on a greatly reduced scale for several more decades, but they would never thrive like before.

Meanwhile, their surrounding communities learned the hard lessons of being one-industry towns. They lost most of their jobs, at least half their people, many of their stores and businesses and churches. Schools were boarded up. Entire neighborhoods disappeared. Rows of houses became rows of cellar holes, surrounded, in summer, by waist-high grass. Generations of children grew up expecting to be poor—or expecting to leave home in order to make it somewhere else.

The costs of economic decline were everywhere apparent. But any benefits of the sharp contraction of mining were fewer and harder to find. When the local industry suddenly shrank, it did have one positive impact on local society: the number of men killed or injured underground fell off dramatically. Also, the slowing and then the ceasing of mining had a positive effect on the environment. The lakes, already damaged enough, no longer received the heavy metals and chemicals, and the air no longer received the smoke from power generation and smelting. Nature started to reclaim much of the ground, covering many scars. The broad hill reaching from the Quincy mine down to Portage Lake was deforested in the 1850s. It stayed barren ground for three-quarters of a century. Now it is a lush green throughout the summer and red and yellow in the fall.

In other parts of the nation, and certainly in other parts of Michigan, deindustrialization is a modern phenomenon that is the cause of much social concern. The old industrial belt running across the base of the Great Lakes—running from Milwaukee and Chicago across Indiana, Michigan, Ohio, and into western Pennsylvania—has seen substantial decline in such important industries as steel and autos. It has become the "Rust Belt." But in fact deindustrialization is not a new process. Other places have seen their day come and then go, leaving behind a depressed economy. The Copper Country is such a place, and it has been for the last 70 years.

Abbreviations Used in Notes

AA	Alexander Agassiz
AIME *Trans.*	American Institute of Mining Engineers *Transactions*
A.R.	*Annual Report*
ASME *Trans.*	American Society of Mechanical Engineers *Transactions*
C&H	Calumet and Hecla
CL	Charles Lawton
CR	Copper Range
DMG	*Daily Mining Gazette*
E&MJ	*Engineering & Mining Journal*
FWD	F.W. Denton
HAER	Historic American Engineering Record
JLH	John L. Harris
JM	James MacNaughton
LSM	*Lake Superior Miner*
LSMI *Proc.*	Lake Superior Mining Institute *Proceedings*
MCJ	*Mining Congress Journal*
MH	*Michigan History*
MTU	Michigan Technological University Archives and Copper Country Historical Collections
NMJ	*Northwestern Mining Journal*
PLMG	*Portage Lake Mining Gazette*
PMC	Pewabic Mining Company
QAS	Quincy A. Shaw, Jr.
QMC	Quincy Mining Company
SBH	Samuel B. Harris

SBW	S.B. Whiting
TFM	Thomas F. Mason
WAP	William A. Paine
WFM	Western Federation of Miners
WPT	William Parsons Todd
WRT	William Rogers Todd

Notes

Chapter 1. Copper Mining on Lake Superior

1. Torch Lake has been listed as an EPA "Superfund" site. Biologists have not isolated a cause for the fish tumors, but many persons suspect the heavy metals and chemicals dumped into the lake by stamp mills and reclamation plants.

2. For descriptions of Keweenaw geology, see B.S. Butler and W.S. Burbank, *The Copper Deposits of Michigan,* U.S. Geological Survey Professional Paper 144 (Washington, 1929); John A. Dorr, Jr., and Donald F. Eschman, *Geology of Michigan* (Ann Arbor, 1977), 70–77; and T.M. Broderick and C.D. Hohl, "Geology and Exploration in the Michigan Copper District," *MCJ* 17 (Oct. 1931): 478–81, 486.

3. John R. Halsey, "Miskwabik—Red Metal: Lake Superior Copper and the Indians of Eastern North America," *MH* (Sept./Oct. 1983): 32–41.

4. This brief summary of European explorations is based on Larry D. Lankton and Charles K. Hyde, *Old Reliable: An Illustrated History of the Quincy Mining Company* (Hancock, Mich., 1982), 1–2; and C. Harry Benedict, *Red Metal: The Calumet and Hecla Story* (Ann Arbor, 1952), 12–17.

5. Robert James Hybels, "The Lake Superior Copper Fever, 1841-7," *MH* 34 (June 1950), 100–105.

6. George N. Fuller, ed., *Geological Reports of Douglass Houghton: First State Geologist of Michigan, 1837–1845* (Lansing, 1928), 557. In "Annual Report of the State Geologist, 1841," pp. 528–59, Houghton gives his conservative views on the commercial value of the copper desposits.

7. Donald Chaput, *The Cliff: America's First Great Copper Mine* (Kalamazoo, 1971), 32–33: Charles K. Hyde, "From 'Subterranean Lotteries' to Orderly Investment: Michigan Copper and Eastern Dollars, 1841–1865," *Mid-America: An Historical Review* 66 (Jan. 1984): 8.

8. Hyde, "Subterranean Lotteries," 5.

9. E.D. Gardner, C.H. Johnson, and B.S. Butler, *Copper Mining in North America,* U.S. Bureau of Mines Bulletin 405 (Washington, 1938), 12–13; Otis E. Young, Jr., "The American Copper Frontier," *The Speculator: A Journal of Butte and Southwest Montana History* 1 (Summer 1984): 4–8.

10. Hyde, "Subterranean Lotteries," 9. For production and dividend records of early companies, also see Butler and Burbank, *Copper Deposits,* 64–98; and William B. Gates, *Michigan Copper and Boston Dollars: An Economic History of the Michigan Copper Mining Industry* (Cambridge, Mass., 1951), 215.

11. Halsey "Miskwabik," 34–35; Hyde, "Subterranean Lotteries," 9.

12. See graph of copper production contributed by the various lodes, Butler and Burbank, *Copper Deposits,* 65.

13. The definitive treatment of the technology of concentrating Lake copper is C. Harry Benedict, *Lake Superior Milling Practice* (Houghton, 1955). Also see Benedict, "Milling at the Calumet and Hecla Consolidated Copper Company," *MCJ* 17 (Oct. 1931): 519–27. For a description and illustration of the earliest milling equipment, see Chaput, *The Cliff*, 41; also see Thomas Egleston, "Copper Dressing in Lake Superior," *The Metallurgical Review*, 2 (May 1878): 227–36; (June 1878): 285–300; and (July 1878): 389–409.

14. Thomas Egleston, "Copper Refining in the United States," AIME *Trans.* 9 (1880–81): 678–730; James B. Cooper, "Historical Sketch of Smelting and Refining Lake Copper," LSMI *Proc.* 7 (1901): 44–49.

15. *PLMG*, 13 May 1869.

16. Butler and Burbank, *Copper Deposits*, 65, 67–69.

17. Ibid., 66.

18. F.L. Collins, "Paine's Career Is a Triumph of Early American Virtues," *American Magazine* (June 1928): 140.

19. This history of the Norwich mine through early 1855 is largely drawn from American Exploring, Mining and Manufacturing Company, *Report* (1847); Norwich Mining Company, *Report* (Aug. 1855); "Charter of the American Mining Company" (New York, 1852); "A Statement of the Condition and Prospects of the Norwich Mine" (1 July 1853); and "To the Shareholders of the American Mining Company" (10 Feb. 1854). Several of these sources can be found in the Norwich mine vertical files, MTU.

20. R.G. Dun & Co. credit records, Michigan, vol. 54 (Ontonagon County), p. 225, Baker Library, Harvard Univ.; Norwich Mining Co., *Report* (1864?), 5.

21. *LSM*, 21 April and 21 July 1866; and from the Norwich vertical file at MTU, "Norwich Mining Company, Year 1865," and material from the *American Journal of Mining*, 18 May 1867.

22. Lankton and Hyde, *Old Reliable*, 1, 10.

23. Regarding the Schoolcraft, Centennial, and Osceola works on the Calumet conglomerate, see Michigan Commissioner of Mineral Statistics, *Mineral Statistics for 1881* (Lansing, 1882), 133 (hereafter cited as *Min. Stats.*); *Min. Stats. for 1899*, 136; Harry Vivian, "Mining Methods Used on the Calumet Conglomerate Lode," *MCJ* 17 (Oct. 1931): 496; and Butler and Burbank, *Copper Deposits*, 92–93.

24. Most of the technical problems that befell the young C&H are detailed in Benedict, *Red Metal*, 60–69.

25. Edwin J. Hulbert provided his side of the story of the discovery of the Calumet conglomerate lode, and of the founding of C&H, in *"Calumet Conglomerate," an Exploration and Discovery Made by Edwin J. Hulbert, 1854 to 1864* (Ontonagon, 1893). Also see Boston *Sunday Globe*, 13 Sept. 1885; Benedict, *Red Metal*, 69–70; and *Min. Stats. for 1882*, 99.

26. Boston *Sunday Globe*, 13 Sept. 1885.

27. See George R. Agassiz, ed., *Letters and Recollections of Alexander Agassiz, with a Sketch of His Life and Work* (Boston, 1913). Briefer treatments of Agassiz's life are found in Benedict, *Red Metal*, 56–58; Richard F. Snow, "Alexander Agassiz: A Reluctant Millionaire," *American Heritage* 34 (April/May 1983): 98–99; and Alvah L. Sawyer, *A History of the Northern Peninsula of Michigan and Its People* (Chicago, 1911), III, 1088–97. Entries for Agassiz are also found in *Who Was Who in America*, *Dictionary of American Biography*, *Dictionary of Scientific Biography* and *Appleton's Cyclopaedia of American Biography*.

28. Benedict, *Red Metal*, 59.

29. Ibid., 62–68; Gates, *Mich. Copper*, 197, 230.

30. Gates, *Mich. Copper*, 71–72.

31. Regarding Stanton's career, see William Barkell, "Honest John Stanton" (Houghton County Historical Society, July 1974); and *DMG*, 7 Aug. 1941. William A. Paine's personal history and investment career is covered in *DMG*, 25 Sept. 1929; *New York Times*, 25 Sept. 1929; Boston *American* and Boston *Evening Globe*, 24 Sept. 1929; Collins, "Paine's Career," 40–41, 139–40; and Paine, Webber, *Paine, Webber & Company, 1880–1930: A National Institution*, (1930), 9–11.

32. Gates, *Mich. Copper*, 72.

33. Ibid., 195, 197–98.

Chapter 2. The Underground: Change and Continuity

1. For descriptions of shafts and drifts, their purpose, size, and spacing, see Robert E. Clarke, "Notes from the Copper Region, " *Harper's New Monthly Magazine* 6 (March 1853): 447; Horace J. Stevens, *The Copper Handbook*, 6th ed. (Houghton, 1906), 841; W.R. Crane, *Mining Methods and Practice in the Michigan Copper Mines*, U.S. Bureau of Mines Bulletin No. 306 (Washington, 1929), 6, 7, 28–29.

2. Crane, *Mining Methods*, 7. For an example of the problems of working a "bunchy," "pockety," "winding" lode, see Quincy Mining Co., *Annual Report for 1872* 18; *A.R. for 1877*, 15; and *A.R. for 1878*, 2.

3. George J. Young, *Elements of Mining* (New York, 1932), 521, 526; Crane, *Mining Methods*, 128.

4. *Min. Stats. for 1881*, 133; T. Egleston comment, AIME *Trans.* 5 (1876–77): 608.

5. *Min. Stats. for 1881*, 132; *Min. Stats. for 1895*, 121.

6. Thomas Egleston, "Copper Mining on Lake Superior," AIME *Trans.* 6 (1879): 290.

7. For C&H, see *Min. Stats. for 1895*, 122–23; and Vivian, "Mining Methods on Conglomerate Lode," 499. For Quincy, see QMC, *A.R.*s for 1927 through 1929, passim. For a more detailed description of advancing and retreating systems, see W.R. Crane, *Mining Methods*, 30–71.

8. Sam W. Hill to TFM, 14 Dec. 1858, QMC records. All manuscript citations relating to the history of the Quincy mine, unless otherwise noted, are from this collection.

9. Egleston, "Copper Mining on Lake Superior," 285–87; Clarke, "Notes from Copper Region," 577; William P. Blake, "The Mass Copper of the Lake Superior Mines, and the Method of Mining It," AIME *Trans.* 4 (1875–76): 110–11; Arthur C. Vivian, "Cutting Mass Copper," *E&MJ* 98 (7 Nov. 1914): 825.

10. Crane, *Mining Methods*, 130–31.

11. PMC, *A.R. for 1866*, 26; and Central Mining Company, *A.R. for 1867*, 16. Both refer to the cutting and use of trip plats.

12. Crane, *Mining Methods*, 7; Egleston, "Copper Mining on Lake Superior," 295.

13. *PLMG*, 4 Oct. 1862 and 3 Dec. 1864; Claude T. Rice, "Copper Mining at Lake Superior," *E&MJ* 94 (27 July 1912), 173. The employment of trammers, bucket-fillers, and loaders can be traced in QMC *Time Books* and *Contract Books* for the 1860s. Similarly, purchases of wheelbarrows, kibbles, tram rails, tramcars, bellwire, and skips are documented in QMC *Invoice Books* from the same period.

14. *Min. Stats. for 1885*, 224.

15. *Mining Magazine* 4 (1855), 189.

16. The QMC *Time Book* for 1876–80 shows only 10 to 12 boys working underground in 1879, just before the advent of the Rand drill. With the widespread adoption of this new machine, more "drill boys" went underground. According to a C&H employment ledger covering 1894 to 1918, in the late 1890s "boys" made up 10 to 12 percent of the company's underground labor force. Some of these "boys" were undoubtedly young men, aged 18 to 20, who happened to work as "drill boys" or "puffer boys."

17. J.W. Rawlings, "Recollections of a Long Life," in Roy Drier, ed., *Copper Country Tales* 1 (Calumet, 1967), 115. For the use of the man-engine in Cornwall, see A.K. Hamilton Jenkin, *The Cornish Miner, an Account of His Life Above and Underground from Early Times* (London, 1927), 183; and Roger Burt, ed., *Cornish Mining, Essays on the Organization of Cornish Mines and the Cornish Mining Economy* (Newton Abbot, 1969), 168–69.

18. QMC, *A.R. for 1866*, 15; PMC, *A.R. for 1868*, 29.

19. Egleston, "Copper Mining on Lake Superior," 294; *Min. Stats. for 1885*, 248; QMC, *A.R. for 1895*, 14; C&H, *Summary of the Operations of the Calumet and Hecla Mining Company for the Year Ending April 30, 1891*, and C&H, *Summary . . . for the Year Ending April 30, 1892*, n.p. (Hereafter, this last source is cited as C&H, *A.R. for 1891–92*.)

20. Arthur W. Thurner, *Calumet Copper and People: History of a Michigan Mining Community, 1864–1970* (pub. by the author, 1974), 40. For discussions of similar sanitary conditions in the West, see Ronald C. Brown, *Hard-Rock Miners: The Intermountain West, 1860–1920* (College Station, 1979), 25; and Mark Wyman, *Hard Rock Epic: Western Miners and the Industrial Revolution, 1860–1910* (Berkeley and Los Angeles, 1979), 114.

21. Egleston, "Copper Mining on Lake Superior," 294; A.J. Corey (QMC agent) to Leopold and Austrian (candle sellers), 25 Sept. 1873; *Min. Stats. for 1885*, 253; S.B. Harris (QMC agent) to WRT, 19 June 1897 (regarding the switch from candles to "moonshine"); and QMC, *A.R. for 1914*, 19.

22. SBH to WRT, 19 June 1897.

23. Rice, "Copper Mining at Lake Superior," 406.

24. For descriptions and illustrations of Cornish hoists and pumps, see Peter Laws, *Cornish Engines and Engine Houses* (London, 1973), and J.H. Trounson, *Mining in Cornwall*, 2 vols. (Redruth, Cornwall, 1985).

25. Carl L. Fichtel, "Underground Power Distribution and Haulage," *MCJ* 17 (Oct. 1931): 503; C&H, *A.R. for 1892–93*, 13; QMC, *A.R. for 1903*, 11–12.

26. Wyman, *Hard Rock Epic*, 110; R.W. Raymond, "The Hygiene of Mines," AIME *Trans.* 8 (1879–80): 115.

27. G.E. McElroy, "Natural Ventilation of Michigan Copper Mines," U.S. Bureau of Mines Technical Paper 516 (Washington, 1932), 1; Crane, *Mining Methods*, 174; *Min. Stats. for 1896*, 125.

28. For underground air and rock temperature data, see Butler and Burbank, *Copper Deposits*, 100; *Min. Stats. for 1897*, 180; *Min. Stats. for 1899*, 270; J. Harris (QMC agent) to R.T. McKeever, 15 March 1904. Also see JM's testimony on air temperatures in U.S. House Committee on Mines and Mining, *Conditions in Copper Mines of Michigan* (63rd Cong., 2nd sess., 1914), Part 4, p. 1429.

29. JLH (QMC agent) to WRT, 6 Sept. 1902 and 6 Jan. 1903.

30. Butler and Burbank, *Copper Deposits*, 100; G.E. McElroy to CL (QMC agent), 5 Sept. 1930; McElroy, "Natural Ventilation," 38.

31. U.S. House, *Conditions in Copper Mines of Michigan*, 443, 446, 1001.

Chapter 3. The Surface: A Celebration of Steam Power

1. For a wonderfully illustrated period piece depicting the centennial, see Frank Norton, ed., *Frank Leslie's Historical Register of the United States Centennial Exposition, 1876* (New York, 1877). For the Corliss engine, see Robert C. Post, ed., *1876: A Centennial Exhibition* (Washington, 1976), 29–33.

2. Nathan Rosenberg, *Technology and American Economic Growth* (New York, 1972), 59–116.

3. *Min. Stats. for 1880,* 42, 112–13; "The Hydraulic Compressed Air Plant at Victoria Mine," *E&MJ* 83 (19 Jan. 1907): 125–30.

4. *LSM,* 24 Jan. 1857.

5. *LSM* of 24 Jan. 1857 provides a detailed breakdown of the steam engines used throughout the district; it gives the size, configuration and function of all 48 engines. This data was also printed in the Detroit *Daily Free Press,* 21 Feb. 1857, and in *Mining Magazine* 8 (1857): 289.

6. Photographs of mine sites taken through the early 1880s show tremendous stockpiles of cordwood located near the engine houses. In addition to burning wood in boilers, the companies cut the local forests for mine supports, fuel for kilnhouses, and for lumber.

7. Quincy, for example, initially relied on eastern-built steam engines. Then, on 16 Nov. 1859, it signed an "Article of Agreement" with J.B. Wayne for a Detroit-built engine, and in the 1860s, it bought at least one Hodge & Christie engine. See QMC, *A.R. for 1867,* 12–13; Clarence M. Burton, *The City of Detroit,* vol. 1 (Chicago, 1922), 545.

8. *Min. Stats. for 1880,* 115.

9. Egleston, "Copper Mining on Lake Superior," 311.

10. For example, see the "Longitudinal Section" included in the Franklin Mining Company's *A.R. for 1879;* this drawing shows a single hoist serving shafts 2, 3, and 5.

11. *Min. Stats. for 1882,* 97; *Min. Stats. for 1899,* 272.

12. C&H, *A.R. for 1892–93,* 7.

13. This summary of Leavitt's life and career is based largely on his obituary, ASME *Trans.* 38 (1916): 1347–49.

14. Ibid., 1348.

15. E.D. Leavitt, "The Superior," ASME *Trans.* 2 (1881): 106–21.

16. *Min. Stats. for 1895,* 124; *Min. Stats. for 1899,* 271–72; C&H, *A.R. for 1892–93,* 10–12.

17. *Min. Stats. for 1891,* 23; *Min. Stats. for 1896,* 124–25; Thomas P. Soddy and Allan Cameron, "Hoisting Equipment and Methods," *MCJ* 13 (Oct. 1931): 509.

18. *Min. Stats. for 1899,* 276. Also see C&H, *A.R. for 1892–93,* 18; *Min. Stats. for 1895,* 124; and *Mining and Scientific Press* (13 Nov. 1897): 459.

19. *Min. Stats. for 1899,* 272.

20. M.C. Ihlseng and Eugene B. Wilson, *A Manual of Mining* (New York, 1911), 127–29.

21. Rice, "Copper Mining at Lake Superior," 121.

22. Charles P. Paulding, "Machinery of the Copper Mines of Michigan," *American Machinist* 20 (8 April 1897): 18.

23. For additional articles showcasing the large hoists used in this district, see "2,500-HP Hoisting Engine for the Quincy Mining Co.," *Engineering News* 33 (18 April 1895): 255 and illus.; O.P. Hood, "Deep Hoisting in the Lake Superior District— Some of the Engines Used to Hoist from a Depth of Over 4,900 Feet," *Mines and*

Minerals (July 1904): 614–17; J.F. Jackson, "Copper Mining in Upper Michigan—A Description of the Region, the Mines, and Some of the Methods and Machinery Used," *Mines and Minerals* (July 1903): 535–40; and Warren O. Rogers, "Tom Hunter, Hoisting Engineer," which is a multi-part "tour" of the copper district's mine and mill equipment, in *Power* 46 (1917).

24. JLH to Wm. B. Mather, 6 April 1905.

25. Clarke, "Notes from the Copper Region," 578; O.W. Robinson, "Recollections of Civil War Conditions in the Copper Country," *MH* 3 (Oct. 1919): 598–99.

26. W.P. Blake, "The Blake Stone- and Ore-Breaker: Its Invention, Forms and Modifications, and the Importance in Engineering Industries," AIME *Trans.* 32 (1902): 1029; and W.P. Blake, "A New Machine for Breaking Stone and Ores," *Mining Magazine and Journal of Geology,* second series, 2:1, 50.

27. *LSM,* 13 June 1864 and 11 Nov. 1865.

28. For descriptions of Quincy's first rockhouse and its initial problems, see *NMJ* (Hancock), 12 Nov. 1893; QMC *A.R. for 1873,* 21. Quincy detailed its calcining and rock-breaking costs in its *Annual Reports.*

29. SBH to N. Daniels, 24 Aug. 1884.

30. *Min. Stats. for 1891,* 22; C&H, *A.R.s for 1890–91, 1891–92,* and *1892–93,* n.p.; *Min. Stats. for 1895,* 124; QMC *A.R. for 1892,* 15; QMC, *A.R. for 1899,* 11; and F.F. Sharpless, "Ore Dressing on Lake Superior," LSMI *Proc.* 2 (1894): 97–104.

31. Sharpless, "Ore Dressing on Lake Superior," 98.

32. JLH to WRT, 22 Dec. 1904 and 26 Sept. 1905; JLH to Gilliland, 13 March 1905; *Copper: A Weekly Review of the Lake Superior Mines* (15 Aug. 1908): 3; Stevens, *Copper Handbook* 10 (1910–11): 1445.

33. L. Hall Goodwin, "Shaft-Rockhouse Practice in the Copper Country—IV," *E&MJ* (10 July 1915): 53–57; T.C. DeSollar, "Rockhouse Practice of the Quincy Mining Company," LSMI *Proc.* 12 (1912): 217–26.

34. See the *Annual Reports* of the respective companies for the early 1860s. These tramroads were indeed gravity systems, operated without steam power. Each incline was double-tracked. A load of rock descended in a car connected to a rope that unwound from a drum. As the loaded car (or cars) went down, the drum wound in a second rope that pulled empty cars back up the incline.

36. Benedict, *Red Metal,* 67; Frank H. Haller, "Transportation System and Coal Dock," *MCJ* 13 (Oct. 1931): 515–16.

37. Gates, *Mich. Copper,* 60–63.

38. The Copper Country, because of its numerous rail lines, is a mecca for rail fans and narrow-gauge railroad buffs. For information on some of the companies, see: Willis Dunbar, *All Aboard!: A History of Railroads in Michigan* (Grand Rapids, 1969), 117; Charles F. O'Connell, Jr., "A History of the Quincy and Torch Lake Railroad Company, 1888–1927," unpublished report, Quincy Mine Recording Project, HAER Collection (Library of Congress), 649–88; John F. Campbell, "The Mineral Range Railroad Company," *The Soo* 2 (Oct. 1980): 12–35; Leon Schaddelee, "The Hancock & Calumet Railroad," *Narrow Gauge and Short-Line Gazette* 8 (Nov./Dec. 1982): 40–45; and Leon Schaddelee, "The Mineral Range Railroad," *Narrow Gauge and Short-Line Gazette* 7 (Jan./Feb. 1982): 54–58.

39. John F. Blandy, "Stamp Mills of Lake Superior," AIME *Trans.* 2 (1874), 208. The use of the Cornish stamp by numerous companies is documented in their *Annual Reports* and in *Min. Stats.* for 1880 and 1881.

40. Benedict, *Red Metal,* 42–45.

41. Frederick G. Coggin, "Notes on the Steam Stamp," *E&MJ* (20 March 1886):

210–11; (27 March 1886): 232–34; and (2 April 1886): 248–50. Also, Benedict, *Lake Superior Milling Practice*, 55–56.

42. Benedict, *Lake Superior Milling Practice*, 57–78; "The Nordberg Compound Steam Stamp," *E&MJ* (24 Aug. 1907): 349–51.

43. T.A. Rickard, "Copper Mines of Lake Superior," *E&MJ* (15 Dec. 1904): 946.

44. *Min. Stats. for 1899*, 274–75; C&H, *A.R. for 1892–93*, 16.

Chapter 4. Men at Work

1. A.J. Corey to WRT, 24 May 1873. Also see Rice, "Copper Mining at Lake Superior," 119–20.

2. Egleston, "Copper Mining on Lake Superior," 280; Ocha Potter and Samuel Richards, "Osceola Lode Operations," *MCJ* 17 (Oct. 1931): 490; U.S. House, *Conditions in the Copper Mines of Michigan*, pt. 4, p. 1393.

3. Egleston, "Copper Mining on Lake Superior," 278–79.

4. *History of the Upper Peninsula of Michigan Containing a Full Account of Its Early Settlement; Its Growth, Development and Resources; an Extended Description of Its Iron and Copper Mines* (Chicago, 1883), 278; "Hon. Samuel Worth Hill," *Michigan Pioneer and Historical Collections* 17 (Lansing, 1892): 51–53.

5. Benedict, *Red Metal*, 87–88; Sawyer, *Northern Peninsula*, 1326–28; *History of Upper Peninsula* (1883), 296, 305.

6. LSMI *Proc.* 13 (1908): 235; S.B. McCracken, *Men of Progress: Embracing Biographical Sketches of Representative Michigan Men* (Detroit, 1900), 202.

7. *History of Upper Peninsula of Michigan* (1883), 293, 302–5, 318; Sawyer, *Northern Peninsula*, 804, 1058–59; 1328–29. For a general treatment of Cornish immigration to the "Copper Frontier in Upper Michigan," see John Rowe, *The Hard Rock Men: Cornish Immigrants and the North American Mining Frontier* (New York, 1974), 62–95; and A.L. Rowse, *The Cousin Jacks: The Cornish in America* (New York, 1969), 161–95.

8. Burt, ed., intro. to *Cornish Mining*, 9.

9. John Taylor, "On the Economy of the Mines of Cornwall and Devon," in Burt, ed., *Cornish Mining*, 21–22, 38.

10. L.L. Price, " 'West Barbary,' or Notes on the Systems of Work and Wages in the Cornish Mines," in Burt, ed., *Cornish Mining*, 136–37.

11. Ibid., 159.

12. Burt, ed., intro. to *Cornish Mining*, 10–11.

13. Sims, "On the Economy of Mining in Cornwall," in Burt, ed., *Cornish Mining*, 98.

14. Ibid., 97–98.

15. Price, " 'West Barbary,' " 152.

16. Chaput, *The Cliff*, 54, 100. For a description of a mine "worked out" by tributors, see PMC, *A.R. for 1875*, 8.

17. For one example, see PMC, *Contract Book for 1862–63*, in QMC records. The contract entries show that the captain gave directions to the miners (such as how wide a stope should be), and he recorded benchmarks to use later in measuring the work done. The benchmark could be a certain stull or timber, a drill hole, a tram rail, or the edge of a shaft or winze.

18. Egleston, "Copper Mining on Lake Superior," 280; and Rice, "Copper Mining at Lake Superior," 219. For the range of supplies sold to miners, see QMC, *Supply Account Books*, 1864–66 and 1866–69.

19. PMC, *Contract Book for 1874–76,* QMC records. Numerous entries show a deduction for using the change house.

20. Taylor, "Economy of Mines of Cornwall," in *Cornish Mining,* 23.

21. See QMC, *Time Book for 1876–80,* 304, 333, 353. This source lists miscellaneous jobs done by miners on the company account, such as sinking wells, cutting copper, and fighting a rockhouse fire. Also, for "stemmers," see U.S. House, *Conditions in the Copper Mines of Michigan,* pt. 4, 1398.

22. Rice, "Copper Mining at Lake Superior," 219.

23. Phoenix Mining Co., "Director's Report" (1 Oct. 1863), 2.

24. *Mining Magazine,* 1 (1853): 294.

25. Numerous entries, QMC, *Contract Book for 1870–72.* Each entry lists partners, places of work, type of work, amount of work done, and monthly earnings.

26. Through much of the late nineteenth century, Quincy's *Annual Reports* recorded the number of contract miners at work and their average monthly earnings.

27. Wm. Tonkin to John Stanton, 24 Feb. 1881, Atlantic Mine *Letter Book,* MTU.

28. Gates, *Mich. Copper,* 113; Benedict, *Red Metal,* 216–19.

29. See Gates, *Mich. Copper,* 195; and QMC, *A.R. for 1874.*

30. See QMC, *Contract Books* for the 1850s and 1860s.

31. "Percentage of Total Shifts Worked by Miners and Trammers at Various Rates/ Wages at C&H and Subsidiary Co's . . . , May, 1913," C&H records, MTU; for employment and wage data covering the Quincy mine in 1865, 1885, and 1905, see Charles K. Hyde, "An Economic and Business History of the Quincy Mining Company," unpublished report, Quincy Mine Recording Project, HAER collection, Library of Congress, 82, 131, 197.

32. Committee of the Copper Country Commercial Club, *Strike Investigation* (Chicago, 1913), 81; U.S. House *Conditions in the Copper Mines of Michigan,* pt. 4, p. 1464.

33. Gardner et al., *Copper Mining in North America,* 11; Gates, *Mich. Copper,* 218.

34. Gardner et al., *Copper Mining in North America,* 4–5, 7; Gates, *Mich. Copper,* 198.

35. Y.S. Leong et al., "Technology, Employment, and Output per Man in Copper Mining," WPA Research Report No. E-12 (Phila., 1940), 18–19, 221–22; Gardner et al., *Copper Mining in North America,* 4–5.

36. Stevens, *Copper Handbook* 10th ed. (Houghton, Mich., 1911), 170.

37. Gates, *Mich. Copper,* 121.

38. Butler and Burbank, *Copper Deposits of Michigan,* 82, 94, 96.

39. Benedict, *Red Metal,* 80, 120; Butler and Burbank, *Copper Deposits of Michigan,* 80.

40. Benedict, *Red Metal,* 122–23; also see C&H *Annual Reports* for years ca. 1900–1910, and *Min. Stats. for 1896,* 125–26.

41. JM, "History of the Calumet and Hecla since 1900," *MCJ* 13 (Oct. 1931): 475; C&H *A.R. for 1907–8,* 7.

42. Lankton and Hyde, *Old Reliable,* 100.

43. Wilfred Erickson collection, *Book 4,* 1529–32, MTU Archives.

44. *DMG,* 26 May 1949.

45. John Lasio, "Copper Range Company: Mine-Mill-Smelter," unpublished manuscript, n.d.

46. Butler and Burbank, *Copper Deposits of Michigan,* 1.

47. M.E. Wadsworth, "The Michigan College of Mines," AIME *Trans.* 17 (1897): 707–8.

48. Ibid., 698–99.

49. See, for example, the membership roster in the AIME *Trans.* vol. 23, p. lxv.

50. *History of the Upper Peninsula* (1883), 318. Invoices covering Emerson's services are found in QMC, *Invoice Book, 1874–76.*

51. JLH to Wm. B. Mather, 6 April 1905.

52. G. Townsend Harley, "A Study of Shoveling as Applied to Mining," AIME *Trans.* 61 (1918–19): 147–87; Glenville A. Collins, "Efficiency-Engineering Applied to Mining," AIME *Trans.* 43 (1912): 649–62.

53. CL to WRT, 10 Aug. 1912.

54. CL to WRT, 21 Nov. 1912.

55. U.S. House, *Conditions in the Copper Mines of Michigan,* pt. 4, p. 1418; George S.S. Playfair to JM, 1 Aug. 1912, and QAS to JM, 12 July 1912, C&H records.

56. Rice, "Copper Mining at Lake Superior," 406.

57. Gates, *Mich. Copper,* 196.

58. See QMC, *Contract Book, 1903–1906,* 359, and *Miners Supplies, 1906,* 253. Quincy juggled the supplies used by the Isaac Peterson team to get the men to $2.50 per day. In January, for instance, the men used 16 boxes of explosives (were charged for 13.5); used 13 coils of fuse (were charged for 10); and used 4 boxes of blasting caps (were charged for 3). In March, when the team used 14 boxes of explosives, it was charged for 15; it used 10 coils of fuse, but paid for 12.

59. U.S. House, *Conditions in the Copper Mines of Michigan,* pt. 4, p. 1398.

Chapter 5. Change and Consensus

1. N. Daniels to SBH, 22 May 1884.

2. Ibid. Also see Daniels to SBH, 28 Jan. 1884.

3. SBH to TFM, 13 Aug. 1885; QMC, *A.R. for 1895,* 14.

4. Lankton and Hyde, *Old Reliable,* 64. Quincy's 1892 duplex hoist at No. 6 had a bore and stroke of 40″ × 84″ and a 21′ diameter drum; its 1894 engine at No. 2 measured 48″ × 84″ with a 26′ drum; and its 1900 No. 7 engine was 52″ × 84″ with a 28′ drum.

5. For example, see WRT to SBH, 15 June 1891, and SBH to WRT, 10 Sept. 1897.

6. *PLMG,* 7 Dec. 1871.

7. Henry S. Drinker, *A Treatise on Explosive Compounds, Machine Rock Drills and Blasting* (New York, 1893), 189–297.

8. Ibid., 202–3, 208, 315–37.

9. For specific reference to the mines' early use of Burleigh drills, see Franklin Mining Co., *A.R. for 1867,* 7; PMC, *A.R. for 1868,* 25; *Min. Stats. for 1889,* 134; *Min. Stats. for 1881,* 63, 87; *LSM,* 1 Aug. 1868 and 17 July 1869; *PLMG,* 18 Aug 1870, 11 May 1871, 7 and 28 Mar. 1872, and 2 May 1872; *Burleigh Rock Drill Co., Manufacturers of Pneumatic Drilling Machines, Air Compressors and Other Machinery* (trade catalog, ca. 1873), 17.

10. *PLMG,* 18 Aug. 1870, 11 May and 7 Dec. 1871, and 28 Mar. and 4 April 1872.

11. QMC, *A.R. for 1872,* 16; and *A.R. for 1873,* 8. Also, QMC, *Invoice Book, 1864–73,* invoices from C.H. Delamater, 12 June 1872, and from Burleigh, 15 July 1872.

12. A.J. Corey to Burleigh Rock Drill Co., 7 Dec. 1872.

13. *PLMG,* 7 Dec. 1871.

14. George G. Andre, *Rock Blasting: A Practical Treatise on the Means Employed in Blasting Rocks* (London, 1878), 142.

15. A.J. Corey to WRT, 12 Aug. and 5 Sept. 1873.
16. QMC, *A.R. for 1872,* 16; and *A.R. for 1873,* 15, 20.
17. *PLMG,* 11 May and 7 Dec. 1871; 7 March, 4 April, and 2 May 1872; and 26 April 1877. Also, J.N. Wright, "The Development of the Copper Industry of Northern Michigan," *Michigan Political Science Assoc.* 3 (Jan. 1899): 6–7.
18. *PLMG,* 7 Dec. 1871.
19. *Mining and Scientific Press,* 22 June 1872, 393; Burleigh, trade catalog (ca. 1873), 14.
20. E. Gybbon Spilsbury, "Rock-drilling Machines," AIME *Trans.* 3 (1874–75): 147.
21. Drinker, *Treatise,* 260–74.
22. George Koether, *The Building of Men, Machines, and a Company* (Ingersoll-Rand, 1971), 58–75.
23. "The Manufacture of Power Drills for Mining, Excavating, etc.," *Scientific American* 43 (25 Dec. 1880): 402. The diffusion of the Rand drill can be followed in *Min. Stats.,* 1880 through 1885, and in *PLMG,* 19 Feb., 12 Aug., 30 Sept., 7 and 21 Oct. 1880; 17 and 24 March, 21 April, 2 and 23 June, 15 Sept., 24 Nov., and 15 and 22 Dec. 1881; and 16 Feb., 6, 13, and 20 April, and 24 Aug. 1882.
24. *Min. Stats. for 1880,* 124, 145; *Min. Stats. for 1881,* 60–61, 117, 120; Gates, *Mich. Copper,* 158.
25. Ingot copper production taken from QMC *Annual Reports;* number of contract miners was calculated from entries in QMC, *Contract Books, 1878–81* and *1881–86.*
26. QMC, *Contract Books, 1878–81,* 17, 32, 40, 50, 65, 160, 166, 178, 181.
27. Ibid., 160, 178, 203.
28. Ibid., 344, 365.
29. QMC, *Contract Book, 1881–86,* 11–12, passim.
30. All figures taken from QMC, *Contract Books, 1878–81* and *1881–86.*
31. A.J. Corey to WRT, 7 Jan. 1873.
32. QMC, *Contract Book, 1878–81,* 451–64.
33. *PLMG,* 11 May 1882; QMC, *Contract Book, 1881–86,* contracts for March 1882 and passim.
34. Raphael Pumpelly, *Report on the Mining Industries of the United States,* 1880 U.S. Bureau of the Census, vol. 15, pt. 2 (Washington, D.C., 1886), 798.
35. QMC, *Contract Book, 1881–86,* contracts for March 1882 and passim.
36. *PLMG,* 11 May 1882.
37. *Min. Stats. for 1881,* 94, 131.
38. Ibid., 344, 365.
39. See Otis E. Young, Jr., *Western Mining* (Norman, Okla., 1970), 204–11; Richard E. Lingenfelter, *The Hardrock Miners* (Berkeley and Los Angeles, 1974), 16–17; Brown, *Hard-Rock Miners,* 80–81; and Wyman, *Hard Rock Epic,* 84–93. The copper mines are not listed as a problem area for lung disease in R.R. Sayers, *Silicosis Among Miners,* U.S. Bureau of Mines Technical Paper 372 (Washington, 1925), 3–4.
40. Drinker, *Treatise,* 135; JLH to R.A. Swain, 29 May 1902. JLH expressed little interest in trying the Durkee electric mining drill, because it would exhaust no air, "a very important factor in ventilating close drifts and stopes."
41. QMC, *Contract Book, 1903–6,* 364–65.
42. QMC, *Contract Book, 1892,* 13.
43. *PLMG,* 2, Feb. 1882; QMC, *Contract Book, 1881–86,* 99–101.
44. Gates, *Mich. Copper,* 208.

45. Sandra Hollingsworth, *The Atlantic: Copper and Community South of Portage Lake* (Hancock, Mich., 1978), 43–45, 142; *Min. Stats. for 1885*, 241–42.

Chapter 6. Change and Conflict

1. Henry S. Drinker, *Tunneling, Explosive Compounds and Rock Drills* (New York, 1878), 56, 59; Charles E. Munroe and Clarence Hall, *A Primer on Explosives for Metal Miners and Quarrymen*, U.S. Bureau of Mines Bulletin No. 80 (Washington, 1915), 20.

2. Arthur Pine Van Gelder and Hugo Schlatter, *History of the Explosives Industry in America* (1927; rep. New York, 1972), 315–16, 321–24.

3. J.G. Jackson, "Report for the Directors," 11 Dec. 1867 (in Phoenix Mining Co. annual report files, MTU).

4. Van Gelder, *History of Explosives Industry*, 325–26.

5. *PLMG*, 5 Aug. 1869 and 24 April 1870.

6. H. Brunswig, *Explosives: A Synoptic and Critical Treatment of the Literature of the Subject as Gathered from Various Sources* (New York, 1912), 295–96.

7. M. Eissler, *A Handbook on Modern Explosives* (London, 1897), 71; Van Gelder, *History of Explosives Industry*, 330–32.

8. Van Gelder, *History of Explosives Industry*, 656–57. For the safety of the dynamites, see Manuel Eissler, *The Modern High Explosives* (New York, 1884), 66.

9. A.J. Corey to WRT, 21 April 1874. Also see Corey to WRT, 15 Feb. 1874, and Corey to Austin, 1 April 1874.

10. A.J. Corey to WRT, 15 Feb. 1874; *NMJ*, 18 Feb. 1874.

11. Egleston, "Copper Mining on Lake Superior," 290.

12. *Min. Stats. for 1880*, 149; *PLMG*, 24 Oct. 1878. For technical details regarding the chemistry of various high explosives, such as Dualin, Giant, and Hercules powders, see Brunswig, *Explosives*; Eissler, *Handbook on Modern Explosives*; and Van Gelder, *History of Explosives Industry*.

13. Quincy's increasing purchases of high explosives and its declining use of black powder can be traced in detail through its *Invoice Books* for 1878–79 and 1879–81. Also, SBH to TFM, 16 Feb. 1884.

14. SBH to Samuel Brady, 27 Jan. 1900.

15. SBH to WRT, 2 March 1901; and to Lake Superior Powder Co., 2 May 1901.

16. WRT to CL, 15 May 1912.

17. CL to WRT, 13 and 18 May 1912.

18. See Ward Royce, "Scraping and Loading in Mines with Small Compressed-Air Hoists," *E&MJ* 112 (1921): 925. Also, U.S. House, *Conditions in the Copper Mines of Michigan*, 1423.

19. U.S. House, *Conditions in the Copper Mines of Michigan*, 1421; Royce, "Scraping and Loading," 925–31; Butler and Burbank, *Copper Deposits*, 100.

20. Rice, "Copper Mining at Lake Superior," 217; Crane, *Mining Methods*, 153.

21. Crane, *Mining Methods*, 92; and T.A. Rickard, *The Copper Mines of Lake Superior* (New York, 1905), 67. For problems of power shovels, see QMC, *A.R. for 1925*, 13.

22. AA to SBW, 28 Dec. 1893, quoted in Benedict, *Red Metal*, 93. And AA to J.P. Channing, 21 Feb. 1894, and to SBW, 2 April 1894, C&H records.

23. JLH, Quincy's agent, wrote to WRT (27 June 1903) that the "successful operating of underground haulage" depended "mostly on the system of operating; that is, distribution of laborers for filling and dumping cars."

24. *Min. Stats. for 1895,* 122; AA to SBW, 2 April 1894; F.C. Stanford, "The Electrification of the Mines of the Cleveland-Cliffs Iron Company, LSMI *Proc.* 19 (1914): 189; U.S. House, *Conditions in Copper Mines of Michigan,* 1423.

25. SBH to WRT, 21 May 1900.

26. Stevens, *Copper Handbook,* 2 (1902), 242; QMC, *A.R. for 1902,* 11; JLH to WRT, 19 June 1902.

27. JLH to L.B. Sutton, 15 Feb. 1903; QMC, *A.R. for 1903,* 12; Stevens, *Copper Handbook,* 4 (1904), 605.

28. U.S. House, *Conditions in Copper Mines of Michigan,* 112; JLH to WRT, 17 Jan. 1904.

29. U.S. House, *Conditions in Copper Mines of Michigan,* 110, 385.

30. Ibid., 443.

31. Treve Holman, "Historical Relationship of Mining, Silicosis, and Rock Removal," *British Journal of Industrial Medicine* 4 (1947): 10–12; J.D. Ditson, "J. George Leyner—Drillmaster," *Compressed Air Magazine* (Nov. 1938): 5735–39.

32. CL to WRT, 18 Aug. 1922 (QMC records); Ingersoll-Rand Co., *Rock Drill Data* (New York, 1960), 14; Eustace M. Weston, *Rock Drills: Design, Construction and Use* (New York, 1910), 49–84; C.E. Nighman and O.E. Kiessling, "Rock Drilling," WPA National Research Project Report No. E-11 (Phila., Feb. 1940), 18–34.

33. Holman, "Mining, Silicosis," 10–12.

34. JLH to J.G. Leyner Co., 24 Oct. 1904.

35. CL to WRT, 30 Nov. 1911.

36. QMC, *A.R. for 1920,* 13–14; CL to WP Todd, 3 Feb. 1921, QMC records.

37. Chief Engineer to CL, 7 Nov. 1913.

38. QMC, *A.R. for 1912,* 14; CL to WRT, 30 Nov. 1911.

39. QMC, "Drilling Machine Record," June 1913; "Machine Drill Inventory," 30 Sept. 1914; and Chief Engineer to CL, 7 Nov. 1913.

40. Chief Engineer to CL, 7 Nov. 1913.

41. U.S. House, *Conditions in Copper Mines of Michigan,* 387; CL to WRT, 10 Sept. 1906.

42. CL to Ingersoll-Rand, 19 Nov. 1912 and 17 March 1913.

43. CL to WRT, 31 Oct. 1913.

44. Telegram, JM to Ingersoll-Rand, 20 April 1912.

45. CL to WRT, 30 Nov. 1912.

46. U.S. House, *Conditions in Copper Mines of Michigan,* 538.

47. CL to WRT, 6 Nov. 1911 and 17 July 1913.

48. U.S. House, *Conditions in Copper Mines of Michigan,* 388.

49. CL to WRT, 22 April 1913. The mining companies, in countering the objection that the drills weighed too much for one man to rig up, said that the men were supposed to pair up to mount the machines, and then separate to run them alone. The men objected to working alone, and the Michigan legislature passed an act that stipulated that drill runners should not be stationed more than 150 feet apart. See Copper Country Commercial Club, *Strike Investigation,* 69.

50. See number of captains employed at C&H, as shown in *Employment Ledger, 1894–1918,* C&H records, MTU.

51. U.S. House, *Conditions in Copper Mines of Michigan,* 615.

52. Although the exact amount varied from mine to mine, a wage increase of some 10 to 20 percent was given to miners willing to work alone on the new machines. See Bureau of Labor Statistics, Bulletin No. 139, *Michigan Copper District Strike* (Washington, 1914), 31.

53. CL to WRT, 26 March 1913. Also, CL to WRT, 17 Jan. and 26 March 1913.
54. QMC's CL to WRT, 12 and 20 Nov., 2 Dec. 1912; and 22 April 1913.
55. CL to WRT, 12 Nov. 1912.

Chapter 7. The Cost of Copper

1. Information on fatal accident victims was drawn from a variety of sources, including the Houghton, Keweenaw, and Ontonagon counties' *Records of Death;* local newspapers, mainly the *PLMG* and the *LSM;* local tombstones; Michigan's *Annual Reports of the Commissioner of Mineral Statistics;* and, most notable, from the annual county *Mine Inspector's Reports* available after 1888. The data collected for each documented fatality (when available) included name, month and year of death, age, marital status, ethnic identity, occupation, type of accident, employing company, and country. Some early results of this research were published in Larry D. Lankton and Jack K. Martin, "Technological Advance, Organizational Structure, and Underground Fatalities in the Upper Michigan Copper Mines, 1860–1929," *Technology and Culture* 28 (Jan. 1987): 42–66.

2. Albert H. Fay, comp., *Metal Mine Accidents in the United States during the Calendar Year 1911,* U.S. Bureau of Mines Technical Paper 40 (Washington, 1913), 14, 17. This source cites 59 "underground" and "shaft" deaths on the Keweenaw and 542 metal-mine deaths across the entire country.

3. "Nationalities Underground at C&H," C&H records, MTU, shows that on 14 July 1913, the company employed 474 Austrians, 383 Italians, 332 Cornishmen, and 291 Finns.

4. The percentage of the underground labor force accounted for by each occupation was calculated from data in C&H, *Employment Ledger, 1894–1918,* MTU.

5. Percentages do not always add up to 100 because some victims died while serving in occupations not discussed here.

6. *NMJ,* 12 Nov. 1873.

7. *History of the Upper Peninsula of Michigan* (1883), 342.

8. Rice, "Copper Mining at Lake Superior," 172.

9. *Min. Stats. for 1881,* 93; *History of the Upper Peninsula of Michigan* (1883), 342.

10. U.S. House, *Conditions in Copper Mines of Michigan,* 523, 1420. Also, *Min. Stats. for 1895,* 123.

11. CL to WRT, 1 April 1921, QMC records.

12. In 1892, a man-engine rod broke while Quincy's day shift was coming up. However, the man-engine was equipped with safety catches that "immediately caught the broken rod." See SBH to WRT, 28 Sept. 1892.

13. See C&H, *A.R.s for 1890–91* and *1891–92,* n.p.

14. Houghton County *Mine Inspector's Report for 1906,* 27; JLH to WRT, 1 Jan. 1903, and to W. Fitch, 27 Dec. 1902, QMC records.

15. For rules for operating and riding man-cars, see QMC, "Notice to Underground Employees," ca. 20 Dec. 1902, and "Instruction to Engineers," ca. 22 Dec. 1902. Also, Thomas P. Soddy and Allan Cameron, "Hoisting Equipment and Methods," *MCJ* 13 (Oct. 1931): 510.

16. *Min. Stats. for 1899,* 167; *Mine Inspector's Report for Houghton County for 1893,* 7–9.

17. Many men who were struck by skips were not supposed to be in the shaft in the first place. Men were not supposed to cross shafts or climb up or down them. How-

ever, companies seem to have done little to enforce rules against being in shafts—and men working under little direct supervision often took deadly shortcuts.

18. Vivian, "Mining Methods in Conglomerate," 501; U.S. House, *Conditions in Copper Mines of Michigan,* 620.

19. Andre, *Rock Blasting,* 124; Rice, "Copper Mining at Lake Superior," 308.

20. Crane, *Mining Methods,* 135; Clarke, "Notes from the Copper Region," 448.

21. CL to WRT, 9 April 1921, QMC records.

22. Eissler, *Modern High Explosives,* 66.

23. The local mines participated in a "Safety First" movement after Michigan passed workmen's compensation legislation. For more details, see Chapter 8, "Weeping Widows, Generous Juries."

24. *Min. Stats. for 1899,* 273.

25. Benedict, *Red Metal,* 87; *Min. Stats. for 1887,* 199.

26. *Min. Stats. for 1888,* 67–68; *PLMG,* 6 Dec. 1888.

27. *Min. Stats. for 1895,* 122, and *for 1899,* 273; C&H, *A.R. for 1890–91,* n.p.

28. AA's 28 Dec. 1893 letter is quoted from Benedict, *Red Metal,* 92, while pp. 90–91 report more details regarding Red Jacket shaft. Also see *Min. Stats. for 1895,* 123.

29. *Mine Inspector's Report for Houghton County for 1895,* 13–16, 18.

30. W.E. Parnall to Thos. Nelson, 15 Sept. 1895, C&H records.

31. W.E. Parnall to Thos. Nelson, 8 Sept. 1895, C&H records.

32. *Mine Inspector's Report for Houghton County for 1895,* 15–16.

33. Lankton and Martin, "Underground Fatalities in the Upper Michigan Copper Mines," 54–57.

34. Fay, "Metal Mine Accidents, 1911," 21. The U.S. Bureau of Mines defined a "serious" accident as one that kept a man off work for 20 days or more.

Chapter 8. Weeping Widows, Generous Juries

1. This narrative account of the undercover work of spy J.A.P. is based on eight reports dated 15 Sept. to 3 Nov. 1906, which the Thiel Detective Service Company mailed to CL. In 1978, the spy reports were discovered in a bottom drawer of CL's rolltop desk in the unused second story of the old Quincy office building.

2. CL to WRT, 9 July 1906.

3. QMC, *A.R. for 1906,* 12–13. For technical treatment of air blasts, see LSMI *Proc.* 12 (1907): 58–67; W.R. Crane, *Subsidence and Ground Movement in the Copper and Iron Mines of the Upper Peninsula, Michigan,* U.S. Bureau of Mines Bulletin 295 (Washington, 1929); and W. Herbert Hobbs, "Earthquakes in Michigan," Michigan Geological and Biological Survey, Publication 5, Geological Series 3 (Lansing, 1911), 82–87.

4. QMC table, "Air Blast Data from March 1914 to April 1920." Also, CL, "Air Blast Repairs and Expenses," table sent in letter to WRT, 11 May 1921. For No. 2 shaft accident, see *Mine Inspector's Report for Houghton County for 1928,* 5–6.

5. In a postscript to a letter of 8 Feb. 1906, addressed to WRT, CL wrote that Quincy's structural problems could be solved by turning to a "safer method of mining" by "driving to the boundaries of the shafts and drawing back." For the company's belated adoption of the retreating method, see QMC, *A.R. for 1927,* 13–14.

6. CL to WRT, 6 Nov. 1911; Suomi College Oral History Collection, transcript of interview with WPT, conducted by R.J. Jalkanen, Oct. 1974, p. 1839.

7. CL to WRT, 5 April 1913.

8. William Graeber, *Coal-Mining Safety in the Progressive Period* (Lexington, 1976), 1-3.

9. See entries for Houghton County in the *Annual Reports of the Secretary of State on the Registration of Births and Deaths, Marriages and Divorces in Michigan* (Lansing).

10. CL to WRT, 9 March 1909 and 24 Jan. 1907, QMC records.

11. CL to WRT, 9 March 1909, QMC records.

12. *Min. Stats. for 1899,* 167; QMC, "Rules and Regulations for the Protection of Employees," 1912, n.p.; JM to AA, 3 Aug. 1905.

13. For the legislative history of the act creating county mine inspectors, see Michigan, *Journal of the Senate, 1887,* II, p. 2423; and *Journal of the House of Representatives, 1887,* III, p. 3033.

14. Michigan, *Public Acts, 1887* (No. 213), 252-54.

15. *PLMG,* 11 Aug. 1887.

16. Houghton County's annual, published *Mine Inspector's Reports* included a brief description of each fatal accident, plus a tabular summary of the year's total fatal accidents. The inspector sometimes included a calculated figure for the death rate per 1,000 employed. But the inspector ignored thousands of less-than-fatal injuries.

17. Lawrence M. Friedman, *A History of American Law* (New York, 1973), 409.

18. Ibid., 412-13; Brown, *Hard-Rock Miner,* 139-40; Wyman, *Hard Rock Epic,* 120-23. For a study of "The Common Law of Work-Related Injuries" in Michigan, see Joseph S. Harrison, "Labor Law and the Michigan Supreme Court, 1890-1930," unpublished M.A. thesis (History, Wayne State University, June 1980), 23-31.

19. Friedman, *American Law,* 422, 426-27.

20. CL to WRT, 30 April 1907 and 21 June 1909.

21. Sawyer, *Northern Peninsula,* 1503-4; *Mine Inspector's Report for Houghton County for 1890,* 12; Arthur W. Thurner, *Rebels on the Range: The Michigan Copper Miners' Strike of 1913-14* (Lake Linden, Mich., 1984), 236.

22. Claim release, signed by Olinto Rocchi, 27 Oct. 1904, QMC records.

23. CL to WRT, 19 June 1912.

24. A.F. Rees to JM, 8 Sept. 1910.

25. CL to WRT, 24 Jan. 1907.

26. CL to WRT, 24 Sept. 1910 and 24 June 1912.

27. JM to A.F. Rees, 10 Sept. 1910.

28. A.F. Rees to JM, 8 Sept. 1910.

29. CL to WRT, 10 Feb. 1910.

30. CL to WRT, 1 Oct. 1912.

31. Ibid.

32. Early in January 1912, just before "Judge Elect O'Brien" took his seat on the bench, he was still representing injured workers. Quincy settled one suit with an O'Brien client for $1,250. The man had been blinded in both eyes by a blast. CL thought it wise to settle his lawsuit out of court, "because the man, in coming into court, would have made a pitiful spectacle" (CL to WRT, 3 Jan. 1912).

33. Quincy's superintendent, CL, on 24 June 1912, wrote WRT that "we will accept the first opportunity to elect to go under the Employer's Liability Act this fall when it becomes possible." On 15 June 1912, CL had written WRT that their attorney was "pretty much at sea as to just what to say in regard to settlements and taking the cases to court. . . . Our cases are nearly all of them very strongly defensive." The act establishing workmen's compensation is found in Michigan, *Public Acts, 1912* (No. 10), 20-39.

34. See Michigan, *Public Acts, 1912,* 25–26.

35. For a study of the political origins of the U.S. Bureau of Mines, see Graebner, *Coal-Mining Safety,* 11–34; also Fred Wilbur Powell, *The Bureau of Mines: Its History, Activities and Organization* (New York, 1922), 3–35.

36. The election of county mine inspectors is stipulated in the act found in Mich., *Public Acts, 1911* (No. 163), 263–67; for the requirement that the inspector visit each mine every 60 days, see Mich., *Public Acts, 1897* (No. 123), 140–41.

37. QMC, "Rules and Regulations," 1912, n.p.; Claude T. Rice, "Labor Conditions at Calumet & Hecla," *E&MJ* 92 (23 Dec. 1911): 1, 238.

38. QMC, *A.R. for 1912,* 7; CL to WRT, 16 Jan. 1913, QMC records; QMC, *A.R. for 1914,* 19–20.

39. CL to WRT, 22 Jan., 3, 5, and 23 April, and 12 July 1913.

40. See CL to WRT, 27 Mar. 1913.

41. CL to WRT, 21 Feb. 1913.

Chapter 9. Homes on the Range

1. *Min. Stats. for 1896,* 121–22.

2. See Michael Brewster Folsom and Steven D. Lubar, eds., *The Philosophy of Manufacturers: Early Debates Over Industrialization in the United States* (Cambridge, Mass., 1982); and John F. Kasson, *Civilizing the Machine: Technology and Republican Values in America, 1776–1900* (New York, 1977).

3. *Mining Magazine,* 1 (1853): 132, 294; Clarke, "Notes from the Copper Region," 581.

4. Robinson, "Recollections of Civil War Conditions," 598–609; John Harrison Forster, "War Time in the Copper Mines," *Michigan Pioneer and Historical Society Collections* 18 (Lansing, 1892), 379.

5. FWD to Editor, *E&MJ,* 20 Dec. 1913, Van Pelt collection.

6. Wright, "Development of the Copper Industry of Northern Michigan," 135, 139–40.

7. "The Upper Peninsula of Michigan," *Harper's New Monthly Magazine* (May 1882): 898–99.

8. See Rice, "Mining Copper at Lake Superior," 217, and "Labor Conditions at C&H," 1, 235.

9. Jensen, *Heritage of Conflict* (Cornell, 1950), 272–73. Also see Arnold R. Alanen, "The Planning of Company Communities in the Lake Superior Mining Region," *Journal of the American Planning Association* 45 (July 1979): 261–64.

10. This brief history of the rise and fall of paternalism at Lowell is largely drawn from Thomas Dublin, *Women at Work: The Transformation of Work and Community in Lowell, Massachusetts, 1826–1860* (New York, 1979).

11. See Stanley Buder, *Pullman: An Experiment in Industrial Order and Community Planning, 1880–1930* (New York, 1967).

12. Stephen Meyer III, *The Five Dollar Day: Labor Management and Social Control in the Ford Motor Company* (Albany, 1981), esp. 123–69.

13. For example, for a negative reaction on the part of industrialists toward the "model town" of Pullman and its paternalism, see Buder, *Pullman,* 131–32, 142.

14. Ely is quoted in Buder, *Pullman,* 103; Meyer, *The Five Dollar Day,* 165.

15. *Min. Stats. for 1880,* 77–78.

16. Bureau of Labor Stats., *Copper District Strike,* 113–17; Copper Country Commercial Club, *Strike Investigation,* 13–29.

17. J.W. Foster and J.D. Whitney, *Report on the Geology and Topography of a Portion of the Lake Superior Land District, Part 1, Copper Lands* (Washington: House of Representatives, 1850), 146–51.

18. *Report of the Pewabic Mining Co.* (Mar. 1858), 7.

19. *Report of the Pewabic Mining Co.* (10 May 1859), 13–14.

20. PMC *A.R. for 1862,* 13.

21. PMC, *A.R. for 1866,* 17, 19.

22. *Mining Magazine,* 8 (1857): 576, 585; *Report of the Directors of the North American Mining Co.* (1857), 9. For additional data on women and children at the early mines, see *Mining Magazine* 1 (1853): 131–32; *Mining Magazine* 2 (1854): 553; *Mining Magazine* 4 (1855): 40. Also, *Min. Stats. for 1880,* 74, 77; and Minesota Mining Co., *A.R. for 1858,* 13.

23. Gates, *Mich. Copper,* 228.

24. Sarah McNear, "Quincy Mining Company: Housing and Community Services, ca. 1860–1931," unpub. report, Quincy Mine Recording Project, HAER collection, Library of Congress, 530.

25. Only one company laid out lots as narrow as 25 feet. Some companies laid out lots 60 to 100 feet wide, and up to 150 feet deep. See Bureau of Labor Stats., *Copper District Strike,* 113.

26. Lankton and Hyde, *Old Reliable,* 85–89.

27. QMC, *Directors' Minutes* for meeting of 15 Jan. 1900.

28. C&H drawing collection, sheets 6344–46, 6348, and 6385–90, drawer 115, MTU.

29. Mendelsohn to WAP, 4 Sept. 1915, Van Pelt Collection, MTU.

30. JM to S. Vertin, 28 June 1917, C&H records.

31. FWD to WAP, 4 April 1916, Van Pelt Collection.

32. Champion Mining Co., "Statement of Employees," 30 Sept. 1905 and 1 Feb. 1912. Also, "Statement of Married and Single Employees," 27 Nov. 1905. All in CR Collection, MTU.

33. Bureau of Labor Stats., *Copper District Strike,* 113.

34. Immigration Commission, *Immigrants in Industries, Part 17: Copper Mining and Smelting,* Senate Doc. No. 633, 61st Cong., 2nd sess. (Washington, 1911), 48–50, 66. For an examination of the effect of discrimination on wages, see Joan Underhill Hannon, "Ethnic Discrimination in a 19th-Century Mining District: Michigan Copper Mines, 1888," *Explorations in Economic History* 15 (1982): 28–50.

35. April Rene Veness-Randle, "The Social-Spatial Lifecycle of a Company Town: Calumet, Michigan," M.A. thesis (Geography, Michigan State University, 1979), 67, 70, 73.

36. JM to M.M. Morrison, 4 May 1915, C&H records.

37. Bureau of Labor Stats., *Copper District Strike,* 114.

38. Copper Country Commercial Club, *Strike Investigation,* 13–29.

39. Ibid., 15, 17.

40. P. West to AA, 24 Sept. 1895.

41. JM to J.S. Cocking, 5 Oct. 1915.

42. FWD to WAP, 14 April 1915, Van Pelt Collection.

43. See, for example, JM to Samuel Jess, 15 June 1915, C&H records. JM wrote that requests for "alterations to barns in order to accommodate an automobile have been so frequent that we have been compelled to refuse in every case."

44. See File 438, Box 20321, C&H records, for ca. 1911 correspondence between workers and C&H relating to electric lights.

45. C&H, *A.R. for 1894–95*, n.p.; U.S. House, *Conditions in Copper Mines of Michigan*, 1441.
47. Bureau of Labor Stats., *Copper District Strike*, 117; Copper Country Commercial Club, *Strike Investigation*, 13.
48. Bureau of Labor Stats., *Copper District Strike*, 117–21; Rice, "Labor Conditions at C&H," 1238.
49. Bureau of Labor Stats., *Copper District Strike*, 119.
50. C&H, *A.R. for 1894–95*, n.p.
51. Ibid. Also see *Min. Stats. for 1895*, 125.
52. C&H, *A.R. for 1892–93*, 9.
53. *Min. Stats. for 1899*, 277.
54. FWD to J. Black, 19 Sept. 1913, Van Pelt Collection.

Chapter 10. Cradle to Grave

1. Minesota Mining Co., *A.R. For 1858*, 13; *Min. Stats. for 1880*, 74; QMC, various entries for "grinding grain" in *Engineer's Time & Day Book, 1863–1870*.
2. Clarke, "Notes from the Copper Region," 581–82.
3. *Mining Magazine*, 1 (1853): 293–94; Minesota Mining Co., *A.R. for 1861*, 15.
4. *Reports of the Directors, Manager and Treasurer of the Ohio Trap Rock Co.* (1854), n.p.; Pittsburgh and Boston Co., *A.R. for 1860*, 21.
5. PMC, *A.R. for 1862*, 19; Tamarack Mining Co., "Surface Map" (Nov. 1913), C&H collection.
6. Bureau of Labor Stats., *Copper District Strike, 126–27*.
7. J. Daniell to H. Bigelow, 16 Sept. 1878, C&H records.
8. *Mining Magazine*, 1 (1853): 295.
9. Robinson, "Recollections of Civil War Conditions," 601; William H. Pyne, "Quincy Mine: The Old Reliable," *MH* 41 (1957): 220; QMC, *A.R. for 1864*, 12, and *for 1866*, 15.
10. J. Daniell to H. Bigelow, 16 Sept. 1878, C&H records.
11. J. Daniell to A.S. Bigelow, 24 Jan. 1878, C&H records.
12. "Trimountain Mining Co's. Stores Reports," Jan.–Dec. 1914, CR collection.
13. FWD to WAP, 14 Aug. 1914, Van Pelt Collection.
14. J. Daniell to A.S. Bigelow, 24 Jan. 1878 and 4 Feb. 1879, C&H records.
15. "Report of the Directors of the North American Mining Company" (1857), 15.
16. C&H, *Directors' Minutes*, meeting of 12 Jan. 1886.
17. Wright, "Development of the Copper Industry of Northern Michigan," 136.
18. Copper Country Commercial Club, *Strike Investigation*, 30–31.
19. Forrest Crissey, "Every Man His Own Merchant," *Saturday Evening Post*, 186 (20 Sept. 1913): 16–17, 57–58; Bureau of Labor Stats., *Copper District Strike*, 123–24.
20. Thurner, *Calumet Copper and People*, 74–75; Arnold Alanen, "The Development and Distribution of Finnish Consumers' Cooperatives in Michigan, Minnesota and Wisconsin," in Michael G. Karni et al., eds., *The Finnish Experience in the Western Great Lakes Region: New Perspectives* (Turku, Finland, 1975), 3.
21. C&H, *A.R. for 1895–96*, n.p.
22. Copper Country Commercial Club, *Strike Investigation*, 41.
23. To trace the rise (and then decline) of schools in the Copper Country, see the entries for Houghton, Keweenaw, and Ontonagon counties in the various *Annual Reports of the Superintendent of Public Instruction of the State of Michigan* (Lansing).
24. Chaput, *The Cliff*, 93.

25. Pittsburgh and Boston Mining Co., *A.R. for 1860*, 21, and *for 1861*, 16.
26. *Mining Magazine* 1 (1853): 132; *Mining Magazine* 2 (1854): 553.
27. PMC, *A.R. for 1861*, 19.
28. *A.R. of the Superintendent of Public Instruction for 1873*, 147.
29. *A.R. of the Superintendent of Public Instruction for 1869*, 90.
30. *A.R. of the Superintendent of Public Instruction for 1877*, 190–91.
31. *A.R. of the Superintendent of Public Instruction for 1870*, 74, and *for 1872*, 71.
32. Arthur Cecil Todd, *The Cornish Miner in America* (Glendale, 1967), 120.
33. *A.R. of the Superintendent of Public Instruction for 1880*, table between 80–81.
34. McNear, HAER report, 547–51; Lankton and Hyde, *Old Reliable*, 37–38.
35. Dorothy J. Klingbeil, "Copper Mining and the Sandwich Kids of Painesdale," unpub. graduate paper for Northern Michigan University, 1973 (MTU), 18, 20.
36. Copper Country Commercial Club, *Strike Investigation*, 42; U.S. House, *Conditions in Copper Mines of Michigan*, 1443–44. Ten schools built by C&H were still used in 1913; some others had been retired by that date. The 1890s had been particularly busy in terms of school construction, due to industrial expansion. C&H built six schools in 1892–93 and two in 1895–96. See C&H, *A.R. for 1892–93*, 9, and *for 1895–96*, n.p.
37. JM to AA, 1, 5, 7, and 30 Sept. and 7 Oct. 1905; and AA to JM, 2 and 4 Sept. and 3 Oct. 1905.
38. *A.R. of the Superintendent of Public Instruction for 1870*, 73.
39. *A.R. of the Superintendent of Public Instruction for 1875*, 96.
40. Thurner, *Calumet Copper and People*, 50.
41. *A.R. of the Superintendent of Public Instruction for 1875*, 96.
42. JM to AA, 7 Sept. 1905.
43. *A.R. of the Superintendent of Public Instruction for 1900*, 37.
44. PMC, *A.R. for 1861*, 19.
45. McNear, HAER report, 562–64; QMC, *A.R. for 1916*, 16, and *for 1918*, 18. Also, WRT to CL, 30 Nov., 9 and 22 Dec. 1915, and 17 March and 7 April 1917.
46. *DMG*, 14 Nov. 1903.
47. Program, "Painesdale Social Club Party, 1904," Van Pelt Collection.
48. Marie F. Grierson to JM, 19 Feb. 1907; Wright, "Development of the Copper Industry of Northern Michigan," 138.
49. Rice, "Labor Conditions at C&H," 1238; Copper Country Commercial Club, *Strike Investigation*, 34; Thurner, *Calumet Copper and People*, 52.
50. C&H, *Annual Report for 1898–99*, n.p.
51. QAS to JM, 12 Sept. 1912; MacNaughton to Shaw, 23 Sept. 1912.
52. Lankton and Hyde, *Old Reliable*, 89, 92; *PLMG*, 18 March 1897; JLH to T. Symington, 27 Nov. 1903, QMC records.
53. Copper Range Band, *Record Book for 1910–13*, CR collection.
54. Tamarack Mining Co., "Surface Map," Nov. 1913; Rice, "Labor Conditions at C&H," 1238; Copper Country Commercial Club, *Strike Investigation*, 34; CR, *A.R. for 1904*, 23; *Min. Stats. for 1899*, 278; QAS to YMCA, 19 Aug. 1908, C&H records.
55. C&H, *A.R. for 1902–3*, n.p.; Miscowaubik Club, "Constitution and Bylaws" (1 Jan. 1910), 7, 14, 21–36.
56. Don Willis Dickerson, "Theatrical Entertainment, Social Halls, Industry and Community: Houghton County, Michigan, 1837–1916," unpub. Ph.d. dissertation (Michigan State University, 1983), 80.
57. Mabel Winnetta Oas, "A History of Legitimate Drama in the Copper Country

of Michigan from 1900 to 1910 with Special Study of the Calumet Theatre," unpub. M.A. Thesis (Michigan State University, 1955), 1, 9–10, 18–21, 30–53.

58. Graham Pope, "Some Early Mining Days at Portage Lake," LSMI *Proc.* 7 (1901): 22.

59. *DMG,* 11 Oct. 1980.

60. Rice, "Mining Copper at Lake Superior," 217.

61. Thurner, *Calumet Copper and People,* 60; *Polk's Houghton County Directory* (1910), 748–52.

62. Lankton and Hyde, *Old Reliable,* 89.

63. FWD to F. Santoni, 26 Aug. 1908, Van Pelt collection.

64. SBH to J. Shuler, 5 May 1888.

65. *Red Jacket Charter and Ordinances* (1898), 74–76, 104.

66. J.R. Ryan to AA, 28 July 1898; Agassiz to S.B. Whiting, 1 Aug. 1898.

67. WRT to CL, 6 June 1913.

68. *Mining Magazine* 1 (1853): 132, 294; *Mining Magazine* 7 (1856): 311; Pittsburgh and Boston Mining Co., *A.R. for 1860,* 21; *Min. Stats. for 1880,* 19.

69. Thurner, *Calumet Copper and People,* 21–24.

70. WRT to CL, 12 Oct. 1908.

71. AA to SBW, 24 June 1898.

72. *Mining and Scientific Press* 77 (20 Aug. 1898): 181.

73. Copper Country Commercial Club, *Strike Investigation,* 41; *Min. Stats. for 1899,* 277.

74. JM to AA, 20 July and 3 Aug. 1905; AA to JM, 28 July and 7 Aug. 1905.

75. JM to Rev. P.W. Pederson, 23 and 26 June 1913.

76. JM to Rev. G.A. Schugran, 21 May 1915.

77. Rev. P.W. Pederson to JM, 21 and 25 June 1913; JM to Pederson, 23 and 26 June 1913.

Chapter 11. The Social Safety Net

1. Burt, ed., *Cornish Mining,* 42.

2. Wright, "Development of the Copper Industry of Northern Michigan," 136.

3. *Mining Magazine* 1 (1853): 132, 294; Pittsburgh and Boston Mining Co., *A.R. for 1860,* 20; *History of the Upper Peninsula of Michigan,* 336; Lankton and Hyde, *Old Reliable,* 36.

4. McNear, HAER report, 540; WRT to J.B. Turner, 17 June 1901, and to CL, 13 Dec. 1910 and 2 Aug. 1912, QMC records.

5. JM to J.P. Channing, 24 March 1913, C&H records.

6. Minesota Mining Co., *A.R. for 1858,* 13; *Mining Magazine* 8 (1857): 576; Lankton and Hyde, *Old Reliable,* 36; PMC, *A.R. for 1862,* 13; *History of the Upper Peninsula of Michigan,* 299; P.D. Bourland, "The Medical Dept. of C&H," *MCJ* 17 (Oct. 1931): 555; and WAP to FWD, 21 April 1906, Van Pelt collection.

7. Dr. A.B. Simonson to JM, 8 Nov. 1916, C&H collection. This letter provides comparable charges for seven hospitals in the area (four were company-owned). The Calumet Public, Lake Superior General, and St. Joseph hospitals charged $15 to $25 per week for rooms; the company hospitals generally charged $7 to $14.

8. JM to M.R. Casanova, 29 June 1910. Also see CM to L.L. Jacquith, 8 Aug. 1910.

9. A.B. Simonson to JM, 24 Sept. 1903.

10. Ibid.

11. A.F. Fischer, "Medical Reminiscence," *MH* 7 (Jan.–April 1923): 31; QMC, *Invoice Book*, entries for May 1874.
12. Fischer, "Medical Reminiscence," 31; SBH to TFM, 26 June 1897, QMC records.
13. *Min. Stats. for 1899*, 276–77; E.H. Pomeroy to SBW, 19 Oct. 1898, C&H records.
14. See extant "Sanitary Bulletins," 1897 through 1921, in C&H records.
15. JM to AA, 7, 12, 15, and 24 June and 6 July 1907; AA to JM, 11 July 1907.
16. JM to T. Lyons, 4 June 1910.
17. R. Wetzel to JM, 16 Sept. 1911.
18. JM to F.C. Kuhn (Attorney General, Michigan), 7 Sept. 1910.
19. A.T. Ellsworth to City Clerk of Calumet, 12 Oct. 1911, C&H records.
20. People of Houghton Co. to "Dear Sir," 14 Dec. 1911, C&H records.
21. J. Daniell to A.S. Bigelow, 6 July 1876, C&H records.
22. See, for example, J. Daniell to A.S. Bigelow, 27 June and 6 July 1878, Osceola Mining Co., *Letter Book, 1878–79*, found in C&H records. A former employee, Michael Sullivan, who had been injured and disabled at Osceola in 1875, returned to the location three years later and wanted a company house. Personally, Daniell wanted to help Sullivan, but he was afraid that if he agreed to house Sullivan, it would "be regarded as a precedent in any case of the kind that might hereafter arise."
23. *LSM*, 9 May 1857.
24. For documentation on the establishment of these societies, their membership requirements, location, and statement of purpose, see Houghton County, *Articles of Association*, vols. 1–5 (held in Clerk's Office).
25. FWD to Judge Murphey, 21 Aug. 1913, Van Pelt collection.
26. W.H. Schacht to WAP, 5 June 1928, Van Pelt collection.
27. Rice, "Labor Conditions at C&H," 1235; Wm. G. Mather, "Some Observations on the Principle of Benefit Funds and Their Place in the Lake Superior Iron District," LSMI *Proc.* 5 (1898): see "Tabulated Statement of Mine Benefit Systems," n.p.; Wright, "Development of the Copper Industry of Northern Michigan," 137.
28. AA to JM, 22 Sept. 1894.
29. See C&H *Annual Reports*, 1898 through 1901, regarding the stopping and then restarting of workers' payments to the aid fund; also, *Min. Stats. for 1899*, 277.
30. F. Valsuano to SBW, 29 March 1899. A C&H official wrote his comments on Valsuano's aid fund record on the back of this letter.
31. Mrs. C. Maier to JM, 22 July 1914; Land Agent to Mrs. Maier, 27 July 1914.
32. Bureau of Labor Stats., *Copper District Strike*, 127; *PLMG*, 26 Jan. 1882.
33. AA to SBW, 7 Aug. 1894.
34. AA to SBW, 21 Sept. 1895. Also, Rice, "Labor Conditions at C&H," 1237.
35. T.H. Soddy to JM, 29 July 1914; and JM to Soddy, 30 July 1914.
36. U.S. House, *Conditions in Copper Mines of Michigan*, 1,457.
37. Bureau of Labor Stats., *Copper District Strike*, 126–27; WRT to CL, 11 Mar. 1915, QMC records.
38. AA to JM, 3 Oct. 1904; and JM to AA, 5 Oct. 1904.
39. AA to SBW, 25 July and 28 Dec. 1898.
40. AA to JM, 6, 8, 11, and 25 June and 1 July 1904; JM to AA, 9, 11, and 28 June and 6 July 1904.
41. R. Dellacqua to JM, 4 April 1913; and JM to Dellacque, 14 April 1913.
42. M. Zagar to W.M. Gibson, 6 July 1913; W.M. Gibson to JM, 17 July 1913; M. Zagar to JM, 21 May 1914; JM to Zagar, 1 June 1914.
43. JM to AA, 1 July 1905.

44. JM to L. Thomas, 4 May 1917.
45. Dr. A.B. Simonson to JM, 24 Sept. 1903, 30 Dec. 1909, and 22 May and 28 Sept. 1914; JM to Simonson, 23 May 1914.
46. Copper Country Commercial Club, *Strike Investigation*, 34; Rice, "Labor Conditions at C&H," 1237.
47. AA to JM, 24 May 1904.
48. JM to R.L. Agassiz, 21 Feb. 1908.
49. See regular correspondence regarding pensions between JM and C&H Treasurer, G.A. Flagg, ca. 1910–11, in File 363, Box 20320, C&H records. Also, JM to QAS, 23 July 1912.
50. JM to H. Fisher, 12 March 1914; U.S. House, *Conditions in the Copper Mines of Michigan*, 1, 448–49.
51. *PLMG*, 25 May 1882.
52. Michigan, *Report of the Board of State Commissioners for the General Supervision of Charitable, Penal, Pauper, and Reformatory Institutions* (Lansing, 1873), 107.
53. *PLMG*, 17 Oct. 1878; 20 Oct. 1881; 19 Oct. 1882; 18 Oct. 1883; 24 Oct. 1884; and 26 Feb. 1885.
54. Michigan, *Sixth Biennial Report of the State Board of Corrections and Charities* (Lansing, 1881–82), 97–100.
55. *PLMG*, 29 Jan. 5 and 12 Feb. 1885; Mrs. G.W. Walker to JM, 18 Sept. 1912, C&H records.
56. "Good Will Farm, Houghton, Michigan, Report for the Years 1910 and 1911," n.p.; Mrs. Walker to JM, 18 Sept. 1912.
57. Lankton and Hyde, *Old Reliable*, 152.

Chapter 12. Social Control at Calumet and Hecla

1. For a brief treatment of the Progressive Era, accompanied by a bibliographical essay, see Arthur S. Link and Richard L. McCormick, *Progressivism* (Arlington Heights, Ill., 1983).
2. JM to QAS, 16 June 1913.
3. *PLMG*, 6 Dec. 1888; *Torch Lake Times*, 11 Dec. 1888; *Min. Stats. for 1888*, 67–68.
4. C&H, *A.R. For 1893–94*, n.p.
5. Bureau of Labor Stats., *Copper District Strike*, 97.
6. Arthur Edwin Puotinen, "Finnish Radicals and Religion in Midwestern Mining Towns, 1865–1914," unpub. Ph.D. dissertation (Univ. of Chicago, 1973), 135; Gates, *Mich. Copper*, 114–15.
7. Quoted from Benedict, *Red Metal*, 217–18.
8. Puotinen, "Finnish Radicals," 103–4
9. Gates, *Mich. Copper*, 113–14. Also see *PLMG*, 8 Jan. 1874.
10. AA to SBW, 15, 16, and 26 July and 7 Aug. 1895, and 30 Nov. 1898.
11. JM to AA, 29 April 1905; and AA to JM, 4 May 1905.
12. Thurner, *Rebels on the Range*, 33.
13. Melvyn Dubofsky, *Industrialism and the American Worker, 1865–1920* (Arlington Heights, Ill., 1975), 54–60; Gerald N. Grob, *Workers and Utopia: A Study of Ideological Conflict in the American Labor Movement* (Evanston, Ill., 1961), 34–59.
14. Dubofsky, *Industrialism*, 34, 42–44.
15. Gates, *Mich. Copper*, 114.

16. This summary of events in Coeur D'Alenes is largely based on Jenson, *Heritage of Conflict,* 25–37, and on Richard E. Lingenfelter, *The Hardrock Miners: A History of the Mining Labor Movement in the American West, 1863–1893* (Berkeley, 1974), 196–215.

17. WFM, "Constitution and By-Laws" (1893), n.p.

18. Jenson, *Heritage of Conflict,* 67.

19. Bureau of Labor Stats., *Copper District Strike,* 36; Thurner, *Rebels on the Range,* 29.

20. Jenson, *Heritage of Conflict,* 119–59; Wyman, *Hard Rock Epic,* 218–20; Dubofsky, *Industrialism,* 107.

21. Thurner, *Rebels on the Range,* 30; Arne Halonen, "The Role of Finnish-Americans in the Political Labor Movement," unpub. M.A. thesis (Univ. of Minnesota, 1945), 129, 135–36.

22. Thurner, *Rebels on the Range,* 28–30. For the rise and fall of the first wave of union locals, see WFM, *Local Records, Receipts, 1903–6* and *Defunct Unions and New Unions, 1894–1933,* WFM records, Western Historical Collection.

23. Charles K. Hyde, "Undercover and Underground: Labor Spies and Mine Management in the Early Twentieth Century," *Business History Review* 60 (Spring 1986): 11–14.

24. Puotinen, "Finnish Radicals," 203–4.

25. JM to AA, 15 April 1904; AA to JM, 25 April 1905.

26. AA to JM, 11 June 1904.

27. AA to JM, 1 July 1905; JM to AA, 6 July 1905.

28. AA to SBW, 17 June 1895.

29. Gates, *Mich. Copper,* 115; *Min. Stats. for 1899,* 278.

30. AA to JM, 19 Nov. 1906.

31. Hyde, "Undercover and Underground," 1–3; W.J. Burns International Detective Agency to JM, 26 May 1919.

32. JM to R.L. Agassiz, 15 March 1915.

33. JM to QAS, 10 July 1913.

34. JM to Charles Nagel, Secretary of Commerce and Labor, 15 June 1912; and to Wm. Williams, Commissioner of Immigration, 20 June 1912. Also, JM to QAS, 19 and 23 June 1913.

35. *PLMG,* 20 Jan. 1881.

36. J.H. Forster, "Life in the Copper Mines of Lake Superior," *Michigan Pioneer Collections* 11 (1887): 182–85.

37. Puotinen, "Finnish Radicals," 20–21, 168–69.

38. Ibid., 156, 163, 171–72.

39. Ibid., 20–33, 63, 88, 178–79. For another study of conservative and leftist Finns, see Al Gedicks, "The Social Origins of Radicalism Among Finnish Immigrants in Midwestern Mining Communities," *Review of Radical Political Economics* 8 (Fall 1976): 2–5, 12–19. Also see Gedicks, "Ethnicity, Class Solidarity, and Labor Radicalism among Finnish Immigrants in Michigan Copper Country," *Politics and Society* 7, No. 2 (1977): 127–56. Finally, see essays by Timo Orta, A. William Hoglund, Matti Kaups, and Arthur Puotinen in Karni et al., eds., *The Finnish Experience in the Western Great Lakes Region.*

40. "Men Working Underground by Nationalities," gives data as of 14 July 1913 and 23 Feb. 1914, C&H records.

41. Champion Mining Co., "Statement of Employees," for 30 Sept. 1905 and 1 Feb. 1912, Copper Range collection.

42. "Impending Labor Changes in Lake Superior Region," *E&MJ* (18 April 1914): 793; and "Proposed Elimination of the Finns," *E&MJ* (2 May 1914): 920.

43. H.A. Gilmartin to JM, numerous letters 1907–8, and return letters from JM, in C&H records, File 250, Box 20319. Joe Grimm, a *Free Press* writer, kindly provided biographical data on Gilmartin in the spring of 1988.

44. W.G. Rice to JM, 8 March 1902. At Quincy, the mine agent asked the company president for $400 to $500 "to recompense [newspaper editors] who have, unsolicited, come over to the Quincy side" during times of trouble (CL to WRT, 27 Feb. 1906).

45. Thurner, *Rebels on the Range,* 22; Thurner, *Calumet Copper and People,* 20–21.

46. W. Rice to JM, 8 March 1902.

47. JM to AA, 23 May 1904; AA to JM, 27 May 1904.

48. Andrew Swaby Lawton, "The Michigan Copper Strike of 1913–14: A Case Study of Industrial Violence during the Age of Reform," unpub. M.A. thesis (Univ. of Wisconsin, 1975), 31–32.

49. AA to SBW, 24 July 1899.

50. AA to SBW, 30 May, 28 July, 3 Aug. 1899, and 2, 5, and 12 July, and 4, 9, and 31 Aug. 1900. Also, James Alain, "The Houghton County Traction Co.," *Headlights: the Magazine of Electric Railways* 39 (May–June 1977): 2–10.

51. Gates, *Mich. Copper,* 115.

Chapter 13. Showdown: The Strike of 1913–14

1. See WFM, *Defunct Unions and New Unions,* 1894–1933, and *Register of Local Union Assessments,* April 1907–22, WFM Records, Western Historical Collection. Unless otherwise noted, all WFM union records are from this one collection.

2. For decisions to reorganize in Michigan, see WFM, *Executive Board Minutes,* meetings of 31 July 1908, 25 Jan. and 4 Aug. 1909, 3 Jan. and 3 Aug. 1910, 12 Jan. 1911, and 2 Feb. and 18 July 1912. For names and numbers of organizers, see month-to-month entries in WFM, *Disbursements,* 1 April 1908–30 June 1912, and 1912–16.

3. WFM, *Register of Local Union Assessments,* April 1907–22.

4. Ibid.

5. Bureau of Labor Stats., *Copper District Strike,* 38.

6. Ibid., 39.

7. Dan Sullivan and C.E. Hietala (president and secretary of the Copper District union) to JM, 14 July 1913, C&H records.

8. JM to QAS, 23, 27, and 31 July 1913.

9. QAS to JM, 24 July 1913.

10. Thurner, *Rebels on the Range,* 43. The membership rolls of the WFM locals have not survived, so it is difficult to prove just how many men belonged at different stages of the strike. In their public statements on membership, the union tended to exaggerate their numbers, while the companies understated union membership. Most likely, WFM membership started at about 7,000, fell to about 4,000 by autumn, and ran about 2,500 at the end of the strike.

11. JM to QAS, 23 and 24 July 1913. The most thorough narrative account of the lengthy strike, one that details all its significant events, is Thurner, *Rebels on the Range.*

12. JM to QAS, 27 July 1913.

13. JM to QAS, telegram of 31 July and letter of 22 Aug. 1913.

14. QAS to JM, 29 and 31 July 1913.
15. JM to QAS, 10 Aug. 1913; and QAS to JM, 13 Aug. 1913; Thurner, *Rebels on the Range*, 63.
16. JM, FWD, CL, et al., to Alfred J. Murphey, 16 Aug. 1913, C&H records.
17. "Joint Statement Respecting the Position of the Calumet and Hecla Mining Company by President Shaw, Vice President Agassiz and General Manager MacNaughton," 3 Jan. 1914, n.p.; QMC, *A.R. for 1912*, 7; JM to QAS, 6 Aug. 1913 and 18 March 1914.
18. Thurner, *Rebels on the Range*, 58–59.
19. Bureau of Labor Stats., *Copper District Strike*, 75–79.
20. U.S. House, *Conditions in Copper Mines of Michigan*, 1482.
21. JM to QAS, 8 Dec. 1913.
22. Thurner, *Rebels on the Range*, 64.
23. Rees, Robinson & Petermann, et al., *Brief of Counsel for the Mining Companies . . . in the Matter of a Hearing Before a Sub-Committee of the Committee on Mines and Mining, House of Representatives . . .* (12 Aug. 1914), 48–49.
24. Thurner, *Rebels on the Range*, 68–74; Bureau of Labor Stats., *Copper District Strike*, 69. Four men were ultimately judged guilty of manslaughter in the Seeberville shootings.
25. Thurner, *Rebels on the Range*, 66.
26. *Calumet Division* ledger, p. 2 of "Strike entry," C&H records, Item 5621, Box 14.
27. Thurner, *Rebels on the Range*, 104; Bureau of Labor Stats., *Copper District Strike*, 46.
28. Thurner, *Rebels on the Range*, 63–64.
29. R.R. Seeber to JM, 12 Feb. 1914, C&H records.
30. JM to QAS, 18 Aug. 1913.
31. WFM, *Michigan Defense Fund Ledger I* (Aug.–Dec. 1913); Thurner, *Rebels on the Range*, 85.
32. Thurner, *Rebels on the Range*, 83.
33. Clarence A. Andrews, " 'Big Annie' and the 1913 Michigan Copper Strike," *MH* 57 (1973): 53-68: Bureau of Labor Stats., *Copper District Strike*, 70.
34. JM to QAS, 2 and 12 Sept. 1913. Also see Thurner, *Rebels on the Range*, 53, 88–89. Out of 263 persons charged with various strike-related offenses through mid-October 1913, 48 were women.
35. Thurner, *Rebels on the Range*, 60–63.
36. A.F. Rees to FWD, 17 Dec. 1914, Van Pelt Collection; and from the C&H collection, see D.L. Robinson to JM, 18 March 1915; A.F. Rees to JM, 9 Oct. 1915; and Rees to WRT, 18 Oct. 1915.
37. Bureau of Labor Stats., *Copper District Strike*, 60–62; Thurner, *Rebels on the Range*, 118.
38. Thurner, *Rebels on the Range*, 47, 107.
39. Rees, Robinson, & Petermann, *Brief of Counsel*, 8–9; C&H, "Men Imported by Calumet and Hecla and Subsidiary Mines, from October 7th to December 18th, 1913," n.p., C&H records.
40. FWD to General Abbey, 30 Oct. 1913, Van Pelt Collection. Also see Rees, Robinson, & Petermann, *Brief of Counsel*, 12–16.
41. Thurner, *Rebels on the Range*, 103; Rees, Robinson, & Petermann, *Brief of Counsel*, 51–52; and JM to QAS, letter and telegram, 22 Jan. 1914.

42. JM to QAS, 2 Nov. 1913.
43. Citizens' Alliance, "Membership Pledge," 1913. Also, see JM to QAS, 9 Nov. 1913.
44. Thurner, *Rebels on the Range*, 112, 114; *Employment Ledger*, 1894–1918, entries for "Deputies and Hotel Men," 1913, C&H records.
45. A.F. Rees to WRT, 18 Oct. 1915, C&H records.
46. QAS to WRT, 18 March 1911, C&H records; JM, FWD, CL, et al., to A.J. Murphey, 16 Aug. 1913, C&H records.
47. "Joint Statement Respecting the Position of the Calumet and Hecla Mining Company . . . ," 3 Jan. 1914, n.p.; JM to QAS, 26 Aug. 1913.
48. Thurner, *Rebels on the Range*, 118–19.
49. Ibid., 120–21, 245–47. Ultimately, Huhta, who confessed to the killings but refused to testify against the others, was convicted of first-degree murder. The others were let off.
50. Ibid., 121–22.
51. FWD to WAP, 9 Dec. 1913, Van Pelt collection; A.F. Rees to WRT, 18 Oct. 1915, C&H records.
52. Thurner, *Rebels on the Range*, 127–28.
53. JM to QAS, 17 Dec. 1913.
54. For the most detailed description of the tragedy, see Thurner, *Rebels on the Range*, 138–53. Also, see Michael F. Wendland, "The Calumet Tragedy," *American Heritage* 37 (April–May 1986): 39–48.
55. Thurner, *Rebels on the Range*, 155; *DMG*, 25 Dec. 1913.
56. *Tyomies*, extra edition, 26 Dec. 1913.
57. Thurner, *Rebels on the Range*, 159–61.
58. *Calumet News*, 6 Jan. 1914; and Thurner, *Rebels on the Range*, 180–81, 189.
59. Thurner, *Rebels on the Range*, 175.
60. H. Goldsmith to Art [?], 15 Jan. 1914, C&H records.
61. Thurner, *Rebels on the Range*, 184.
62. WFM, *Executive Board Minutes*, meeting of 3 April 1914; Thurner, *Rebels on the Range*, 229.
63. Thurner, *Rebels on the Range*, 255–56.
64. Ibid., 248; also, Arthur W. Thurner, "Western Federation of Miners in Two Copper Camps: The Impact of the Michigan Copper Miners' Strike on Butte's Local No. 1," *Montana: The Magazine of Western History* (Spring 1983): 30–45.
65. Thurner, *Rebels on the Range*, 236.
66. Ibid., 47; Puotinen, "Finnish Radicals," 5–6, 286.
67. FWD to Messrs. Rees, Robinson & Petermann, 6 Jan. 1915, and to A.F. Rees, 12 Jan. 1915, Van Pelt collection; A.F. Rees to JM, 9 Oct. 1915, and to WRT, 18 Oct. 1915, C&H records.
68. U.S. House, *Conditions in the Copper Mines of Michigan*, 11.
69. Production figures used in calculations are taken from tables in Gates, *Mich. Copper;* C&H employment data are drawn from *Employment Ledger*, 1894–1918, C&H records; Quincy productivity data drawn from company report, "Cost of Producing Copper . . . Should be Less Regardless of . . . Increasing Depth," ca. 1921, p.2.
70. Ingersoll-Rand to JM, 27 Feb. 1914.
71. MacNaughton, "History of C&H Since 1900," 477; FWD to WAP, 14 Aug. 1914, Van Pelt collection; R.R. Seeber to W.H. Schacht, 18 Aug. 1914, CR collection.
72. Gates, *Mich. Copper*, 147.

73. A. Hill to JM, 10 June 1915; JM to A. Beck, 10 June 1915; Beck to JM, 14 and 30 June 1915; JM to Hill, 15 June 1915.

74. FWD to WAP, 14 Aug. 1914, Van Pelt collection.

75. R.R. Seeber to W.H. Schacht, 18 Aug. 1914, CR collection.

Chapter 14. Shutdown: Death of an Industry

1. Gates, *Mich. Copper,* 199, 207, 210, 220.

2. Ibid., 147–51.

3. Leong et al., "Technology, Employment and Output per Man," 84; Gardner et al., *Copper Mining in North America,* 120–22, 140–44, 271–74.

4. "Group Insurance," 1 Dec. 1919, QMC records; and CL to WRT, 29 Oct. 1919.

5. WAP to W.H. Schacht, n.d., and Schacht to WAP, 12 June 1922, Van Pelt collection.

6. Mendelsohn, untitled reflection on wages, labor situation at Copper Range mines, ca. 1922, Van Pelt collection.

7. Suomi College Oral History collection, transcript of Heritage Line broadcast, "Quincy No. 2," 4518; "Statement of German Importation Costs," n.d., CR collection; Schacht to WAP, 27 Dec. 1923, Van Pelt collection.

8. Gates, *Mich. Copper,* 155.

9. Mendelsohn to Schacht, 22 Jan. 1924, Van Pelt collection.

10. A.H. Wohlrab, "Shrinkage Stoping in Amygdaloid Lodes," *MCJ* 17 (Oct. 1931): 493, 495; Potter and Richards, "Osceola Lode Operations," 490; CL to C. Kendall, 11 Oct. 1918, QMC records.

11. CL to WPT, 15 and 29 July 1926.

12. See Crane, *Mining Methods and Practice,* 132–34, 153–59, 161–62; Lankton and Hyde, *Old Reliable,* 112.

13. From 1871 through 1900, C&H recovered only 78.8 percent of the copper contained in its stamp rock. See Benedict, *Red Metal,* 80.

14. Benedict, *Lake Superior Milling Practice,* 79–89, 98–106, 107–22; Butler and Burbank, *Copper Deposits of Michigan,* 68, 81–82, 94; H.S. Munroe, "The Losses in Copper Dressing at Lake Superior," AIME *Trans.* 8 (1879–80): 409–51; C. Harry Benedict, "Methods and Costs of Treatment at the Calumet and Hecla Reclamation Plant," U.S. Bureau of Mines Information Circular 6357 (Sept. 1930), 1–11.

15. Gates, *Mich. Copper,* 146.

16. Ibid., 153; Lankton and Hyde, *Old Reliable,* 127.

17. C&H, "To the Stockholders of Calumet and Hecla Mining Company" (Dec. 1910), 1; Benedict, *Red Metal,* 125; also see C&H *Annual Reports,* 1905 through 1910, for details on the company's expanding interests.

18. MacNaughton, "History of C&H Since 1900," 475.

19. Gates, *Mich. Copper,* 230.

20. Ibid., 154.

21. Ibid., 157–62; Gardner et al., *Copper Mining in North America,* 11.

22. CL to WPT, 28 April 1933, QMC records; R.L. Agassiz, "Review of the Copper and Brass Research Association," *MCJ* 17 (Oct. 1931): 467.

23. Gates, *Mich. Copper,* 166.

24. Letter is in QMC records. See Lankton and Hyde, *Old Reliable,* 141–43.

25. C&H Consolidated, *A.R. for 1932,* 15; Mohawk Mining Company, *A.R. for 1932,* 5; CR, *A.R. for 1932,* 6.

26. CR, *A.R. for 1929*, 7; *A.R. for 1930*, 4; *A.R. for 1932*, 3.
27. C&H, "Operations—Past, Present and Future" (25 Nov. 1950), 2–3, C&H records.
28. CR, *A.R. for 1931*, 5; C&H Consolidated, *A.R. For 1931*, 7.
29. Lankton and Hyde, *Old Reliable*, 142.
30. Gates, *Mich. Copper*, 210. Also see employment figures for these years as reported in the Houghton County *Mine Inspector's Report*, and in the U.S. Bureau of Mines, *Metal Mine Accidents in the United States*.
31. Gates, *Mich. Copper*, 167–68, 232.
32. Ibid., 164.
33. CR, *A.R. for 1931*, 5.
34. CR, *A.R. for 1930*, 6; *A.R. for 1931*, 6–7; Benedict, *Lake Superior Milling Practices*, 66–69.
35. Lucien Eaton, "Reopening and Rehabilitating the Isle Royale Copper Mine," *MCJ* 24 (Oct. 1938): 39–43; Lankton and Hyde, *Old Reliable*, 143; Gates, *Mich. Copper*, 162; Benedict, *Lake Superior Milling Practices*, 95–97.
36. Gates, *Mich. Copper*, 199, 210, 229.
37. C&H Consolidated, *A.R. for 1939*, 4–5.
38. "Scrap Recovery Campaign in Michigan Iron and Copper Country a Model," *Mining and Metallurgy* 24 (July 1943): 317.
39. Gates, *Mich. Copper*, 173–77.
40. Ibid., 199–200, 211, 221.
41. C&H Consolidated, *A.R. for 1941*, 9.
42. CR, *A.R. for 1944*, 2; Lankton and Hyde, *Old Reliable*, 146; *DMG*, 30 Sept. 1944.
43. Gates, *Mich. Copper*, 171–72; C&H Consolidated, *A.R. for 1942*, 5.
44. C&H Consolidated, *A.R. for 1942*, 3; *A.R. for 1943*, 3; C&H, "Operations—Past, Present and Future" (25 Nov. 1950), 3–4.
45. Underground employment figures for Houghton County were taken from annual *Mine Inspector's Reports*, 1942–70.
46. CR, *A.R. for 1945*, 3; *A.R. for 1946*, 3–4; *A.R. for 1948*, 4; *A.R. for 1950*, 4, 7–8.
47. For the closing of mine operations, see *Milwaukee Journal* article, "Old Quincy Mine at Hancock Closes," n.d., in Quincy vertical file, MTU; for its reclamation plant, see C. Harry Benedict, "Reclaiming Quincy Tailings from Torch Lake," *E&MJ* 145 (April 1944): 74–78.
48. Butler and Burbank, *Copper Deposits of Michigan*, 172–74; Harold B. Ewoldt, "Mining and Milling at White Pine," *MCJ* 41 (Mar. 1955), 25.
49. CR, *A.R. for 1929*, 8; *A.R. for 1940*, 5; Richard F. Moe, "White Pine Mine Development," *Mining Engineering* 6 (April 1954): 381–86; "White Pine Uses New Methods and Equipment for Mine Development," *Mining World* 16 (April 1954): 35–38; "Constructing $70 Million Michigan Copper Mine," *Excavating Engineer* 48 (Jan. 1954): 16–23.
50. CR, *A.R. for 1966*, 4. Also see *DMG*, 28 Jan., 4 March, and 19 Oct. 1967.
51. See C&H, *Annual Reports*, 1950–57.
52. A.S. Kromer et al., "Unwatering the Osceola Lode," *Mining Engineering* 8 (April 1956): 375–81.
53. Quote is found inside the front cover.
54. C&H, *A.R. for 1949*, 3; *A.R. for 1952*, 6; *A.R. for 1955*, 9–10; *A.R. for 1965*, 2.
55. Universal Oil Products, *A.R. for 1968*, 5.

56. Ibid., 14; *DMG,* 1 May 1968; *New York Times,* 22 Dec. 1967; *Wall Street Journal,* 22 Dec. 1967 and 23 Feb. 1968.

57. *DMG,* 9 April 1969; also see UOP, *A.R. for 1969,* 4.

58. *DMG,* 4 April 1969 (also see 5, 9, 10, and 15 April).

59. For example, see the efforts of a Citizens Committee and Congressman Philip Ruppe to resolve the strike, *DMG,* 9 April 1969, and a strikers' vote on a mediated settlement, *DMG,* 23 April 1969.

60. UOP, *A.R. for 1970,* 4; *DMG,* 28 and 30 Dec. 1970.

61. Issue for May 1882, pp. 892–902.

Bibliography

Archival/Public Records

Baker Library, Harvard University. R.G. Dun and Co. credit record collection.
Houghton County, Michigan. Records of Death, and Articles of Association.
Keweenaw County, Michigan. Records of Death.
Michigan Technological University Archives and Copper Country Historical Collections. Calumet and Hecla Mining Company, Copper Range Consolidated Mining Company, and Quincy Mining Company collections; Wilfred Erickson and J. R. Van Pelt collections.
Ontonagon County, Michigan. Records of Death.
Suomi College Archives, Hancock, Michigan. Oral history collection.
Western Historical Collection, Norlin Library, University of Colorado. Western Federation of Miners collection.

Newspapers/Periodicals

Boston American
Boston Evening Globe
Boston Sunday Globe
Calumet News
Copper: A Weekly Review of the Lake Superior Mines
Detroit Free Press
Miners' Bulletin
Lake Superior Miner
New York Times
Northwestern Mining Journal
Tyomies
Wall Street Journal

Mining Company Annual Reports

American Exploring, Mining and Manufacturing Co.
Calumet and Hecla Mining Co.
Central Mining Co.
Copper Range Consolidated Mining Co.
Franklin Mining Co.
Minesota Mining Co.
Mohawk Mining Co.
Norwich Mining Co.

Ohio Trap Rock Co.
Pewabic Mining Co.
Phoenix Mining Co.
Pittsburgh and Boston Mining Co.
Quincy Mining Co.
Universal Oil Products

Articles, Books, Government Documents, and Unpublished Works

Agassiz, George R., ed. *Letters and Recollections of Alexander Agassiz, with a Sketch of His Life and Work* (Boston, 1913).
Agassiz, R. L. "Review of the Copper and Brass Research Association," *Mining Congress Journal* 17 (Oct. 1931): 467
Alain, James. "The Houghton County Traction Co.," *Headlights: The Magazine of Electric Railways* 39 (May–June 1977): 2–10.
Alanen, Arnold R. "The Planning of Company Communities in the Lake Superior Mining Region," *Journal of the American Planning Association* 45 (July 1979): 256–78.
Andre, George G. *Rock Blasting, A Practical Treatise on the Means Employed in Blasting Rocks* (London, 1878).
Andrew, Clarence A. " 'Big Annie' and the 1913 Michigan Copper Strike," *Michigan History* 57 (Spring 1973): 53–68.
Barkell, William. "Honest John Stanton." Pamphlet, Houghton County Historical Society, July 1974.
Benedict, C. Harry. *Lake Superior Milling Practice* (Houghton, 1955).
———. *Red Metal: The Calumet & Hecla Story* (Ann Arbor, 1952).
———. "Methods and Costs of Treatment at the Calumet and Hecla Reclamation Plant." U.S. Bureau of Mines Information Circular 6357. Washington, 1930.
———. "Milling at the Calumet & Hecla Consolidated Copper Company," *Mining Congress Journal* 17 (Oct. 1931): 519–27.
———. "Reclaiming Quincy Tailings from Torch Lake," *Engineering & Mining Journal* 145 (April 1944): 74–78.
Blake, William P. "The Blake Stone- and Ore-Breaker: Its Invention, Forms and Modifications, and Its Importance in Engineering Industries," American Institute of Mining Engineers *Transactions* 33 (1902): 988–1031.
———. "The Mass Copper of the Lake Superior Mines, and the Method of Mining It," American Institute of Mining Engineers *Transactions* 4 (1875–76): 110–11.
———. "A New Machine for Breaking Stone and Ores," *Mining Magazine and Journal of Geology* (1:2, second series): 50.
Blandy, John F. "Stamp Mills of Lake Superior," American Institute of Mining Engineers *Transactions* 2 (1874): 208–15.
Board of Corrections and Charities, Michigan. *Biennial Reports.* Lansing, 1873–1900.
Bourland, P. D. "The Medical Dept. of C&H," *Mining Congress Journal* 17 (Oct. 1931): 555–57.
Broderick, T. M., and C. D. Hohl. "Geology and Exploration in the Michigan Copper District," *Mining Congress Journal* 17 (Oct. 1931): 478–81, 486.
Brown, Ronald C. *Hard-Rock Miners: The Intermountain West, 1860–1920* (College Station, 1979).
Brunswig, H. *Explosives: A Synoptic and Critical Treatment of the Literature of the Subject as Gathered from Various Sources* (New York, 1912).

Buder, Stanley. *Pullman: An Experiment in Industrial Order and Community Planning, 1880–1930.* (New York, 1967).
Burleigh Rock-Drill Company. *Burleigh Rock-Drill Company, Manufacturers of Pneumatic Drilling Machines, Air Compressors, and Other Machinery.* Trade cat. (Fitchburg, Mass, ca. 1873–76).
Burt, Roger, ed. *Cornish Mining: Essays on the Organisation of Cornish Mines and the Cornish Mining Economy* (Newton Abbot, 1969).
Burton, Clarence M. *The City of Detroit, Michigan, 1701–1922* (Chicago, 1922).
Butler, B. S., and W. S. Burbank. *The Copper Deposits of Michigan.* U.S. Geological Survey Professional paper 144. (Washington, 1929).
Calumet and Hecla Mining Company. "To the Stockholders of Calumet and Hecla Mining Company" (Dec. 1910).
Campbell, John F. "The Mineral Range Railroad Company," *The Soo* 2 (Oct. 1980): 12–35.
Chaput, Donald. *The Cliff: America's First Great Copper Mine.* (Kalamazoo, 1971).
Clarke, Robert E. "Notes from the Copper Region," *Harper's New Monthly Magazine* 4 (March 1853): 433–48; and (April 1853): 577–88.
Coggin, Frederick G. "Notes on the Steam Stamp," *Engineering & Mining Journal* (20 March 1886): 210–11; (27 March): 232–34; and (3 April): 248–50.
Collins, Frederick L. "Paine's Career Is a Triumph of Early American Virtues," *American Magazine* (June 1928): 40–41, 139–43.
Collins, Glenville A. "Efficiency-Engineering Applied to Mining," American Institute of Mining Engineers *Transactions* 43 (1912): 649–62.
Commissioner of Mineral Statistics, Michigan. *Annual Reports* (Lansing, ca. 1880–1915).
Committee of the Copper Country Commercial Club of Michigan. *Strike Investigation* (Chicago, 1913).
Conant, H.D. "The Historical Development of Smelting and Refining Native Copper," *Mining Congress Journal* 17 (Oct. 1931): 531–32.
"Constructing $70 Million Michigan Copper Mine," *Excavating Engineer* 48 (Jan. 1954): 16–23.
Cooper, James B. "Historical Sketch of Smelting and Refining Lake Copper," Lake Superior Mining Institute *Proceedings* 7 (1901): 44–49.
Crane, W. R. *Mining Methods and Practice in the Michigan Copper Mines.* U.S. Bureau of Mines Bulletin 306. (Washington, 1929).
———. *Subsidence and Ground Movement in the Copper and Iron Mines of the Upper Peninsula, Michigan.* U.S. Bureau of Mines Bulletin 295. (Washington, 1929).
Crissey, Forrest. "Every Man His Own Merchant," *Saturday Evening Post* 186 (20 Sept. 1913): 16–17, 57–58.
DeSollar, T. C. "Rockhouse Practice of the Quincy Mining Company," Lake Superior Mining Institute *Proceedings* 13 (1912): 217–26.
Dickerson, Don Willis. "Theatrical Entertainment, Social Halls, Industry and Community: Houghton County, Michigan, 1837–1916." Ph.D. diss., Michigan State University, 1983.
Dorr, John A., Jr., and Donald F. Eshman. *Geology of Michigan* (Ann Arbor, 1977).
Drinker, Henry S. *A Treatise on Explosive Compounds, Machine Rock Drills and Blasting* (New York, 1883).
———. *Tunneling, Explosive Compounds and Rock Drills* (New York, 1878).
Dublin, Thomas. *Women at Work: The Transformation of Work and Community in Lowell, Massachusetts, 1826–1860* (New York, 1979).

Dubofsky, Melvyn. *Industrialism and the American Worker, 1865–1920* (Arlington Heights, Ill., 1975).
Dunbar, Willis. *All Aboard! A History of Railroads in Michigan* (Grand Rapids, 1969).
Eaton, Lucien. "Reopening and Rehabilitating the Isle Royale Copper Mine," *Mining Congress Journal* 24 (Oct. 1938): 39–43.
Egleston, Thomas. "Copper Dressing in Lake Superior," *Metallurgical Review* 2 (May 1878): 227–36; (June): 285–300; and (July): 389–409.
———. "Copper Mining on Lake Superior," American Institute of Mining Engineers *Transactions* 6 (1879): 275–312.
———. "Copper Refining in the United States," American Institute of Mining Engineers *Transactions* 9 (1880–81): 678–730.
Eissler, Manuel. *A Handbook on Modern Explosives* (London, 1897).
———. *The Modern High Explosives* (New York, 1884).
Ewoldt, Harold B. "Mining and Milling at White Pine," *Mining Congress Journal* 41 (March 1955): 24–26.
Fay, Albert H., comp. *Metal-Mine Accidents in the United States during the Calendar Year 1911*. U.S. Bureau of Mines Technical Paper 40. (Washington, 1913).
Fichtel, Carl L. "Underground Power Distribution and Haulage," *Mining Congress Journal* 17 (Oct. 1931): 503–6, 508.
Fischer, A. F. "Medical Reminiscence," *Michigan History Magazine* 7 (Jan.–April 1923): 27–33.
Folsom, Michael Brewster, and Steven D. Lubar. *The Philosophy of Manufacturers: Early Debates over Industrialization in the United States* (Cambridge, Mass., 1982).
Forster, John H. "Life in the Copper Mines of Lake Superior," *Michigan Pioneer and Historical Collections* 11 (1887): 175–86.
———. "War Time in the Copper Mines," *Michigan Pioneer and Historical Collections* 18 (1892): 375–82.
Foster, J. W., and J. D. Whitney. *Report on the Geology and Topography of a Portion of the Lake Superior Land District in the State of Michigan, Part 1, Copper Lands*. U.S. House, Ex. Doc. 69 (31st Cong., 1st sess.), 1850.
Fuller, George N., ed. *Geological Reports of Douglass Houghton, First State Geologist of Michigan, 1837–1845* (Lansing, 1928).
Friedman, Lawrence M. *A History of American Law* (New York, 1973).
Gardner, E. D., C. H. Johnson, and B. S. Butler. *Copper Mining in North America*. U.S. Bureau of Mines Bulletin 405. (Washington, 1938).
Gates, William B. *Michigan Copper and Boston Dollars: An Economic History of the Michigan Copper Mining Industry* (Cambridge, Mass., 1951).
Gedicks, Al. "Ethnicity, Class Solidarity, and Labor Radicalism among Finnish Immigrants in Michigan Copper Country," *Politics and Society* 7, no. 2 (1977): 127–56.
———. "The Social Origins of Radicalism among Finnish Immigrants in the Midwestern Mining Communities," *The Review of Radical Political Economics* 8 (Fall 1976): 1–31.
Goodwin, L. Hall. "Shaft-Rockhouse Practice in the Copper Country," *Engineering & Mining Journal* 99 (19 June 1915): 1061–6; (27 June): 1107–10; 100 (3 July 1915): 7–12; and (10 July): 53–7.
Graebner, William. *Coal-Mining Safety in the Progressive Period* (Lexington, Ky., 1976).

Grob, Gerald N. *Workers and Utopia: A Study of Ideological Conflict in the American Labor Movement* (Evanston, 1961).
Haller, Frank H. "Transportation System and Coal Dock." *Mining Congress Journal* 17 (Oct. 1931): 515–18.
Halonen, Arne. "The Role of Finnish-Americans in the Political Labor Movement." Master's thesis, University of Minnesota, 1945.
Halsey, John R. "Miskwabik—Red Metal: Lake Superior Copper and the Indians of Eastern North America," *Michigan History* (Sept.–Oct. 1983): 32–41.
Hannon, Joan Underhill. "Ethnic Discrimination in a 19th-Century Mining District: Michigan Copper Mines, 1888," *Explorations in Economic History* 19 (1982): 28–50.
Harley, G. Townsend. "A Study of Shoveling as Applied to Mining," American Institute of Mining Engineers *Transactions* 61 (1918–19): 147–87.
Harrison, Joseph S. "Labor Law and the Michigan Supreme Court, 1890–1930." Master's thesis, Wayne State University, 1980.
Hobbs, William Herbert. *Earthquakes in Michigan*. Michigan Geological and Biological Survey, Pub. 5, Geol. Series 3. (Lansing, 1911).
Hollingsworth, Sandra. *The Atlantic: Copper and Community South of Portage Lake* (Hancock, 1978).
Holman, Treve. "Historical Relationship of Mining, Silicosis, and Rock Removal," *British Journal of Industrial Medicine* 4 (Jan. 1947): 1–29.
"Hon. Samuel Worth Hill," *Michigan Pioneer and Historical Collections* 17 (1890): 51–53.
Hood, O. P. "Deep Hoisting in the Lake Superior District—Some of the Engines Used to Hoist from a Depth of Over 4,900 Feet," *Mines and Minerals* (July 1904): 614–17.
Hulbert, Edwin J. *"Calumet Conglomerate," an Exploration and Discovery Made by Edwin J. Hulbert, 1854 to 1864* (Ontonagon, 1893).
Hybels, Robert James. "The Lake Superior Copper Fever, 1841–47," *Michigan History* 34 (1950): 97–119, 224–44, and 309–26.
Hyde, Charles K. "An Economic and Business History of the Quincy Mining Company." Historic American Engineering Record report, Library of Congress, 1978.
———. "From 'Subterranean Lotteries' to Orderly Investment: Michigan Copper and Eastern Dollars, 1841–1865," *Mid-America: An Historical Review* 66 (Jan. 1984): 3–20.
———. "Undercover and Underground: Labor Spies and Mine Management in the Early Twentieth Century." *Business History Review* 60 (Spring 1986): 1–27.
"The Hydraulic Compressed Air Plant at Victoria Mine," *Engineering & Mining Journal* 83 (19 Jan. 1907): 125–30.
Ihlseng, M. C., and Eugene B. Wilson. *A Manual of Mining* (New York, 1911).
Ingersoll-Rand Co. *Rock Drill Data* (New York, 1960).
Jackson, J. F. "Copper Mining in Upper Michigan—A Description of the Region, the Mines, and Some of the Methods and Machinery Used," *Mines and Minerals* (July 1903): 535–40.
Jenkin, A. K. Hamilton. *The Cornish Miner, An Account of His Life Above and Underground from Early Times* (London, 1927).
Jensen, Vernon H. *Heritage of Conflict: Labor Relations in the Nonferrous Metals Industry up to 1930* (Cornell, 1950).

Karni, Michael G., Matti E. Kaups, and Douglas J. Ollila, eds. *The Finnish Experience in the Western Great Lakes Region: New Perspectives* (Turku, Finland, 1975).

Kasson, John F. *Civilizing the Machine: Technology and Republican Values in America, 1776–1900* (New York, 1977).

Klingbeil, Dorothy J. "Copper Mining and the Sandwich Kids of Painesdale." Unpub. graduate paper, Northern Michigan University, 1973.

Koether, George. *The Building of Men, Machines and a Company* (Ingersoll-Rand, 1971).

Kromer, A. S., et al. "Unwatering the Osceola Lode," *Mining Engineering* 8 (April 1956): 375–81.

Lankton, Larry D. "The Machine *under* the Garden: Rock Drills Arrive at the Lake Superior Copper Mines, 1868–1883," *Technology and Culture* 24 (Jan. 1983): 1–37.

Lankton, Larry D., and Charles K. Hyde. *Old Reliable: An Illustrated History of the Quincy Mining Company* (Hancock, 1982).

Lankton, Larry D., and Jack K. Martin. "Technological Advance, Organizational Structure, and Underground Fatalities in the Upper Michigan Copper Mines, 1860–1929," *Technology and Culture* 28 (Jan. 1987): 42–66.

Lasio, John. "Copper Range Company: Mine—Mill—Smelter." Unpub. manuscript, n. d.

Laws, Peter. *Cornish Engines and Engine Houses* (London, 1973).

Lawton, Andrew Swaby. "The Michigan Copper Strike of 1913–14: A Case Study of Industrial Violence during the Age of Reform." Master's thesis, University of Wisconsin, 1975.

Leavitt, E. D. "The Superior," American Society of Mechanical Engineers *Transactions* 2 (1881): 106–21.

Leong, Y. S., et al. "Technology, Employment, and Output per Man in Copper Mining." Work Projects Administration National Research Project Report E-12. (Philadelphia, 1940).

Lingenfelter, Richard E. *The Hardrock Miners: A History of the Mining Labor Movement in the American West, 1863–1893* (Berkeley, 1974).

Lovell, Endicott R., and Herman C. Kenny. "Present Smelting Practice," *Mining Congress Journal* 17 (Oct. 1931): 533–38.

McCracken, S. B. *Men of Progress: Embracing Biographical Sketches of Representative Michigan Men* (Detroit, 1900).

McElroy, G. E. "Natural Ventilation of Michigan Copper Mines." U.S. Bureau of Mines Technical Paper 516. (Washington, 1932).

MacNaughton, James. "History of the Calumet and Hecla since 1900," *Mining Congress Journal* 17 (Oct. 1931): 474–77.

McNear, Sarah. "Quincy Mining Company: Housing and Community Services, ca. 1860–1931." Historic American Engineering Record report, Library of Congress, 1978.

The Manufacture of Power Drills for Mining, Excavating, Etc.," *Scientific American* 43 (25 Dec. 1880): 399, 402.

Mather, William G. "Some Observations on the Principle of Benefit Funds and Their Place in the Lake Superior Iron Mining District," Lake Superior Mining Institute *Proceedings* 5 (1898): 10–20.

Michigan. *Journal of the House* (Lansing, 1887).

———. *Journal of the Senate* (Lansing, 1887).

———. *Public Acts* (Lansing, 1887, 1897 and 1911).

———. *Report of the Board of State Commissioners for the General Supervision of Charitable, Penal, Pauper, and Reformatory Institutions* (Lansing, 1873).
Miscowaubik Club. "Constitution and By-Laws." 1 Jan. 1910.
Moe, Richard F. "White Pine Mine Development." *Mining Engineering* 6 (April 1954): 381–86.
Munroe, Charles E., and Clarence Hall. *A Primer on Explosives for Metal Miners and Quarrymen.* U.S. Bureau of Mines Bulletin 80. (Washington, 1915).
Munroe, H. S. "The Losses in Copper Dressing at Lake Superior," American Institute of Mining Engineers *Transactions* 8 (1879–80): 409–51.
Nighman, C. E., and O. E. Kiessling. "Rock Drilling." Work Projects Administration National Research Project Report No. E-11. (Philadephia, 1940).
Oas, Mabel Winnetta. "A History of Legitimate Drama in the Copper Country of Michigan from 1900 to 1910 with Special Study of the Calumet Theatre." Master's Thesis, Michigan State University, 1955.
O'Connell, Charles F. "A History of the Quincy and Torch Lake Railroad Company, 1888–1927." Historic American Engineering Record Report, Library of Congress, 1978.
Paulding, Charles P. "Machinery of the Copper Mines of Michigan," *American Machinist* 20 (8 April 1897): 17–20.
Pope, Graham. "Some Early Mining Days at Portage Lake," Lake Superior Mining Institute *Proceedings* 7 (1901): 17–31.
Post, Robert C., ed. *1876: A Centennial Exhibition* (Washington, 1976).
Potter, Ocha, and Samuel Richards. "Osceola Lode Operations," *Mining Congress Journal* 17 (Oct. 1931): 487–90.
Powell, Fred Wilbur. *The Bureau of Mines: Its History, Activities and Organization* (New York, 1922).
Pumpelly, Raphael. *Report on the Mining Industries of the United States.* U.S. Bureau of the Census, 1880 census, vol. 15, pt. 2. (Washington, 1886).
Puotinen, Arthur Edwin. "Finnish Radicals and Religion in Midwestern Mining Towns, 1865–1914." Ph.D. diss., University of Chicago, 1973.
Pyne, William. "Quincy Mine: The Old Reliable," *Michigan History* 41 (1957): 219–42.
Rawlins, J.W. "Recollections of a Long Life," in Drier, Roy, ed., *Copper Country Tales* 1 (Calumet, 1967): 75–127.
Raymond, R. W. "The Hygiene of Mines," American Institute of Mining Engineers *Transactions* 8 (1879–80): 97–120.
Red Jacket Village, Michigan. *Charter and Ordinance* (Calumet, 1898).
Rees, Robinson, & Petermann, et al. "Brief of Counsel for the Mining Companies, the Sheriff of Houghton County, Michigan, and the Michigan National Guard in the Matter of the Hearing Before a Sub-Committee of the Committee of Mines and Mining. . . . " 12 August 1914.
Rice, Claude T. "Copper Mining at Lake Superior," *Engineering & Mining Journal* 94 (20 and 27 July; 3, 10, 17, 24, and 31 Aug. 1912): 119–24, 171–75, 217–21, 267–70, 307–10, 365–68, and 405–7.
———. "Labor Conditions at Calumet & Hecla," *Engineering & Mining Journal* 92 (23 Dec. 1911): 1235–39.
Rickard, T. A. *The Copper Mines of Lake Superior* (New York, 1905).
Robinson, O. W. "Recollections of Civil War Conditions in the Copper Country," *Michigan History Magazine* 3 (Oct. 1919): 598–609.
Rosenberg, Nathan. *Technology and American Economic Growth* (New York, 1972).

Rowe, John. *The Hard Rock Men: Cornish Immigrants and the North American Mining Frontier* (New York, 1974).
Rowse, A. L. *The Cousin Jacks: The Cornish in America* (New York, 1969).
Royce, Ward. "Scraping and Loading in Mines with Small Compressed-Air Hoists," *Engineering & Mining Journal* 112 (1921): 925–31, 973–77, and 1014–19.
Sawyer, Alvah L. *A History of the Northern Peninsula and Its People* (Chicago, 1911).
Sayers, R. R. *Silicosis Among Miners.* U.S. Bureau of Mines Technical Paper 372. (Washington, 1925).
Schaddelee, Leon. "The Hancock & Calumet Railroad," *Narrow Gauge and Short-Line Gazette* 8 (Nov.–Dec. 1982): 40–45.
———. "The Mineral Range Railroad," *Narrow Gauge and Short-Line Gazette* 7 (Jan.–Feb. 1882): 54–58.
"Scrap Recovery Campaign in Michigan Iron and Copper Country a Model," *Mining and Metallurgy* 24 (July 1943): 317.
Secretary of State, Michigan. *Annual Reports on the Registration of Births and Deaths, Marriages and Divorces in Michigan* (Lansing, ca. 1865–1910).
Sharpless, F. F. "Ore Dressing on Lake Superior," Lake Superior Mining Institute *Proceedings* 2 (1894): 97–104.
Snow, Richard F. "Alexander Agassiz: A Reluctant Millionaire," *American Heritage* 34 (April–May 1983): 98–99.
Soddy, Thos. P., and Allan Cameron. "Hoisting Equipment and Methods," *Mining Congress Journal* 17 (Oct. 1931): 509–14.
Spilsbury, E. Gybbon. "Rock-Drilling Machinery," American Institute of Mining Engineers *Transactions* 3 (1874–75): 144–50.
Stanford, F. C. "The Electrification of the Mines of the Cleveland-Cliffs Iron Company," Lake Superior Mining Institute *Proceedings* 19 (1914): 189–222.
Stevens, Horace C. *The Copper Handbook: A Manual of the Copper Industry of the World,* 10 (Houghton, 1911).
Superintendents of the Poor, Houghton County, Michigan. *Annual Reports* (Houghton, 1917–26 and 1929–34).
Superintendent of Public Instruction, Michigan. *Annual Reports* (Lansing, ca. 1855–1910).
Swinford, Alfred P. *History and Review of the Material Resources of the South Shore of Lake Superior* (Marquette, 1877).
Thurner, Arthur W. *Calumet Copper and People: History of a Michigan Mining Community* (By the author, 1974).
———. *Rebels on the Range: The Michigan Copper Miners' Strike of 1913–1914.* (Lake Linden, Mich., 1984).
———. "The Western Federation of Miners in Two Copper Camps: The Impact of the Michigan Copper Miners' Strike on Butte's Local No. 1," *Montana: The Magazine of Western History* (Spring 1983): 30–45.
Todd, Arthur Cecil. *The Cornish Miner in America* (Glendale, 1967).
Trounsen, J. H. *Mining in Cornwall.* 2 vols. (Redruth, Cornwall, 1985).
"The Upper Peninsula of Michigan," *Harper's New Monthly Magazine* (May 1882): 892–902.
U.S. Bureau of Labor Statistics. *Michigan Copper District Strike.* Bulletin No. 139. (Washington, 1914).
U.S. House. Committee on Mines and Mining. *Conditions in Copper Mines of Michigan.* Hearings before subcommittee pursuant to House Resolution 387. 63rd Cong., 2nd sess. 7 pts. 1914.

U.S. Immigration Commission. *Immigrants in Industries.* Part 17: Copper Mining and Smelting. (Washington, 1911).

Van Gelder, Arthur Pine, and Hugo Schlatter. *History of the Explosives Industry in America* (New York, 1972 reprint of 1927 ed.).

Veness-Randle, April Rene. "The Social-Spatial Lifecycle of A Company Town: Calumet, Michigan." Master's thesis, Michigan State University, 1979.

Vivian, Arthur C. "Cutting Mass Copper," *Engineering & Mining Journal* 98 (7 Nov. 1914): 825.

Vivian, Harry. "Mining Methods used on the Calumet Conglomerate Lode," *Mining Congress Journal* 17 (Oct. 1931): 496–502, 508.

Wadsworth, M. E. "The Michigan College of Mines," American Institute of Mining Engineers *Transactions* 17 (1897): 696–711.

Wendland, Michael F. "The Calumet Tragedy," *American Heritage* 37 (April–May 1986): 39–48.

Western Historical Society. *History of the Upper Peninsula of Michigan Containing a Full Account of Its Early Settlement; Its Growth, Development and Resources* (Chicago, 1883).

Weston, Eustace M. *Rock Drills: Design, Construction and Use* (New York, 1910).

"White Pine Uses New Methods and Equipment for Mine Development," *Mining World* 16 (April 1954): 34–38.

Wohlrab, A. H. "Shrinkage Stoping in Amygdaloid Lodes," *Mining Congress Journal* 17 (Oct. 1931): 491–95.

Wright, J.N. "The Development of the Copper Industry of Northern Michigan," *Michigan Political Science Association* 3 (Jan. 1899): 127–41.

Wyman, Mark. *Hard Rock Epic: Western Miners and the Industrial Revolution* (Berkeley and Los Angeles, 1979).

Young, George J. *Elements of Mining* (New York, 1932).

Young, Otis E., Jr. *Western Mining* (Norman, Okla., 1970).

———. "The American Copper Frontier, 1640–1893," *The Speculator: A Journal of Butte and Southwest Montana History* 1 (Summer 1984): 4–15.

Index

Abbey, General, 226
Accidents, fatal, 24, 29, 96, 128, 132, 135, 265; by age, 111, 113; causes of, 113–22, 129, 132, 141, 187; by company, 111; deaths per thousand employed, 124; by decade, 110–11; by ethnic group, 112; law-suits over, 134–39, 141; by occupation, 113, 115; Osceola fire, 114, 122–24, 132, 237; in other districts, 125, 130; out-of-court settlements, 136–39; preventability of, 132, 140; reporting of, 131, 133. *See also* Mine safety; Mutual aid and benefit societies; Safety First movement; Workmen's compensation
Accidents, less-than-fatal, 124–25, 134, 140; charity work for accident victims, 190, 196–97
Agassiz, Alexander, 21, 23, 122, 170, 173, 177–78, 215–16, 257; on aid fund, 188–89; anti-unionism of, 202–4, 205, 206, 209; as autocrat, 200–201, 210; career as scientist, 20, 218; on charity, 190–93; on churches, 178–79; on elimination of trammers, 101–2, 103; family background and education, 19–20; on health services, 183; named C&H president, 20; opposes streetcars, 210, 216–17; on pension plan, 194–95; resides at Calumet, 20, 44, 157; technological style of, 44
Agent, mining company: conflicts with eastern officers, 79–81, 94, 96, 97–99; engineers assume post, 72–73; role of, 59, 151, 168, 187; routes taken to agent's post, 60–61
Ahmeek mine, 10, 72, 101; C&H acquires, operates, 251, 254, 256, 258
Allis, E. P., & Co., 47, 48, 56, 80
Allouez conglomerate lode, 10, 14

Allouez mine, 10, 21, 54, 254; C&H acquires, 251
American Federation of Labor, 211, 229
American Institute of Mining Engineers, 74, 85
Americanization (of immigrants), 143, 144, 151, 171, 174, 211
American Mining Co., 15–16
Amerikan Uutiset, 215
Amygdaloid deposits, 5, 10, 14
Anaconda copper, 245
Animals, use in mines, 32, 101
Apostolic Lutheran Movement in America, 212
Arcadian mine, 72
Arizona, as copper producer, 22, 71, 76, 245
Armenians, 112, 154
Ash Bed lode, 67
Assumption-of-risk rule, 134, 136
Athletics, workers participation in, 175–76
Atlantic amygdaloid lode, 14
Atlantic mine, 10, 21; drilling machines at, 78, 86, 89, 91; dynamite at, 96–97; hanging-wall problems, 72
Austrians, 112, 155, 161, 211, 213, 214
Automobile industry, attracts mine workers, 240–41, 246
Aztec mine, 82, 84, 95

Ball, Edwin P., 56
Baltic amygdaloid lode, 11, 14, 73
Baltic mine, 11, 21, 38, 71, 72
Bands, company, 174–75
Bars (saloons), 142, 176–77, 210
Bathhouses, company, 172, 173, 191
Beck, August, 242
Benedict, C. Harry, 249
Bennett, Robert, 167

311

Bigelow, Albert S., 21
Bigelow, Horatio, 18, 21
Bird, George M., & Co. 43
Black powder. *See* Explosives
Blake, Eli Whitney, 49
Brockway, Daniel, 60–61
Burleigh, Charles, 82
Burleigh Rock Drill Co., 81, 85
Butte, Mont.: labor conditions in, 23; as leading copper producer, 22, 71, 76, 245; and support for strike of 1913–14, 229–30; unions in, 206, 207, 219, 239

Cages, 35
Calumet and Hecla Consolidated Mining Co.: creation of, 251; diversifies, 258–59, 261; and Great Depression, 253–54, 256
Calumet and Hecla, Inc.: and closing of C&H mine, 262–63; formed, 261; strikes at, 261–62
Calumet and Hecla mine, 35; fires in, 120–22, 202; high explosives in, 97; location of, 10, 17; mechanized tramming in, 101–2; one-man drills in, 105–6; Red Jacket shaft, 46, 122; rockhouses at, 50; shaft-rockhouses at, 51; shut-down of, 24, 256; steam engines at, 44–46, 47; surface plant at, 41, 43–44, 46; timbering within, 30; two-man rock drills at, 82, 84, 86, 88; underground temperatures, 38
Calumet and Hecla Mining Co.: acquires stock in other companies, 72, 251; blacklists WFM leaders, 228, 242; and churches, 178–80; consolidation efforts, 251; difficult start, 18–20; dominance in region, 14–15, 143–44, 201, 251; employs efficiency engineers, 75–76; employs geologists, 73; founding of, 20; funding for 1913–14 strike, 225; library of, 172–74; reclamation plants, 250, 254, 259; and schools, 170–72; smelter of, 13, 18; stamp mills of, 18, 56, 248–49; strikes against, 69, 202–4; wages paid, 70, 204
Calumet conglomerate lode, 10; declining copper values, 72; discovery of, 18; limited profitability along, 17, 19; operations closed at C&H, 256; richest in district, 14, 20
Calumet Mining Co., 18, 19, 20
Calumet News, 226
Cass, Lewis, 7
Centennial mine, 17, 251, 254, 258, 263
Central mine, 21, 35, 82, 84, 190
Champion mine, 35, 71, 72; on Baltic lode, 73; closed, 260; company housing at, 154–56; elimination of Finns at, 214; and Stanton-Paine group, 21; underground temperatures, 38
Charities: company-sponsored, 135, 184, 186–87, 190, 194; private, 197, 255; public, 196, 255
Chicago, Ill.: labor unrest and strikes at, 205, 206; home of labor radicals, 209, 211
Child labor: boys' deaths underground, 110, 111, 112–13, 124; drill boys, 88–89; legislation regarding, 133, 190–91; and machinery, 33, 102
Children: and Italian Hall disaster, 237, 239; orphans and needy, 197; in school, 168–69
Churches, 14, 22, 188, 196; company support of, 177–80; effect of 1913–14 strike on, 240
Citizens' Alliance: condemns WFM, 233–35; and Italian Hall disaster, 237–38; publishes *Truth*, 233
Civil War, 9, 14, 16–17, 24, 67, 143, 165
Clark mine, 82
Clemenc, "Big Annie," 230
Cleveland-Cliffs iron mine, 102
Cliff, Capt. John, 59, 61
Cliff mine, 21, 33, 34, 65, 182; early development of, 9–10, 149; steam power at, 42
Coeur d'Alene Executive Miners' Union, 206
Colorado: labor conditions, 23; labor wars in Cripple Creek and Telluride, 208–9; unions in, 207
Colorado School of Mines, 74, 75
Columbia School of Mines, 30, 73, 74
Conglomerate lodes, 5, 10, 14
Conglomerate mine, 89, 169
Connecticut, as copper producer, 9
Contract system of mining: in Cornwall, 62–63, 66; decline of, 76–77, 247; effects of one-man drill on, 107, 221; effects of Rand drill on, 87–92; inequities of, 63–65; on Lake Superior, 59, 62, 65–69, 70
Contributory negligence, 134, 136
Cooper, James, 217
Copper: costs of producing, 23, 71, 73, 76, 83, 86, 105, 245, 247, 252; foreign producers of, 245, 259; geological formation on Keweenaw, 5–6; importance of amygdaloid and conglomerate, 10, 14, 39; Keweenaw production of, 9, 14, 20, 22, 23, 71, 242, 244, 256, 257, 263, 265; markets for, 8, 23, 245; mass copper, 9–10, 31, 104; native copper, 5–6, 11; producers' cartel, 245, 252; western U.S. produc-

tion of, 22, 71, 76, 200, 241, 245; yields from lodes, 72
Copper Falls mine, 32, 42, 56, 114; rock drills at 82, 83, 84, 85
Copper Harbor, Mich., 4, 8
Copper Range Consolidated Mining Co., 71; controls Baltic, Trimountain, and Champion mines, 72; develops White Pine mine, 259–60; formation of, 21; and Great Depression, 253, 254, 255–56; library of, 172, 173; operates stores, 166; reclamation plant of, 250, 256, 259; and schools, 170; shutdown of, 24; smelter of, 13, 259
Corey, A. J., 59, 83, 96, 97
Cornishmen: accept risks, 131, 140; and churches, 177; emigration of, 22, 61, 70, 246; favor contract mining, 63; fill top mine posts, 60, 61–62; killed underground, 112; leave for auto plants, 246; paternalism favors, 151, 155, 156; pride in mining, 30, 131, 212; reputation as independent, 62; as working-class elite, 227
Cornish mines: and contract mining, 59, 62–65; doctors at, 181; machinery at, 33, 36–37, 55; produce sought-after experts, 61–62, 63
Cost of living, workers', 161, 165, 167
Croatians, 112, 154, 156, 214, 228, 229, 230
Cruse, Sheriff James, 223, 224, 233, 235

Daily Mining Gazette: ally of companies, 214, 215; and 1913–14 strike, 219, 226, 235
D'Aligny, Henry, 19
Danes, 112
Daniell, John, 61, 187; and company stores, 165, 167
Daniell, William, 61
Daniels, Nathan, 79–80
Darrow, Clarence, 230
Debs, Eugene, 216
Defense Materials Procurement Agency, 260
Deindustrialization, 267
Delaware mine, 54
Denton, Frederick W., 73, 144, 154, 166, 176, 242–43; and strike of 1913–14, 227, 232
Depression, Great: effect on companies, 252–55; effect on local communities, 252–53, 255; and organized labor, 257–58; relief programs, 255
Detective service companies, 126, 207, 211, 228
Detroit and Lake Superior Copper Co., 13

Detroit *Free Press,* 214
Doctors, company, 22, 66, 151, 182; at C&H, 182, 183, 184; at Quincy, 182–83; tradition of, in Cornwall, 181; treat widows, 194
Douglas, Columbus C., 60
Downer, Dr., 183–84
Drills. *See* Rock drilling
Duncan, Capt. John, 192, 217
Dynamite. *See* Explosives

Eagle Harbor, Mich., 4
Eagle River, Mich., 4, 42
Economic decline, regional, 24–25, 242, 244–46, 266; companies combat, 245, 247, 248, 251, 252, 255–56, 258–59, 261; loss of jobs, 244, 255, 256, 258; mine closures, 250–51, 252, 253, 259, 260, 263. *See also* Out-migration of workers; Unemployment
Edwards, Richard, 122, 123
Egleston, Thomas, 30, 96
Eight-hour day, 133, 225, 233–34
Electric Park, 176
Eloheimo, Pastor J. W., 212–13
Ely, Richard, 147
Emerson, Luther G., 74
Engineers: as company agents, 61; education of, 74; as efficiency experts, 75–76, 77, 140, 266; evolving role of, 74–75; few employed, 19th century, 60, 74; replace "practical" men, 22, 58–59, 72–74, 75, 76, 107
Environmental impact of mining, 5, 12, 43, 266
Ethnic groups, 22, 23–24, 60, 71, 144, 161, 174, 208, 213; and agrarian traditions, 164; benefit societies, 188; and churches, 177–78, 212–13; discrimination in hiring, 203, 211–14, 265; and housing, 148, 154, 155–56; importation programs, 246; killed at Italian Hall, 237; killed underground, 112, 131; newspapers of, 214–15; and paternal benefits, 197–98; roles in strike of 1913–14, 227, 229, 235; social halls, 175, 236; stereotypes of, 212. *See also* Armenians; Austrians; Cornishmen; Croatians; Danes; Finns; French Canadians; Germans; Greeks; Hungarians; Irish immigrants; Italians; Lithuanians; Mexicans; Norwegians; Polish immigrants; Russians; Scotsmen; Slovenians; Swedes
Evergreen Bluff mine, 49
Explosives: black powder, 30, 62, 94, 95, 96, 97; dynamites and high explosives, 31,

Explosives (*continued*)
    39–40, 93–99; fear of nitro, 94–95; safety of, 93–94, 96, 98, 99, 113, 118–20; sold to miners, 66. *See also* Accidents, fatal; Mine safety

Farms, company, 15–16, 22, 163–64
Fellow servant rule, 134, 136, 137
Ferris, Governor Woodbridge, 224, 226, 230, 231, 238
Finlander Workmen's Society, 127
Finnish Evangelical Lutheran Church in America, 212
Finnish Temperance Society, 127
Finns: and company housing, 154, 155–56; and cooperative stores, 167; discriminated against, 23, 161, 212–13; and farming, 164; killed underground, 112; migration to U.S., 71, 212–13; as miners, 70, 155; and newspapers, 214–15; numbers reduced, 213–14; and politics, 213, 215, 235; and religion, 178, 212–13, 215, 240; and strike of 1913–14, 227, 229, 230, 240
Fires, 114; at C&H, 120–22; at Osceola, 122–24, 126, 132
First-aid training, 140
First World War, 24, 146, 154, 266; effect on district, 242, 244–45, 247
Fischer, Dr., 183
"Five Dollar Day," 146, 147, 240–41
Flotation process, 249–50
Forest mine, 42
Ford, Henry, 145–46, 147, 240
Forster, John H., 61
Franklin mine, 10, 54, 65, 72, 182; library at, 172; mechanized drilling, 82, 84, 86
French Canadians, 22, 43, 112, 155, 156

Gardens, workers', 142, 164, 244, 254–55
General Electric Co., 102
Geologists, 11; early surveys and assessments, 3, 7–8; as mine managers, 60–61; role in industry, 59–60, 73, 74
Germans, 22, 24, 60, 112, 155, 156, 178, 211, 212
Gilmartin, H. A., 214
Gompers, Samuel, 229
Goodwill Farm, 197
Graton, Prof. L. C., 73
Greeks, 112, 211
Greeley, Horace, 8
Gundry, John, 61

Hall, Josiah, 133
Hancock *Evening Journal*, 214

Harris, John L., 48, 72, 104
Harris, Samuel B., 61, 72, 102, 176; disputes with eastern officers, 79–81, 97–98
Hecla Mining Co., 18, 19, 20
Hill, Samuel W., 60
Hodge and Christie (engine builders), 43
Hoisting technology: balanced hoisting, 47; C&H engines for, 45–46, 117; by hand, with windlasses, 46–47; horsewhims, 47; kibbles and skips, 32, 39, 47, 118; ropes and chains, 47, 117; specialized engines for, 47–48
Holy Rosary Catholic Church, 179
Hoosac Tunnel, 82, 83
Hospitals, company, 182–83, 184
Houghton, Douglass, 7–8, 9, 60, 264
Houghton Club, 175
Houghton County Charitable Assoc., 197
Houghton County Poorhouse and Farm, 196
Houghton County Traction Co., 175, 217
Housing, company-owned, 22; and community planning, 149–51; discriminatory distribution of, 149, 154–57; evictions from during strike, 232; hierarchy of, 152–53, 154, 156; improvements to, modernization of, 151–52, 157–58, 161; profitability of, 148, 158, 160; reasons for building, 147–48, 160–61; rents, 148, 157, 160; types of, 16, 148–51, 153; widows and, 192–94
Housing, private (on leased company ground), 148, 149, 150, 158–60
Howe, Thomas, 21
Hubbard, Dr. Lucius L., 73
Hubbell, Jay A., 133
Hulbert, Edwin, 18, 19, 20
Hulbert Mining Co., 18
Hungarian Magyar Reformed Church, 240
Hungarians, 23, 112, 156, 229
Huron mine, 10, 95
Hussey, C. G., & Co., 257, 259
Hussey, Curtis, 21

Idaho: labor conditions in, 23; unions in, 206–7
Industrialization: general, 143, 145, 199; importance of steam power to, 42; of mining, 22, 58
Ingersoll-Rand Co., 104, 106, 241, 248
International Union of Mine, Mill and Smelter Workers, 240, 258
International Workers of the World, 208
International Workingmen's Assoc., 205
Irish immigrants, 22, 24, 60, 112, 155, 156, 212, 229
Iroquois lode, 258

Isle Royale, 6, 8, 9, 12
Isle Royale amygdaloid lode, 14
Isle Royale mine, 10, 72, 101, 252, 253, 256
Italian Hall, disaster at, 236–39, 240
Italians, 22, 155, 161, 213, 214, 229; and company housing, 154, 156; discrimination in hiring, 211; killed underground, 112; as unskilled mine workers, 23, 70, 71, 161

Jackson, J. G., 94
Jane, Arthur, 234–35
Jane, Harry, 234–35
Jensen, Marelius, 184–86
Jensen, Vernon, 145
Jones, Mother, 225

Kearsarge amygdaloid lode, 14, 72; C&H works on, 251, 254
Kearsarge mines, 251, 254, 256, 258
Kendall, Captain, 128, 132
Kennecott copper, 245
Keweenaw Peninsula: appearance today, 4–5; early explorations of, 6–7; geological formation of, 5–6; natural environment of 3, 5; population of, 9, 22, 150–51, 246, 255, 256; remoteness and isolation of, 4, 13, 14–15, 142, 145, 151, 253, 263
Kingston mine, 261, 263
Knights of Labor, 133, 205–6, 211
Korean War, effect on district, 259–60
Krone, Rev. Father, 177
Krupp Works, 45

Labor unrest, 77, 199–200, 204; and ethnic groups, 200, 213; explosives-related, 93–95, 97–99; "Great Upheaval" of 1886, 205; grievance policies, 202, 208, 234; paternalism erodes, 197–98; trammers' alienation, 101, 103, 198; in western U.S. mines, 23, 200. *See also* Strike of 1913–14; Strikes (other than 1913–14)
Ladders, 33, 120
Lake Chemical Co., 258
Lake Superior, 3, 13, 42; coastline and ports, 4; as mill site, 11, 53, 54
Lake Superior Iron Works, 43
Lake Superior mine, 42
Lake Superior Powder Co., 96, 97
Larson, O. J., 173, 215
Lawton, Charles, 132, 135, 173, 208, 253; attitude regarding safety, 130, 132, 136, 141; background and education, 72–73; death of, 257; and labor spies, 126–29; as modern practitioner, 74–75, 98, 247; and one-man drill, 104–9; selects explosives, 98–99; and strike of 1913–14, 223
Lawton, Swaby, 137, 138
Leaching process, 249
Leavitt, Erasmus Darwin, 44–47, 56
LeGendre, Edward F. 135, 137
Leprosy, at C&H, 184–86
Ley, Samuel, 68–69
Leyner, J. George, 104
Libraries, company, 172–74
Lighting, underground, 36
Lithuanians, 155
Lowell, Mass., 44–45, 145–46

McCormick Harvester, strike at, 206
MacDonald, William J., 199, 241
MacNaughton, James: on accident victims and safety, 132–33, 137; as administrator of social safety net and paternal programs, 157–58, 170, 173, 174, 183, 184–86, 190–95, 218; as autocratic leader, 201, 218; on Calumet and Hecla's consolidation 251; career of, 201; on church support, 179–80; death of, 257; education of, 73; and ethnic discrimination, 211, 213; and one-man drill, 105; seeks control over press, 214–16; and strike of 1913–14, 222–34, 236, 242; on trammers and tramming, 102, 154; on WFM's initial union drive, 209
Mahoney, C. E., 221, 226, 232
Man-cars, 35, 116–18
Man-engine, 33–35
Marriage and family: companies' prefer married workers, 150–51, 161, 191, 201; husband as breadwinner, 190. *See also* Widows, treatment of; Women
Mason, Thomas F., 21
Massachusetts Institute of Technology, 75
Mechanization, on surface, 42, 58; of rock-breaking, 49–53. *See also* Hoisting technology; Stamp mills; Steam power; Surface plants
Mechanization, underground, 22, 26; and child labor, 33, 88, 102; of copper cutting, 31; of drilling shot holes, 31, 39–40, 81–92; of mucking and tramming, 32, 99–103, 248; and safety, 113, 115–18; transporting men, 33–35; of ventilation, 38. *See also* Rock drilling; Tramming technology; Transport, human
Medical programs, company, 181, 182–83, 202; insurance, 182; public health services, 184–86

Mentally ill, treatment of, 196
Mesnard mine, 72
Messimer, Dr. 192
Metals Reserve Co., 257, 258, 259
Mexicans, 112, 246
Michigan College of Mines, 4, 73, 75; role in industry, 74
Michigan mine, 21, 208
Michigan Technological University, 4
Miller, Guy, 220
Milling machinery. *See* Stamp mills
Milwaukee, Wisc.: labor strike in, 206; as home of radical labor, 209
Mine inspectors, county, 133–34, 139
Miners: on company account, 66; on contracts, 62–66; definition of, 30; earnings of, 68–70, 76, 87–88, 89, 90, 108; skills of, 30, 40, 85; status of, 30, 69–70, 76–77
Mine safety: safety practices, companies, 117–18, 120, 121–22, 127, 132, 139–40; safety practices, workers, 114–15, 119, 132; as social issue, 125, 128–30, 134–35, 140–41; workers' acceptance of risk, 131, 133, 134, 139, 140. *See also* Accidents, fatal; Accidents, less-than-fatal; Safety First movement; Workmen's compensation
Minesota mine, 32, 65, 177, 182, 187; early development, 9–10, 149–50; steam engines at, 42
Mining captains, 58–61, 107, 151; Cornishmen as, 61–62; lose responsibilities, 74, 76; role in contract mining, 62–63, 64, 66, 68
Mining companies: capital investment in, 8–9, 15; collusion during 1913–14 strike, 225, 240; dividends paid, 9, 14–15, 70, 242, 244, 252, 257; domination over local society, 23, 168, 172, 177, 216, 252; "families" of investors, 20–21; financial risks of starting, 8–9, 14; increase in size of, 58, 143, 197; management structure, 58–60, 79–81
Mining methods: advancing system, 29, 39; drifts, drifting, 27–28, 32, 66; retreating system, 29, 30, 39; rock pillars, leaving, 29–30, 39, 129; shafts, shaft-sinking, 26–27, 32, 66, 114; stopes, stoping, 28–29, 66; timbering, 30, 32, 114; winzes, 28
Miscowaubik Club, 175, 193
Mohawk mine, 21, 54, 156, 252, 253
Morris, I. P., 45, 46, 48
Moyer, Charles, 221, 229, 230, 239; and Italian Hall disaster, 237–38; knows strike is lost, 232
Mucking rock, 100–101, 248

Mutual aid and benefit societies, 196; set up by C&H, 188–90; workers' own, 187–88

National Labor Relations Board, 258
National mine, 49
Native American Indians: mine copper, 6, 10, 18; treaties with, 7, 8
New Jersey, as copper producer, 9
Newspapers: control over by companies, 214–16, 226; English-language, 34, 82, 83, 89, 197, 212, 214, 215, 219, 226, 235; foreign-language, 208, 215, 237
Nicholas II, Czar, 213
Nitroglycerine oil. *See* Explosives
Nobel, Alfred, 94, 95
Nobel, Emmanuel, 94
Nonesuch lode, 259
Nonesuch mine, 42
Nordberg Manufacturing Co., 47, 48, 56, 248
North American mine, 42, 149, 150, 164, 166
Northwest mine, 86, 149
Norwegian-Danish Methodist Episcopal Church, 179
Norwegians, 112, 156, 178
Norwich mine, 15–17

O'Brien, Patrick H.: as judge during 1913–14 strike, 231–32, 233, 234, 235, 240; as personal injury lawyer, 135, 137, 138
Ogima mine, 49, 82
Ohio Trap Rock Co., 164
One-man drill. *See* Rock drilling
O'Neill, Capt. Jerry, 247–48
"Ontonagon Boulder," 7, 8
Osceola amygdaloid lode, 14, 17, 72; C&H works on, 251, 254, 261
Osceola mine, 21, 35, 72, 86, 89, 187; becomes profitable, 17; C&H acquires, 251; fire at, 122–24, 126, 132; location of, 10; store at, 165–66
Out-migration of workers: 240–41, 245–47, 255

Paine, Ward, 244
Paine, William A., 21, 154, 182
Painesdale High School, 170
Paivalehti, 215
Parnall, W. E., 123–24, 126
Pasty, Cornish, 35
Paternalism: breadth of, 22, 145, 172; criticisms of, 146–47, 149, 163, 264–65; and Great Depression, 254; perquisites for

managers, 151–52, 156, 157, 182; as practiced elsewhere, 145–46; praised, 144, 159; and social control, 23, 172, 200, 202, 218; after strike of 1913–14, 242–43; weakening bonds of, 197–98, 202, 217. *See also* Bands, company; Bathhouses, company; Charities; Churches; Doctors, company; Farms, company; Gardens, workers'; Hospitals, company; Housing, company-owned; Housing, private (on leased company ground); Libraries, company; Medical programs, company; Mutual aid and benefit societies; Pensions, workers'; Recreational activities; Schools; Stores; Widows, treatment of

Pederson, Reverend P. W., 179–80

Pennsylvania Mining Co., 8, 49

Pensions, workers', 194–95

Perkins, T. Harry, 21

Pewabic amygdaloid lode, 14, 17, 72, 149, 253

Pewabic mine, 33, 34, 54, 65, 182; acquired by Quincy, 72; finds lode, 17; first library, 172; housing at, 149–51; location of, 10; mechanized drilling at, 82, 84, 86

Phelps-Dodge copper, 245

Phoenix mine, 35, 67, 94

Pittsburgh and Boston Mining Co. *See* Cliff mine

Playfair & Hurd (efficiency engineers), 76

Polish immigrants, 112, 156, 211

Pollard, Joy, 206

Pontiac mine, 72

Portage Lake, 42, 54; milling and smelting at, 11–12, 13, 21, 53; mines near, 10–11

Portage Lake Foundry and Machine Works, 43

Portage Lake Mining Gazette, 34; ally of companies, 214; criticizes use of drill boys, 88–89; on Finlanders, 212; supports home for disabled, 196–97; on two-man drill, 82–83

Powderly, Terrance, 205

Productivity, 40, 71, 92, 252; improvements necessary, 76, 105; losses due to Cornish contract system, 64; man-engine as time-saving device, 33; and one-man drill, 241; and trammers, 103; and two-man drill, 86; and wage-plus-bonus system, 247. *See also* Scientific management; Taylor, Frederick W.

Progressive era and reforms, 24, 140, 199–200, 234

Prostitution, 176

Pullman, George, 145–46, 147

Quincy mine, 21, 27, 33, 34, 35; copper yields, 11; Cornishmen strike at, 30; and Great Depression, 253, 254–55; hanging-wall problems and "air blasts," 72, 127–29; high explosives at, 97–99; location, 10; one-man drill in, 104–9; retreating system at, 30, 130; rock-breaking technology, 48–53; rock drills introduced, 82–91; shuts down, 259; steam engines at, 47, 248; tramming at, 32, 100, 101–3; underground temperatures, 38

Quincy Mining Company: acquires other properties, 72; churches supported by, 177, 178; company store, 165; difficult beginning, 17; engineering staff, 72, 74–75; library, 172–73; manipulates miners' earnings, 68–69; reclamation plant, 250, 257; school controlled by, 169–70; smelter, 13; wages paid, miners and trammers, 69

Railroads, 14, 18, 27, 51, 53–54

Rand Drill Co., 82, 85

Rawlings, John, 34

Reclamation plants, 250, 254, 256, 257, 259. *See also* Stamp mills

Recreational activities, 142, 173, 174–76, 210. *See also* Athletics; Bands, company; Bars; Electric Park; Prostitution; Theatres

Rees, Allen F., 137; and 1913–14 strike, 231, 235

Rees, Robinson & Petermann, 231, 235, 240

Rice, Claude, 144–45, 173

Rice, W. G. 214–15

Roberts, John, 68–69

Robinson, S. S., 60

Rock breaking (on surface): with Blake crushers, 49–53; by hand, 48; in rockhouses, 49–51; in shaft-rockhouses, 51–53; with steam and drop hammers, 49, 50, 51, 52

Rock drilling: Burleigh drills, 82–88, 92; dust and health hazards, 89–90; by hand, 29, 30, 31, 81, 86, 90–91; by machine, 31, 39–40, 78–92; one-man drills, 93, 104–9, 241, 248; Rand two-man drills, 82–92, 96, 103–4, 105

Rockland mine, 150

Rock receiving (on surface), 48–52

Roosevelt, Franklin D., 257

Russians, 112

Ryan, John, 177

Safety First movement, 139, 141

Saint Joseph's Catholic Church, 179

Salvation Army, 210

Sanitation, underground, 35–36
Sarah Sargeant Paine Memorial Library, 173
Saunders, W. L., 106
Schact, William, 244, 257
Schoolcraft, Henry, 7
Schoolcraft mine, 17
Schools, 14, 168–72
Scientific management, 75, 107
Scotsmen, 60, 112, 155, 229
Sears, Roebuck and Co. houses, 151, 156
Second World War, 252, 257, 258
Seeber, R. R., 229, 243
Seeberville, 154; murders at, 228, 231
Seneca mine, 252, 258
Shaw, Quincy A., 19, 20, 21, 257
Shaw, Quincy A., Jr., 174, 218, 251; and 1913–14 strike, 222, 223, 225, 226, 228, 230, 233, 236
Shea, Bartholomew, 203
Shea, John, 128, 136
Shepard, Dr. A. A., 183
Shifts: bosses of, 60, 89; length of, 31, 202; rotate day and night, 70, 247
Simonson, Dr. A. B., 181, 183
Sims, James, 64–65
Slovenians, 156, 214
Smelting, 11, 12, 13, 21, 54, 250
Sobrero, Ascanio, 94
Socialism, 188, 207–8, 213, 215, 225, 230, 234
Societa Mutua Beneficenza Italiana, 236
South Dakota, miners' unions in, 207
Spies, labor: at C&H, 203, 211; at Quincy, 126–29. *See also* Detective service companies
Stamp mills: Cornish drop stamps, 55–56; crushing rollers, 55; description of, 11–12; pumps at, 56; recovery of fine copper, 248–49; sites of, 53; steam stamps, 55–56, 256. *See also* Flotation process; Leaching process; Reclamation plants
Stanton, John, 21, 78
Steam power; boiler fuels, 42–43, 80–81; at Calumet and Hecla, 45–46, 257; in Cornwall, 62; to drive generators, 248; early engines, 42, 43, 47; hoisting applications, 45–48; importance to district, 42, 56–57; at Quincy, 79–80, 248
Stores: company-owned, 14, 16, 22, 165; co-operatives, 165, 167; independent merchants, 166; leased, on company ground, 166
Strike of 1913–14, 24, 69, 109; back-to-work movement, 227, 228, 232; Christmas eve disaster, 236–39; citizen rallies against WFM, 235; congressional investigation of, 226, 238; effect on companies, 240–41; ends, 239; imported scabs, 230, 231, 232, 233, 243; indictments during, 239; and Michigan National Guard, 224–25, 226, 227, 230, 231, 233; mines reopen during, 229, 231; and one-man drill, 107–9, 110, 221, 225, 266; start of, 222–24; union demands, 225; use Waddell men, 228; Western Federation of Miners' strategy, 223–24, 230; and women, 230–31. *See also* Western Federation of Miners
Strikes (other than 1913–14), 202–5, 208, 213, 260, 261
Suomi College, 212, 215
Superintendent of the Poor, 184, 185, 196, 255
Superior mine, 110
Supplies, charged to miners, 63, 66, 76
Surface plants, 27, 41, 43; at C&H, 46
Swedes, 60, 112, 156, 178, 211, 214
Swedish Methodist Episcopal Church, 179
Synagogues, 178
Syrians, 112, 211

Tailings, mill, 248, 250, 259
Tamarack Co-operative Assoc., 167
Tamarack mine, 21, 72, 164; C&H acquires, 250; location of, 10; steam engines at, 47; underground temperatures at, 38; vertical shafts at, 35, 91
Taylor, Frederick W., 75
Teachers, 169
Temperatures, mine, 37–38, 39, 40
Tennessee, as copper producer, 9
Timbermen, 101, 115, 155; definition, 32; work description, 32–33, 137
Theatres, 14, 175
Thiel Detective Service Co., 126
Todd, William Parsons, 247, 253
Todd, William R., 102, 129, 130, 135, 173, 177, 178; disputes with Quincy agents, 80–81, 96, 97–99; and eight-hour day, 233
Toltec mine, 42
Tonkin, William, 78
Torch Lake, 42; milling and smelting site, 11, 13, 53, 248, 250, 259
Trammers: as "beasts of burden," 32, 103; and company housing, 154–55, 162; definition, 31; elimination of, 101, 102–3; hard labor of, 40, 70, 103; pay, 69–70, 103; relations with miners, 77, 221; work longer shifts, 70
Tramming technology: by animal, 32, 101; by cable-car system, 32, 102; car filling

and dumping, 100–103; by locomotive, 32, 102–3, 248; with wheelbarrows, 31
Tramroads, surface, 53–54
Transport, human: by cages, 35; by ladders, 33; by man-cars, 35, 116–17; by man-engine, 33–35; safety of devices, 116–17
Trembath, Richard, 122
Tributing, 62, 63–64, 65
Trimountain mine, 11, 21, 38, 71, 72
Tyomies, 208, 215, 237

Underground, conditions in: dampness, 36–37; darkness, 36; depths, 21–22, 26, 30, 33, 37, 72, 245; sanitation, 35–36; temperatures, 37–38, 39, 40; ventilation and air quality, 27, 28, 37–39, 40, 93, 98–99, 119
Unemployment, 247, 254, 255. *See also* Economic decline, regional; Out-migration of workers
United Brewery Workmen, 229
United Mine Workers, 229
United Steel Workers of America, 258, 261–62
Universal Oil Products (UOP), 262–63
Uren, Richard, 61, 82
U.S. Bureau of Immigration, 231
U.S. Bureau of Labor Statistics, 159, 205
U.S. Bureau of Mines, 124, 139, 140
U.S. Commissioner of Immigration, 212
U.S. Department of Labor, 231
U.S. Geological Survey, 73
U.S. Secretary of Commerce and Labor, 212
Utah: as copper producer, 245; miners' unions in, 207

Venereal disease, 183–84
Ventilation. *See* Underground, conditions in
Vermont, as copper producer, 9
Victoria hydroelectric plant, 255–56
Victoria mine, 42
Vivian, Johnson, 61, 190
Vocational training, 171–72

Waddell-Mahon Corporation, 228
Wages, 14, 246, 247; of contract miners, 67–68, 70, 76, 87–88, 108, 241–42; differences between companies, 38, 166; of trammers, 70, 103, 104
Wagner Act, 257
War Industries Board, 242
Washington School, 170–71
Water: for milling, 11, 12, 53, 56; ridding from mines, 11, 36–37; for transport, 11, 13
Waterpower, 42
Wayne, J. B., & Co., 43
Western Federation of Miners, 211; benefits paid during strike, 227, 229, 233; Copper Country membership, 220, 223; effects of strike on union, 230; ends strike, 239; executive board and officers, 219, 221, 222, 232; finances for strike, 219, 221, 229, 239; history and western U.S. origins, 206–7, 209, 219, 220; ignored by mine managers, 226; and Italian Hall disaster, 237–38; leftist politics, 207–8; organizational drives in Michigan, 205, 208, 219–20; publishes *Miners' Bulletin,* 225, 233. *See also* Strike of 1913–14
White Pine mine, 259–60, 263
Whiting, Stephen B., 61, 73, 189, 192, 201, 210, 217
Widows, treatment of, 189–94
Wilfley, Arthur R., 248–49
Wills, Thomas, 61
Wilson, William B., 231
Winona mine, 21, 71
Winter weather, 13, 39, 142, 167; and labor unrest, 108–9, 221
Wolverine mine, 156
Wolverine Tube Co., 258, 261
Women: and company jobs, 150, 191; percentage of population, 150–51, 168; as positive influence on men, 150–51, 161; in strike of 1913–14, 230–31; when widowed, 189–94
Wood, R. J., 203
Workmen's compensation, 138–39, 141
Wright, James North, 61, 144, 152, 166, 167, 169, 203

YMCA, 175